Occurrence and Behavior of Emerging Contaminants in Organic Wastes and Their Control Strategies

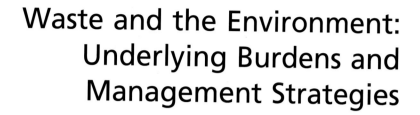

Waste and the Environment: Underlying Burdens and Management Strategies

Occurrence and Behavior of Emerging Contaminants in Organic Wastes and Their Control Strategies

Series Editor

Sunil Kumar

Volume Editors

Kui Huang
School of Environmental and Municipal Engineering, Lanzhou Jiaotong University, Lanzhou, China

Sartaj Ahmad Bhat
River Basin Research Center, Gifu University, Gifu, Japan

ELSEVIER

Elsevier
Radarweg 29, PO Box 211, 1000 AE Amsterdam, Netherlands
125 London Wall, London EC2Y 5AS, United Kingdom
50 Hampshire Street, 5th Floor, Cambridge, MA 02139, United States

ISBN: 978-0-443-13585-9

For information on all Elsevier publications visit our website at
https://www.elsevier.com/books-and-journals

Publisher: Candice Janco
Acquisitions Editor: Jessica Mack
Editorial Project Manager: Helena Beauchamp
Production Project Manager: Paul Prasad Chandramohan
Cover Designer: Miles Hitchen

Typeset by TNQ Technologies

Contents

List of contributors

Muhammad Adil Pharmacology & Toxicology Section, University of Veterinary & Animal Sciences, Lahore, Jhang Campus, Jhang, Pakistan

Sartaj Ahmad Bhat River Basin Research Center, Gifu University, Gifu, Japan

Christian Sekomo Birame National Industrial Research and Development Agency, Kigali, Rwanda

Muniyandi Biruntha Vermiculture Technology Laboratory, Department of Animal Health and Management, Alagappa University, Karaikudi, Tamil Nadu, India

Susanta Kumar Chakraborty Department of Zoology, Vidyasagar University, Midnapore, West Bengal, India

Jin Chen School of Environmental and Municipal Engineering, Lanzhou Jiaotong University, Lanzhou, China

Hongguang Cheng State Key Laboratory of Environmental Geochemistry, Institute of Geochemistry, Chinese Academy of Sciences, Guiyang, Guizhou, China

Antonio Cristaldi Department of Medical, Surgical Sciences, and Advanced Technologies "G.F. Ingrassia", University of Catania, Catania, Italy

Guangyu Cui School of Environment and Energy, Peking University Shenzhen Graduate School, Shenzhen, China

Shiamita Kusuma Dewi United Graduated School of Agricultural Science, Gifu University, Gifu, Japan

Margherita Ferrante Department of Medical, Surgical Sciences, and Advanced Technologies "G.F. Ingrassia", University of Catania, Catania, Italy; CRIAB—Centro di Ricerca Interdipartimentale per L'implementazione dei Processi di Monitoraggio Fisico, Chimico e Biologico nei Sistemi di Biorisanamento e di Acquacoltura e di Biorisanamento, of Department "G.F. Ingrassia", University of Catania, Catania, Italy

Ram Kumar Ganguly Department of Zoology, Vidyasagar University, Midnapore, West Bengal, India

Gloria Gómez Engineering and Environmental Biotechnology Group, Environmental Science Faculty, Universidad de Concepción, Concepción, Chile; Water Research Center for Agriculture and Mining (CRHIAM), ANID Fondap Center, Concepción, Chile

Zaw Min Han Graduate School of Engineering, Gifu University, Gifu, Japan

Kui Huang School of Environmental and Municipal Engineering, Lanzhou Jiaotong University, Lanzhou, China; Key Laboratory of Yellow River Water Environment in Gansu Province, Lanzhou, China

Olivier Irumva School of Science and Engineering, Tongji University, Shanghai, China

Natchimuthu Karmegam PG and Research Department of Botany, Government Arts College (Autonomous), Salem, Tamil Nadu, India

Ana María Leiva Engineering and Environmental Biotechnology Group, Environmental Science Faculty, Universidad de Concepción, Concepción, Chile; Water Research Center for Agriculture and Mining (CRHIAM), ANID Fondap Center, Concepción, Chile

Fusheng Li River Basin Research Center, Gifu University, Gifu, Japan

Jingyang Luo Key Laboratory of Integrated Regulation and Resource Development on Shallow Lakes, Ministry of Education, Hohai University, Nanjing, China; College of Environment, Hohai University, Nanjing, China

Naomi Monsalves Engineering and Environmental Biotechnology Group, Environmental Science Faculty, Universidad de Concepción, Concepción, Chile; Water Research Center for Agriculture and Mining (CRHIAM), ANID Fondap Center, Concepción, Chile

Gea Oliveri Conti Department of Medical, Surgical Sciences, and Advanced Technologies "G.F. Ingrassia", University of Catania, Catania, Italy

J. Christina Oviya Department of Biotechnology, St. Joseph's College of Engineering, Chennai, Tamil Nadu, India

Eloise Pulvirenti Department of Medical, Surgical Sciences, and Advanced Technologies "G.F. Ingrassia", University of Catania, Catania, Italy; Department of Biological, Geological and Environmental Sciences, University of Catania, Catania, Italy

Balasubramani Ravindran Department of Environmental Energy and Engineering, Kyonggi University Yeongtong-Gu, Suwon, Gyeonggi-Do, Republic of Korea

Sirat Sandil National Laboratory for Water Science and Water Security, Institute of Aquatic Ecology, Centre for Ecological Research, Budapest, Hungary

Jiwei Shi School of Environmental and Municipal Engineering, Lanzhou Jiaotong University, Lanzhou, China

Pragya Tiwari Department of Biotechnology, Yeungnam University, Gyeongsan, Gyeongbuk, Republic of Korea

Gratien Twagirayezu State Key Laboratory of Environmental Geochemistry, Institute of Geochemistry, Chinese Academy of Sciences, Guiyang, Guizhou, China

Abias Uwimana College of Science and Technology, University of Rwanda, Kigali, Rwanda

Gladys Vidal Engineering and Environmental Biotechnology Group, Environmental Science Faculty, Universidad de Concepción, Concepción, Chile; Water Research Center for Agriculture and Mining (CRHIAM), ANID Fondap Center, Concepción, Chile

Bangchi Wang School of Environmental and Municipal Engineering, Lanzhou Jiaotong University, Lanzhou, China

Chengchen Wei School of Environmental and Municipal Engineering, Lanzhou Jiaotong University, Lanzhou, China

Yongfen Wei River Basin Research Center, Gifu University, Gifu, Japan

Yang Wu State Key Laboratory of Pollution Control and Resource Reuse, School of Environmental Science and Engineering, Tongji University, Shanghai, China

Zilong Wu School of Environmental and Municipal Engineering, Lanzhou Jiaotong University, Lanzhou, China

Jiachen Xie School of Environmental and Municipal Engineering, Lanzhou Jiaotong University, Lanzhou, China

Zhiquan Yan School of Environmental and Municipal Engineering, Lanzhou Jiaotong University, Lanzhou, China

Jin Yang School of Environmental and Municipal Engineering, Lanzhou Jiaotong University, Lanzhou, China

Ananthanarayanan Yuvaraj Vermitechnology and Ecotoxicology Laboratory, Department of Zoology, School of Life Sciences, Periyar University, Salem, Tamil Nadu, India

Licheng Zhu School of Environmental and Municipal Engineering, Lanzhou Jiaotong University, Lanzhou, China

Preface

Occurrence and Behavior of Emerging Contaminants in Organic Wastes and Their Control Strategies is a comprehensive book that covers the occurrences of microplastics, antibiotic resistance genes, antimicrobial resistance genes, pharmaceutical and personal care products, and other emerging contaminants discharged in different types of organic wastes. This book addresses the challenges and recent advances in understanding and mitigating the impact of emerging contaminants in organic wastes. Emerging contaminants have gained significant attention due to their widespread occurrence and potential environmental and health hazards. This book provides valuable insights into the occurrence and behavior of these contaminants, showing contributions from top experts in the field. This book promotes the development of green and eco-friendly technologies for removing emerging contaminants in organic wastes. This book is spread into 13 chapters covering the occurrence, monitoring, and behavior of emerging contaminants and their control strategies.

Chapter 1 discusses the fate of antibiotic resistance genes in organic wastes such as sewage sludge from sewage treatment plants in the framework of the circular economy. This chapter also examines the risks associated with the reuse of organic waste in agriculture. Chapter 2 evaluates the fate and behavior microplastics in biosolid, and its effects on agricultural soils, plants, and earthworms. Chapter 3 focuses on the occurrence, fate, detection, ecological impact, and mitigation of antimicrobial resistance genes derived from animal wastes. Chapter 4 discusses E-waste legislation framework and green and sustainable E-waste management. Chapter 5 reviews the research on microplastics in excess sludge. This chapter also discussed the sources, content, and potential hazards of microplastics, and a review of sampling, preparation, and identification methods for microplastic sample detection. Chapter 6 covers the fate of personal care products and pharmaceuticals in sewage sludge. Chapter 7 addresses the environmental behavior of exogenous emerging contaminants on the anaerobic digestion of waste activated sludge. This chapter also discusses the impacts of exogenous emerging contaminants on waste activated sludge digestion efficiency, disclosed the underlying mechanisms, and also provides the corresponding promising mitigation approach. Chapter 8 investigated the potential applications of biochar in mitigating environmental pollutants and also discussed the insights toward the framework of circular bioeconomy. Chapter 9 discussed the common synthetic composting additives in reducing antibiotics and antibiotic resistance genes in the sludge composting process. This chapter also reviewed the current laws and factors influencing the reduction of antibiotics and antibiotic resistance genes in the sludge composting process. Chapter 10 focuses on the change in species and abundance of antibiotic resistance genes across different vermicomposting products and also discusses the antibiotic resistance genes removal efficiency, including biochar and herbal medicine residues. Chapter 11 discusses the critical influence factors for decreasing the antibiotic resistance genes during anaerobic digestion of organic wastes. This chapter also addresses the various factors that have significant impacts on the digestion process of organic waste. Chapter 12 reviews the occurrence, behavior, and fate of microplastics in agricultural and livestock wastes and their impact on agricultural fields. Finally, Chapter 13 covers organic waste management and health-related problems. As waste management is strongly linked to health, if it is not managed correctly, it can damage the environment and human health.

This book gathers the expertise of leading scientists in the field and offers insights into the fate, occurrence, monitoring, and control strategies for these pollutants. By using a comprehensive approach and considering various aspects of the issue, it aims to contribute to the development of effective strategies for mitigating emerging contaminants in our environment.

PART I

Occurrence of emerging contaminants in different organic wastes

1

Fate of antibiotic resistance genes in organic wastes from sewage treatment plants in the framework of circular economy

Ana María Leiva[1,2], Naomi Monsalves[1,2], Gloria Gómez[1,2] and Gladys Vidal[1,2]

[1]ENGINEERING AND ENVIRONMENTAL BIOTECHNOLOGY GROUP, ENVIRONMENTAL SCIENCE FACULTY, UNIVERSIDAD DE CONCEPCIÓN, CONCEPCIÓN, CHILE; [2]WATER RESEARCH CENTER FOR AGRICULTURE AND MINING (CRHIAM), ANID FONDAP CENTER, CONCEPCIÓN, CHILE

1. Introduction

Circular economy (CE) is a framework that considers the efficient use of resources through waste minimization, long-term value retention, reduction of primary resources, and closed loops of products to achieve sustainable development (Morseletto, 2020). This concept can be applied in sewage treatment plants (STPs) for reducing their environmental impacts and for revaluing treated wastewater and organic wastes (Guerra-Rodríguez et al., 2020). Specifically, the production of organic wastes such as sewage sludge (SS) is considered a problem in STPs, and its disposal is a challenge. For this reason, SS must be treated using aerobic or anaerobic digestion for stabilizing and ensuring safe disposal. The stabilized SS is known as treated SS or biosolids (Buta et al., 2021). The concept of CE offers a new perspective on treated SS, considering it as a product and not a residue. Due to the physicochemical properties of SS, they represent a good alternative of biofertilizers for agricultural purposes (Kacprzak et al., 2017). Treated SS contains concentrations of organic matter and nutrients, which can improve the soil properties (Venegas et al., 2021). However, the presence of emerging contaminants such as ARGs is a challenge for the reuse of treated SS in agriculture (Leiva et al., 2021). The dissemination of antibiotic resistance (AR) is a global concern related to the increase in the difficulty of treating infectious diseases using traditional antibiotics (WHO, 2022). SS is one of the principal pathways of spreading AR into the environment, reporting ARG abundances between 10^6–10^{18} copies/g·dry weight (DW) (Jang et al., 2018; Xu et al., 2018; Zhang et al., 2019). Many studies have determined that AR elements such as

Occurrence and Behavior of Emerging Contaminants in Organic Wastes and Their Control Strategies
https://doi.org/10.1016/B978-0-443-13585-9.00010-0

antibiotics, ARGs, and antibiotic-resistant bacteria (ARB) can enter the food chain when treated SS is used as a soil amendment. Moreover, these elements can be translocated from the soil to the different parts of the plant (Piña et al., 2020). Despite the multiple studies identifying the potential risks to human health and the environment of the SS land application, the risk assessment of AR transmission is still unclear (Manaia, 2017).

Under this context, the main objective of this book chapter is to determine the fate of ARGs in organic wastes from STPs in the framework of the CE. The occurrence and performance of ARGs in different processes for treating organic wastes in STPs as aerobic and anaerobic digestion are discussed. Moreover, the risks associated to the reuse of organic wastes in agriculture related to the ARG's fate are also discussed in this book chapter.

2. Antibiotic resistance genes in sewage treatment plants

As mentioned above, STPs present essential characteristics for increasing AR, and this fact limits the possible reuse of treated SS in a CE context. For this reason, this section will discuss the concept of AR, what it is, where it comes from, how it is disseminated in the environment, and its associated risks.

2.1 Antibiotic resistance genes

The AR corresponds to a natural phenomenon used by bacteria, which gives them adaptive advantages in the environment over other species competing for resources (Alonso et al., 2001; Martínez et al., 2014). This phenomenon may occur because the antibiotics present in the medium generate a selective pressure that inhibits the growth of susceptible bacteria while selecting for bacteria that are intrinsically resistant or have acquired resistance over time. The AR is carried out by ARGs, which are nucleic acids that encode proteins that participate in different resistance mechanisms such as alteration of the antibiotic, modification of the target site, and alteration in the bacterial membrane (Monsalves et al., 2022). The acquisition of ARGs by previously susceptible bacteria can be due to vertical gene transfer, point gene mutation, or the fact that they accumulate in mobile genetic elements such as plasmids, integrons, or transposons, which are able to capture genes and mediate their movement (De oliveira et al., 2020; Nguyen et al., 2021). This gene movement can occur within the bacterial genome or between different cells through horizontal gene transfer (HGT), which can occur by three main mechanisms: Transduction, transformation, and conjugation (Amarasiri et al., 2019; Jiang et al., 2022). The dissemination of ARGs into the environment has been recognized as one of the main concerns of the 21st century. According to the World Health Organization (WHO), it is expected that by 2050, this AR will be the main cause of death in the world population (WHO, 2017). This is due to the impossibility of treatment for resistant infections caused mainly by *Enterococcus faecium, Staphylococcus aureus, Klebsiella pneumoniae, Acinetobacter baumannii, Pseudomonas aeruginosa,* and *Enterobacter* species that have been designated as "priority status" (WHO, 2017; De oliveira et al., 2020).

2.2 Development of antibiotic resistance genes for dissemination in sewage treatment plants

The human microbiota is recognized as an important reservoir of ARGs, where antibiotic consumption can provide selective pressure within the gut, leading to the development of AR in enteric bacteria (Amarasiri et al., 2019). Furthermore, considering that 30%–70% of antibiotics are partially metabolized in the human body, both antibiotics and ARB are continuously excreted into the STPs (Bouki et al., 2013). For this reason, ARGs have been detected in sewage and SS with abundances ranging from 1.7×10^3–1.0×10^9 copies/mL and 4.26×10^9–2.76×10^{10} copies/g·DW, respectively (Liao and Chen, 2018; García et al., 2020; Sabri et al., 2021). Fig. 1.1 shows the contribution of STPs to AR dissemination. These concentrations of ARGs and trace antibiotics (30–34,500 ng/L) in a medium containing a significant supply of nutrients, temperature, stable pH, and high cell proximity in the flocs generate the right environment for the proliferation and HGT of ARGs (Manaia et al., 2018; Leiva et al., 2021; Markowicz et al., 2021). Moreover, many studies indicate that the selection of ARB by other types of stressors, such as pesticides, disinfectants, and heavy metals (He et al., 2018). Due to this, it is indicated that STPs act as primary reactors, and they are considered a key point to control the spread of AR in the environment (Wang et al., 2020; Tang et al., 2022).

Table 1.1 shows the concentrations of ARGs detected in wastewater (copies/mL) and in SS (copies/g·DW) and the resistance mechanisms that ARGs exhibit once they enter

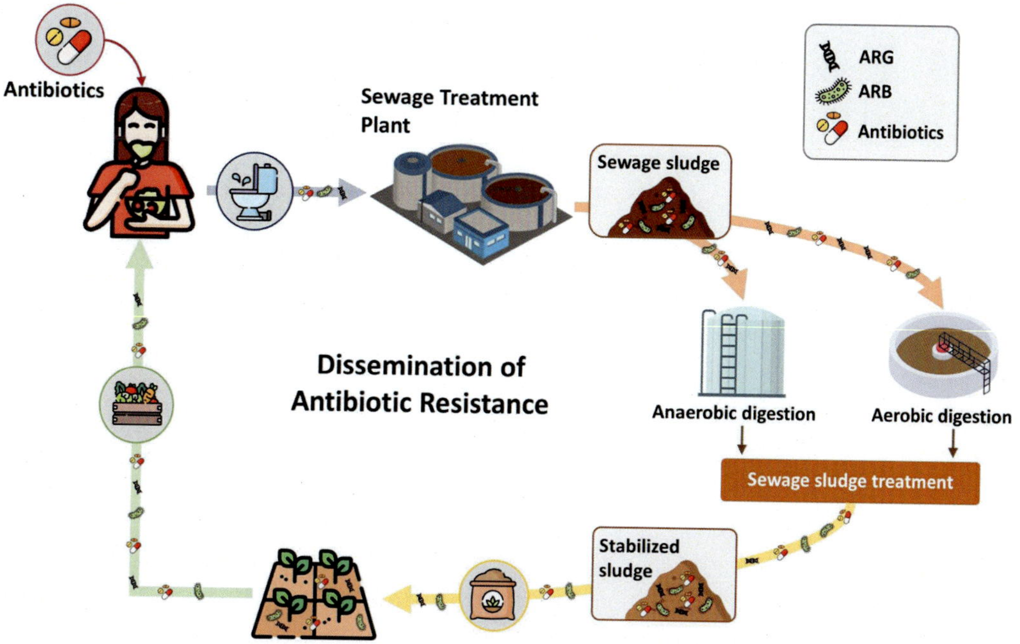

FIGURE 1.1 The contribution of STPs to the antibiotic resistance dissemination.

Table 1.1 The concentrations of ARGs detected in wastewater (copies/mL) and in SS (copies/g·DW), and their resistance mechanisms.

ARG class	Resistance mechanism	ARGs	Abundances: Wastewater (copies/mL)	Abundances: Sewage sludge (copies/g·DW)	References
Beta-lactams	Alteration of bacterial outer membrane by decreasing the number of porin channels. Alteration of the antibiotic by cleavage of the beta-lactam ring by *amp*, *bla*, and *oxa* genes.	*bla*TEM *bla*OXA-48	1.0×10^4–1.0×10^7 1.0×10^4–1.0×10^8	4.3×10^2–1.2×10^4 3.5×10^3–4.7×10^5	Cacace et al. (2019), Zhang et al. (2019). Pazda et al. (2019)
Sulfonamides	Modification of the target site and changes in membrane permeability affect proper drug uptake.	*sul*1 *sul*2	1.0×10^5–1.0×10^9 1.7×10^3–1.6×10^8	2.8×10^7–5.4×10^9 5.0×10^7–1.7×10^9	Cacace et al. (2019), Zhang et al. (2019), Pazda et al. (2019), Zhang et al. (2021), Liu et al. (2019)
Tetracyclines	Modification of the target site by ribosomal protection and alteration of the bacterial membrane decrease the intracellular concentration of the drug. Enzymatic inactivation of the antibiotic.	*tet*O *tet*X	3.8×10^5–1.4×10^6 5.0×10^6–7.7×10^7	1.1×10^7–5.1×10^7 1.2×10^7–1.6×10^8	Zhang et al. (2019), Pazda et al. (2019), Zhang et al. (2018), Chen et al. (2016a,b)
Macrolides	Modification of the target site by methylation of 23S rRNA. Alteration of the antibiotic by cleavage of the lactone ring.	*erm*B *erm*C	9.0×10^4–2.8×10^6 6.5×10^5–3.4×10^6	2.9×10^4–3.6×10^7 6.3×10^4–9.8×10^7	Zhang et al. (2019), Pazda et al. (2019), Zhang et al. (2018), Ávila et al. (2021), He et al. (2018)
Quinolones	Modifications of the target enzyme and decreased outer membrane permeability through changes in porins and overexpression of the exit pump.	*qnr*S *qnr*D	8.4×10^5–2.3×10^6 3.8×10^4–5.1×10^6	5.3×10^4–0.54×10^7 1.1×10^5–8.1×10^6	Zhang et al. (2019), Pazda et al. (2019), Zhang et al. (2021), Chen et al. (2019)

Notes: *ARGs*, antibiotic resistance genes.

the bacterial cell. In general, this table is a sample of the high anthropogenic contamination to which STPs and SS microorganisms are exposed. The high abundances detected in SS (4.26×10^2–5.35×10^9 copies/g·DW) can be explained by their structure, which allows them to accumulate significant amounts of emerging contaminants, including antibiotics and microorganisms that contribute to the rapid appearance of ARGs (Buta et al., 2021).

3. Organic wastes from sewage treatment as a source of antibiotic resistance genes

The effectiveness of STPs in reducing ARB and ARGs in effluents discharged into the aquatic environment through disinfection processes such as chlorination, ultraviolet light, and ozone treatment has been documented. However, it is estimated that 90%–95% of ARBs and ARGs in STPs are associated with the sludge fraction (Redhead et al., 2020), resulting in high concentrations of ARGs in sludge and biosolids (up to 10^9 copies/g·DW) (Nguyen et al., 2021). The loading rate of ARGs into the environment from sludge is estimated to be 1000 times that of effluent; as a consequence, sludge treatment is a key process in controlling the release of ARB and ARGs into the environment (Syafiuddin and Boopathy, 2021).

3.1 Processes for treating organic waste in sewage treatment plants

SS is first separated from the liquid fraction by primary and secondary sedimentation and then combined for incineration, disposal, or treatment (Nguyen et al., 2021). SS treatment aims on the one hand to sanitize and remove toxic compounds and, on the other hand, to reduce the volume of sludge (Grobelak et al., 2019) and can represent up to 60% of the total operating costs of STPs (Liu et al., 2011). For sludge treatment, in most STPs, anaerobic and aerobic digestion processes are usually used for stabilization. Then sludge is dewatered to a dry matter content of the order of 20%–30% (Ignatowicz, 2017).

In general, anaerobic digestion is used in large-size STPs, thus allowing methane recovery, and aerobic digestion is implemented in medium- and small-size STPs due to its low capital cost, high energy demand, and simple control requirements, as anaerobic digestion is a costly process (Liu et al., 2010, 2011). Fig. 1.2 shows a diagram of different SS treatments.

3.1.1 Aerobic digestion

Aerobic digestion is a bacterial process that occurs in the presence of oxygen. Under aerobic conditions, organic matter is oxidized, and products such as carbon dioxide, nitrate, or phosphate are generated (Martin et al., 2015). Aerobic SS treatment is similar to activated sludge treatment, its objective being to stabilize solids through long-term diffused aeration or mechanical surface aeration, thereby reducing biological oxygen

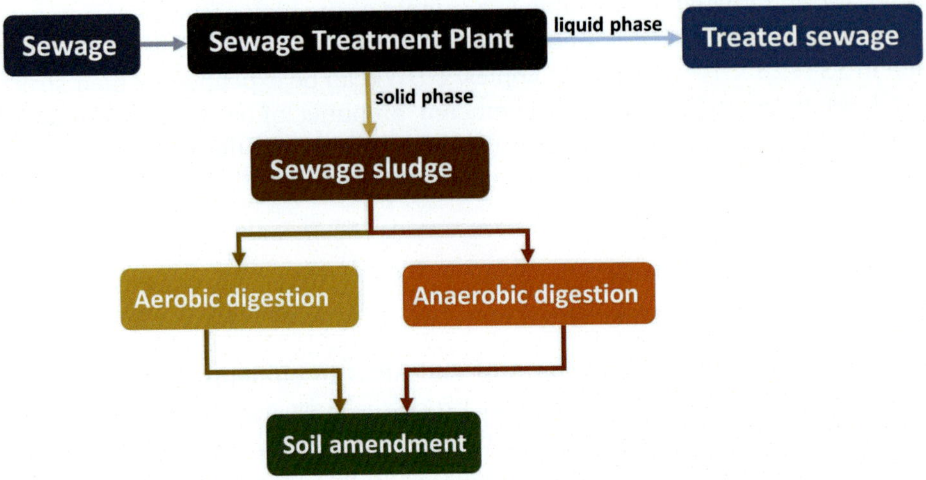

FIGURE 1.2 Diagram of different sewage sludge treatments.

demand and volatile solids. The reduction of volatile solids is usually around 40% (Demirbas et al., 2017; Grobelak et al., 2019).

Aerobic thermophilic autothermal aerobic digestion is a promising technology in the aerobic treatment of SS, achieving the production of Class A biosolids from SS (Liu et al., 2011). According to USEPA, aerobic thermophilic digestion is defined as liquid sludge agitated with air or oxygen to maintain aerobic conditions at a residence time of 10 days and a temperature of 55–60 °C with a volatile solids reduction of at least 38%. The process is called autothermal if the concentrated waste is aerated in isolated reactors, raising the temperature to the thermophilic range (45–65°C) without supplemental heat. Other benefits of this technology are rapid biomass degradation, short sludge retention time, and effective inactivation of pathogens compared to anaerobic digestion (Layden et al., 2007; Lloret et al., 2012).

Another aerobic process is composting, in which the decomposition of organic matter occurs naturally through the use of aerobic microorganisms capable of transforming organic matter into heat, carbon dioxide, and ammonium (Martin et al., 2015; Grobelak et al., 2019). During composting putrescible organic compounds are biologically stabilized, pathogenic organisms are eliminated, and waste volume is reduced. As a result of composting volatile solids are reduced by 20%–30% (Demirbas et al., 2017).

3.1.2 Anaerobic digestion

Anaerobic digestion is one of the most widely used technologies in SS treatment due to its effectiveness in stabilizing sludge, producing renewable energy, and reducing pathogen load to levels suitable for reuse (Redhead et al., 2020). Anaerobic digestion is considered a complex microbial process that is divided into the following stages: Hydrolysis, acidogenesis, acetogenesis, and methanogenesis. Anaerobic digestion requires stringent anaerobic conditions (oxidation-reduction potential < -200 mV) and is dependent on the coordinated activity of a complex microbial association to transform organic matter (Appels et al., 2008). In the absence of oxygen, organic matter is biodegraded and converted to biogas and other energy-rich organic compounds as end products. The biogas, which generally consists of 48%–65% methane, could be used for energy generation (Hanum et al., 2019).

The anaerobic digestion process can occur under both mesophilic (30–42°C) and thermophilic (50–60°C) conditions, with the latter being better at reducing pathogens and digestion time. However, thermophilic anaerobic digestion is more sensitive to small changes in process parameters and requires more energy to heat the SS to the required temperature compared to mesophilic anaerobic digestion (Grobelak et al., 2019).

A disadvantage of anaerobic digestion is the requirement of high reactor volumes and long retention times to achieve adequate stabilization of SS before disposal or reuse. This is due to the presence of high molecular weight compounds and complex organic matter in the sludge, limiting the hydrolysis stage in anaerobic digestion (Neumann et al., 2016). In most cases, hydrolysis is the slowest or rate-limiting stage due to the formation of volatile fatty acids and other toxic by-products. Hydrolysis is often accelerated through the implementation of various sludge pretreatment strategies (Kumar and Samadder, 2020).

3.2 Performance of sewage treatment plant for reducing antibiotic resistance genes

As mentioned above, organic wastes such as SS in STPs are usually treated by aerobic and anaerobic digestion (Kacprzak et al., 2017). In these types of treatments, the reduction of ARGs is carried out by physical, chemical, and biological processes, where biodegradation is the principal pathway for removing these compounds (Pärnänen et al., 2019). Generally, the performances of ARG reductions in aerobic and anaerobic digesters fluctuates between 1–3 log units (Haffiez et al., 2022). However, these values depend on different factors, such as operational and design factors and sewage quality (Leiva et al., 2021). Moreover, many studies have reported the occurrence of ARGs in treated SS (Aziz et al., 2022; Burch et al., 2017; Chen et al., 2016a,b; Markowicz et al., 2021). Under the context of CE, the presence of these compounds in organic wastes after applying a treatment represents a risk for the reuse of SS as a soil amendment in agriculture (Bondarczuk et al., 2016). In this section, the occurrence and reduction of ARGs by aerobic and anaerobic treatments in STPs will be discussed.

3.3 Occurrence of antibiotic resistance genes in sewage sludge

The occurrence of ARGs has been reported in different environmental matrixes, and SS is not the exception (Larsson and Flach, 2022). Table 1.2 shows the occurrence of ARGs in SS treated by aerobic and anaerobic digestion. In general, the abundance of ARGs in aerobic and anaerobic digestion varied between 3.2×10^6–3.4×10^{18} copies/g·DW and 1.7×10^7–1.7×10^{12} copies/g·DW, respectively. The most common ARGs found in SS

Table 1.2 The occurrence and performance of ARGs in SS treating by aerobic and anaerobic digestion.

SS treatment	Pretreatment	Operational conditions: SRT (d)	HRT (d)	OLR (gVS/m3·d)	Temperature (°C)	ARGs	SS (copies/g)	ARG removal (%)	ARG reduction (ulog)	References
Anaerobic digestion	Microwave irradiation	15–20	–	2700–2900	37	*tet*X, *tet*M, *tet*G, *sul*1, *sul*2, *mef*A, *erm*F, *erm*B, *ere*A, *bla*TEM, *bla*oxa-1	3.2×10^{6}–3.2×10^{7}	18.7–92.7	0.9–2.2	Zhang et al. (2019)
Anaerobic digestion	–	–	15	28,440	35–55	*tet*A, *tet*L, *tet*M, *tet*O, *tet*W, *tet*X, *sul*1, *sul*2	3.5×10^{-1}–$5.8 \times 10^{\circ}$*	81.7–99.3	0.9–2.1	Xu et al. (2018)
Anaerobic digestion	–	–	–	–	25–55	*tet*A, *tet*C, *tet*O, *tet*W, *sul*1, *sul*2	3.0×10^{7}–3.2×10^{10}	64.5–91.9	(–0.3)–1.1	Zhang et al. (2021)
Aerobic digestion	–	–	1	–	55	*tet*A, *tet*B, *tet*D, *tet*E, *tet*G, *tet*H, *tet*M, *tet*Q, *tet*X, *tet*Z, *tet*B, *sul*1, *sul*2, *qnr*S, *qnr*D, *erm*B, *erm*C, *bla*Tem, *bla*CTX, *bla*SHV, *oqx*A, *flo*R	2.8×10^{10}–4.4×10^{11}	20.4–100.0	0.1–1.0	Jang et al. (2018)
Anaerobic digestion	Ozonation and thermal hydrolysis	–	15	–	35	*tet*A, *tet*G, *tet*Q, *tet*W, *tet*X	7.4×10^{10}–$1,7 \times 10^{12}$	84.3–99.3	0.8–2.2	Pei et al. (2016)
Anaerobic digestion	Microwave irradiation and peroxide oxidation	30	–	–	38	*tet*A, *tet*C, *tet*M, *tet*O, *tet*X, *bla*SHV, *bla*CTX-M, *ampC*, *intI*1	5.4×10^{5}–3.4×10^{18}	70.5–100.0	0.5–12.8	Tong et al. (2016)
Aerobic digestion	–	1.5	–	–	28	46 ARGs*	1.0×10^{10}–1.7×10^{10}	(-69.8)–(–28.8)	(-0.2)–(–0.1)	Zheng et al. (2019)
Aerobic digestion	–	15	–	–	–	aac(6′)-Ib-cr, *erm*B, *erm*F, dfrA1, *sul*1, *sul*2, *tet*A, *tet*X and *mef*A	1.7×10^{9}–1.7×10^{12}	59.6–96.7	0.8–3.7	Zhang et al. (2021)

Notes: *ARGs*, antibiotic resistance genes; *HRT*, hydraulic retention time; *OLR*, organic loading rate; *SRT*, solid retention time; *SS*, sewage sludge.

are genes associated to tetracycline (*tet*) and sulfonamide (*sul*) resistance in aerobic and anaerobic digesters. The tetracycline resistance genes encode ribosomal protection proteins and efflux pumps. In the case of sulfonamide resistance, these genes encode a sulfonamide-resistant dihydropteroate synthase. Tetracycline and sulfonamide resistance genes are reported widely and these resistances are related to the extensive use of veterinary antibiotics in livestock farming to control diseases and promote growth (Cheng et al., 2019).

3.4 Reduction of antibiotic resistance genes in sewage treatment plants

Table 1.2 also shows the performance of aerobic and anaerobic digestion in reducing ARGs in SS. Regarding the reduction of ARGs in both types of treatments, these values varied between 0.1–3.7 log units and 0.5–12.8 log units, respectively. In the case of removal efficiencies, aerobic digestion achieved a percentage between (−69.8)−100%, while anaerobic digestion reported values between 18.7%– 100%. In general, the reduction of ARGs in anaerobic digestion was 200%–210% higher than values achieved in aerobic digestion. This behavior can be related to the use of pretreatment in anaerobic digestion (Tong et al., 2019). This strategy consists of improving the hydrolysis by adding a prior physical, chemical, and/or biological step to anaerobic digestion. Within the types of pretreatment, there are thermal, ultrasound, and irradiation processes (Neumann et al., 2017). Pei et al. (2016) studied the use of ozone and thermal hydrolysis pretreatment to improve the removal of ARGs in anaerobic digestion. In this study, ozonation and thermal hydrolysis reduced more *tet* genes (*tet*A, *tet*G, *tet*Q, *tet*W, and *tet*X) than anaerobic digestion without pretreatment achieving reductions between 2.01–3.79 log units. The advantage of using a pretreatment is that this process allows the destruction of the bacterial cell wall and then the release of different cellular components, such as ARGs. Therefore, the biodegradation of ARGs will be favored (Aziz et al., 2022).

Moreover, the reduction of ARGs in STPs depends on the operational and design parameters of technologies such as solid retention time (SRT), hydraulic retention time, organic loading rate, and antibiotic concentration in SS (Haffiez et al., 2022). Few studies analyze the effect of these parameters on the performance of anaerobic and aerobic digestion in reducing ARGs. Zhang et al. (2019) analyzed the effects of SRT on the fate of ARGs in the anaerobic digestion of SS. This study showed that the reduction of ARGs depends on the ARG types and reactor configuration. In the case of *erm*B and *tet*M, their reductions were favored using anaerobic digestion with a short SRT. On the contrary, the removal of *bla*OXA-1 was lower in the short SRT. These results demonstrated that the influence of operational parameters on ARG reduction is still unknown. For this reason, more investigations into this topic should be developed.

Despite anaerobic and aerobic digestion achieving ARG reductions between 1 and 2 log units, treated SS still contains ARG abundances. The reuse of organic wastes from

STPs in agricultural soil represents a risk for the environment and human health related to the spread of AR. The presence of these compounds is a challenge for the safe implementation of CE in STPs (Bondarczuk et al., 2016).

4. Risks associated to the reuse of organic wastes in agriculture in the framework of circular economy

Despite the major advances in ensuring the safety of organic wastes from STPs, treated SS may contain antibiotics, ARGs, and ARBs. This section will discuss the risks of the reuse of organic wastes in agriculture for human health and the environment focusing on the framework of EC.

4.1 Health risks

Despite the fact that the spread of AR is a global concern, the risk assessment of the transmission of AR elements (ARB and ARGs) from the environment to humans is poorly understood (Manaia, 2017). Fig. 1.3 shows a schematic diagram of human exposure to AR through the ingestion of crops contaminated with ARGs and ARB. Under the scenario of the CE, people will consume food that has been grown with treated SS as fertilizer. In

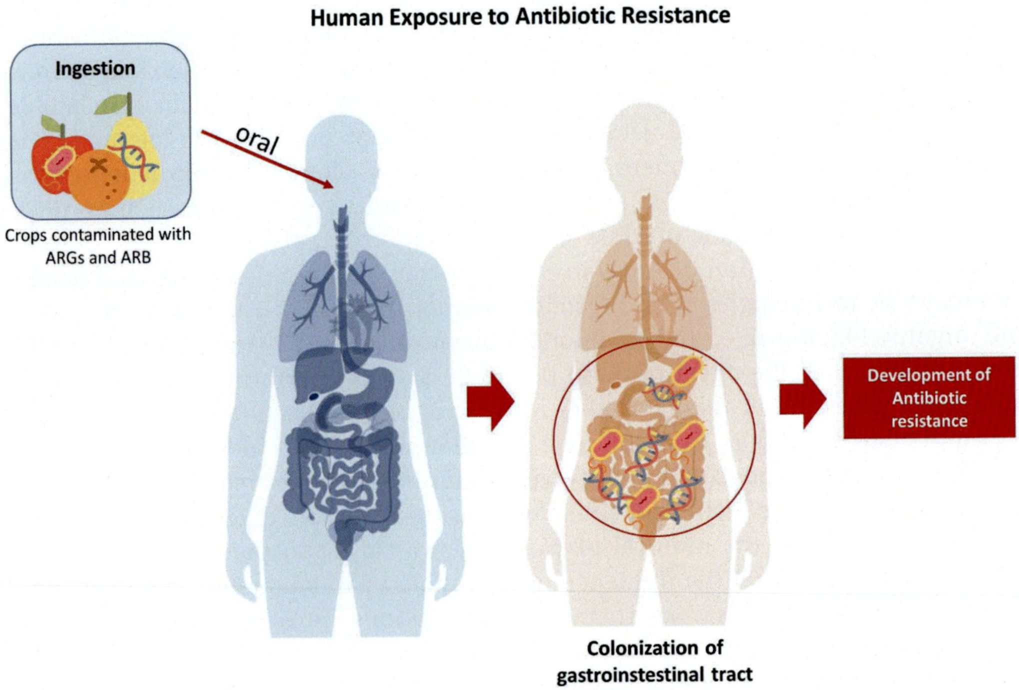

FIGURE 1.3 Schematic diagram of the human exposure to AR through the ingestion of crops contaminated with ARGs and ARB.

turn, these foods contain an abundance of ARGs and ARBs, and therefore, they will come into contact with the human body (Wengenroth et al., 2021). In this case, bacteria play a key role because they are responsible for colonizing and infecting the human microbiome. It is important to mention that the effects of the bacteria on the host will depend on the health conditions of the host and also on the infectious capacity of the bacteria (Chen et al., 2021). The greatest risk of AR to health is the increase in infectious diseases that are difficult to treat (WHO, 2022). In this line, it is estimated that 4.95 million deaths were associated to AR in 2019 (Murray et al., 2022). Moreover, high rates of AR have been reported in low and middle-income countries compared to rates achieved by high-income countries (WHO, 2022). Thus, AR will affect poorer countries, being not only a public health problem but also a socioeconomic one.

Regarding the human health risk assessment of AR, few studies propose a quantitative model. Ben et al. (2019) characterize this risk associated with antibiotic residues in the environment. This model summarizes the available data and realizes hazard identification, exposure assessment, dose-response assessment, and risk characterization. The principal conclusion of this study is that the information required for this analysis is limited, and future studies could focus on three determinant parameters: The establishment of a standardized monitoring guide of AR elements; determining the relation between antibiotic concentration and pathogenic ARB; and finally, establishing a relationship between dose-response, ARB, and infection diseases. To ensure the safe reuse of treated SS in agriculture, the development of a human health risk is fundamental. Moreover, this evaluation is an interesting tool for applying the CE to STPs.

4.2 Environmental risks

In addition to the risks to human health, the extensive use of treated SS in agricultural soil generates also negative effects on the environment. The reuse of treated SS as a soil amendment improves soil quality by enriching it with amounts of organic matter, nutrients, and biogenic compounds. Moreover, this practice contributes to minimizing the amount of generated waste and promoting the development of CE (Aleisa et al., 2021). However, the presence of AR elements such as antibiotics, ARGs, and ARB in treated SS represents a risk for the environment due to the dissemination of AR (Buta et al., 2021). Recently, different authors have focused their research in this area, and they have investigated the occurrence of ARGs in different parts of plant species and on soil used for agricultural purposes (Rahube et al., 2016; Liao and Chen, 2018; Murray et al., 2019).

Table 1.3 shows the ARGs detected in different parts of plant species and soil using treated SS as a soil amendment. In general, the most common ARGs detected in crops and soil are *intI*1, *sul*1, and genes associated to tetracycline resistance (*tet*). The association between the occurrence of *sul*1 and *intI*1 is very often described in the literature (Cerqueira et al., 2019). The class 1 integron (*intI*1) is an important form of mobile gene elements that transfer ARGs through HGT, and within its structure, *sul*1 is associated

Table 1.3 The ARGs detected in different part of plant species and soil using treated sewage sludge as a soil amendment.

Plant species	Sample types	ARG detection									References
		*intI*1	*bla*TEM	*sul*1	*sul*2	*tet*M	*tet*O	*tet*W	*qnr*S	*bla*OXA-58	
Lactuca sativa	Leaves	+	−	+	−	−	−	−	−	−	Lau et al. (2017)
	Soil	+	−	+	−	−	−	−	−	−	
Raphanus sativus	Fruit	+	−	+	−	−	−	−	−	−	
	Soil	+	−	+	−	−	−	−	−	−	
Daucus carota	Leaves	+	−	+	−	−	−	−	−	−	
	Soil	+	−	+	−	−	−	−	−	−	
Lactuca sativa	Roots	+	−	+	+	+	+	+	−	−	Wang et al. (2015)
	Soil	+	−	+	+	+	+	+	−	−	
Cichorum endivia	Roots	+	−	+	+	+	+	+	−	−	
	Soil	+	−	+	+	+	+	+	−	−	
Lactuca sativa	Leaves	−	−	+	−	+	−	−	−	−	Murray et al. (2019)
	Soil	+	−	+	−	−	−	−	−	+	
Raphanus sativus	Fruit	+	+	+	−	+	−	−	−	+	
	Soil	+	−	+	−	−	−	−	−	+	
Daucus carota	Leaves	−	−	+	+	+	−	−	−	+	
	Soil	+	−	+	−	−	−	−	−	+	
Lactuca sativa	Leaves	−	+	−	−	−	+	−	+	−	Yang et al. (2018)
	Soil	−	+	+	−	+	−	−	−	+	

Notes: (+) indicates the target is detected; (−) indicates the target was not detected.

with this type of integron (Liao and Chen, 2018). The application of treated SS to agricultural soil may also raise environmental exposure to ARGs affecting plants and soil. These resistant genetic elements can change the soil and plant microbiomes, making them potentially pathogenic (Urra et al., 2019). Moreover, bacteria are capable of entering the plant vascular system, translocating to the aerial parts (leaves, fruits), and colonizing the endophytic or periphytic microbiome (Mansilla et al., 2021).

5. Final remarks, future perspectives and recommendations

Under the context of CE, one of the main challenges in the reuse of organic wastes in STPs is to avoid the dissemination of AR when they are used for agricultural purposes. This book chapter demonstrates that treated SS contains an abundance of ARGs, which can be transferred to crops when treated SS is used as a soil amendment. Despite the fact that SS treatments such as anaerobic and aerobic digestion allow ARG reductions between 1 and 4 log units, pretreatment and advanced technologies should be implemented to decrease the ARG load in the treated SS. Along the same lines, the optimization and implementation of new technologies must be accompanied by a new risk assessment model that allows for reducing the environmental and human health impacts of treated SS reuse. Likewise, for facing the dissemination of AR and ensuring a safe CE application, updates to regulations and guidelines for the reuse of treated SS are mandatory combining different approaches such as metagenomic, epidemiological, environmental, and microbiological data.

Acknowledgments

This study was funded by the following Grants: ANID/FONDAP/15130015. A.M. Leiva thanks to National for Research and Development (ANID) for their Scholarship Program National for Research and Development (ANID)/Scholarship Program/Doctorado Nacional/2019-21191116 for supporting her Ph.D. studies at the University of Concepción.

References

Aleisa, E., Alsulaili, A., Almuzaini, Y., 2021. Recirculating treated sewage sludge for agricultural use: life cycle assessment for a circular economy. Waste Management 135, 79–89. https://doi.org/10.1016/j.wasman.2021.08.035.

Alonso, A., Sánchez, P., Martinez, J., 2001. Environmental selection of antibiotic resistance genes. Environmental Microbiology 3, 1–9. https://doi.org/10.1046/j.1462-2920.2001.00161.x.

Amarasiri, M., Sano, D., Suzuki, S., 2019. Understanding human health risks caused by antibiotic resistant bacteria (ARB) and antibiotic resistance genes (ARG) in water environments: current knowledge and questions to be answered. Critical Reviews in Environmental Science and Technology 50, 2016–2059. https://doi.org/10.1080/10643389.2019.1692611.

Appels, L., Baeyens, J., Degrève, J., Dewil, R., 2008. Principles and potential of the anaerobic digestion of waste-activated sludge. Progress in Energy and Combustion Science 34 (6), 755–781. https://doi.org/10.1016/j.pecs.2008.06.002.

Ávila, C., García-Galán, M.J., Borrego, C.M., Rodríguez-Mozaz, S., García, J., Barceló, D., 2021. New insights on the combined removal of antibiotics and ARGs in urban wastewater through the use of two configurations of vertical subsurface flow constructed wetlands. Science of the Total Environment 755, 142554. https://doi.org/10.1016/j.scitotenv.2020.142554.

Aziz, A., Sengar, A., Basheer, F., Farooqi, I.H., Isa, M.H., 2022. Anaerobic digestion in the elimination of antibiotics and antibiotic-resistant genes from the environment—a comprehensive review. Journal of Environmental Chemical Engineering 10 (1), 106423. https://doi.org/10.1016/j.jece.2021.106423.

Ben, Y., Fu, C., Hu, M., Liu, L., Wong, M.H., Zheng, C., 2019. Human health risk assessment of antibiotic resistance associated with antibiotic residues in the environment: a review. Environmental Research 169, 483–493. https://doi.org/10.1016/j.envres.2018.11.040.

Bondarczuk, K., Markowicz, A., Piotrowska-Seget, Z., 2016. The urgent need for risk assessment on the antibiotic resistance spread via sewage sludge land application. Environment International 87, 49–55. https://doi.org/10.1016/j.envint.2015.11.011.

Bouki, C., Venieri, D., Diamadopoulos, E., 2013. Detection and fate of antibiotic resistant bacteria in wastewater treatment plants: a review. Ecotoxicology and Environmental Safety 91, 1–9. https://doi.org/10.1016/j.ecoenv.2013.01.016.

Burch, T.R., Sadowsky, M.J., LaPara, T.M., 2017. Effect of different treatment technologies on the fate of antibiotic resistance genes and class 1 integrons when residual municipal wastewater solids are applied to soil. Environmental Science and Technology 51 (24), 14225–14232. https://doi.org/10.1021/acs.est.7b04760.

Buta, M., Hubeny, J., Zieliński, W., Harnisz, M., Korzeniewska, E., 2021. Sewage sludge in agriculture—the effects of selected chemical pollutants and emerging genetic resistance determinants on the quality of soil and crops—a review. Ecotoxicology and Environmental Safety 214, 112070. https://doi.org/10.1016/j.ecoenv.2021.112070.

Cacace, D., Fatta-Kassinos, D., Manaia, C.M., Cytryn, E., Kreuzinger, N., Rizzo, L., Karaolia, P., Schwartz, T., Alexander, J., Merlin, C., Garelick, H., Schmitt, H., de Vries, D., Schwermer, C.U., Meric, S., Ozkal, C.B., Pons, M.N., Kneis, D., Berendonk, T.U., 2019. Antibiotic resistance genes in treated wastewater and in the receiving water bodies: a pan-European survey of urban settings. Water Research 162, 320–330. https://doi.org/10.1016/j.watres.2019.06.039.

Cerqueira, F., Matamoros, V., Bayona, J., Piña, B., 2019. Antibiotic resistance genes distribution in microbiomes from the soil-plant-fruit continuum in commercial Lycopersicon esculentum fields under different agricultural practices. Science of the Total Environment 652, 660–670. https://doi.org/10.1016/j.scitotenv.2018.10.268.

Chen, J., Wei, X.D., Liu, Y.S., Ying, G.G., Liu, S.S., He, L.Y., Su, H.C., Hu, L.X., Chen, F.R., Yang, Y.Q., 2016b. Removal of antibiotics and antibiotic resistance genes from domestic sewage by constructed wetlands: optimization of wetland substrates and hydraulic loading. Science of the Total Environment 565, 240–248. https://doi.org/10.1016/j.scitotenv.2016.04.176.

Chen, Q., An, X., Li, H., Su, J., Ma, Y., Zhu, Y.G., 2016a. Long-term field application of sewage sludge increases the abundance of antibiotic resistance genes in soil. Environment International 92, 1–10. https://doi.org/10.1016/j.envint.2016.03.026.

Chen, J., Deng, W.J., Liu, Y.S., Hu, L.X., He, L.Y., Zhao, J.L., Wang, T.T., Ying, G.G., 2019. Fate and removal of antibiotics and antibiotic resistance genes in hybrid constructed wetlands. Environmental Pollution 249, 894–903. https://doi.org/10.1016/j.envpol.2019.03.111.

Chen, Z., Li, Y., Ye, C., He, X., Zhang, S., 2021. Fate of antibiotics and antibiotic resistance genes during aerobic co-composting of food waste with sewage sludge. Science of the Total Environment 784, 146950. https://doi.org/10.1016/j.scitotenv.2021.146950.

Cheng, D., Feng, Y., Liu, Y., Xue, J., Li, Z., 2019. Dynamics of oxytetracycline, sulfamerazine, and ciprofloxacin and related antibiotic resistance genes during swine manure composting. Journal of Environmental Management 230, 102–109. https://doi.org/10.1016/j.jenvman.2018.09.074.

De Oliveira, D., Forde, B., Kidd, T., Harris, P., Schembri, M., Beatson, S., Paterson, D., Walker, M., 2020. Antimicrobial resistance in ESKAPE pathogens. Clinical Microbiology Reviews 33 (3). https://doi.org/10.1128/CMR.00181-19.

Demirbas, A., Edris, G., Alalayah, W.M., 2017. Sludge production from municipal wastewater treatment in sewage treatment plant. Energy Sources, Part A: Recovery, Utilization, and Environmental Effects 39 (10), 999–1006. https://doi.org/10.1080/15567036.2017.1283551.

García, J., García-Galán, M., Day, J., Boopathy, R., White, J., Wallace, S., Hunter, R., 2020. A review of emerging organic contaminants (EOCs), antibiotic resistant bacteria (ARB), and antibiotic resistance genes (ARGs) in the environment: increasing removal with wetlands and reducing environmental impacts. Bioresource Technology 307, 123–228. https://doi.org/10.1016/j.biortech.2020.123228.

Grobelak, A., Czerwińska, K., Murtaś, A., 2019. General considerations on sludge disposal, industrial and municipal sludge. In: Industrial and Municipal Sludge. Butterworth-Heinemann, pp. 135–153. https://doi.org/10.5937/aaser2152117j.

Guerra-Rodríguez, S., Oulego, P., Rodríguez, E., Singh, D.N., Rodríguez-Chueca, J., 2020. Towards the implementation of circular economy in the wastewater sector: challenges and opportunities. Water 12 (5), 1431. https://doi.org/10.3390/w12051431.

Haffiez, N., Chung, T.H., Zakaria, B.S., Shahidi, M., Mezbahuddin, S., Hai, F.I., Dhar, B.R., 2022. A critical review of process parameters influencing the fate of antibiotic resistance genes in the anaerobic digestion of organic waste. Bioresource Technology 127189. https://doi.org/10.1016/j.biortech.2022.127189.

Hanum, F., Yuan, L.C., Kamahara, H., Aziz, H.A., Atsuta, Y., Yamada, T., Daimon, H., 2019. Treatment of sewage sludge using anaerobic digestion in Malaysia: current state and challenges. Frontiers in Energy Research 7, 19. https://doi.org/10.3389/fenrg.2019.00019.

He, Y., Nurul, S., Schmitt, H., Sutton, N.B., Murk, T.A., Blokland, M.H., Rijnaarts, H.H., Langenhoff, A.A., 2018. Evaluation of attenuation of pharmaceuticals, toxic potency, and antibiotic resistance genes in constructed wetlands treating wastewater effluents. Science of the Total Environment 631–632, 1572–1581. https://doi.org/10.1016/j.scitotenv.2018.03.083.

Ignatowicz, K., 2017. The impact of sewage sludge treatment on the content of selected heavy metals and their fractions. Environmental Research 156, 19–22. https://doi.org/10.1016/j.envres.2017.02.0355.

Jang, H.M., Lee, J., Kim, Y.B., Jeon, J.H., Shin, J., Park, M.R., Kim, Y.M., 2018. Fate of antibiotic resistance genes and metal resistance genes during thermophilic aerobic digestion of sewage sludge. Bioresource Technology 249, 635–643. https://doi.org/10.1016/j.biortech.2017.10.073.

Jiang, Q., Feng, M., Ye, C., Yu, X., 2022. Effects and relevant mechanisms of non-antibiotic factors on the horizontal transfer of antibiotic resistance genes 60 in water environments: a review. Science of the Total Environment 806, 150568. https://doi.org/10.1016/j.scitotenv.2021.150568.

Kacprzak, M., Neczaj, E., Fijałkowski, K., Grobelak, A., Grosser, A., Worwag, M., Rorat, A., Brattebo, H., Singh, B.R., 2017. Sewage sludge disposal strategies for sustainable development. Environmental Research 156, 39–46. https://doi.org/10.1016/j.envres.2017.03.010.

Kumar, A., Samadder, S.R., 2020. Performance evaluation of anaerobic digestion technology for energy recovery from organic fraction of municipal solid waste: a review. Energy 197, 117253. https://doi.org/10.1016/j.energy.2020.117253.

Larsson, D.G., Flach, C.F., 2022. Antibiotic resistance in the environment. Nature Reviews Microbiology 20 (5), 257–269. https://doi.org/10.1038/s41579-021-00649-x.

Lau, C.H.F., Li, B., Zhang, T., Tien, Y.C., Scott, A., Murray, R., Sabourin, L., Lapen, D.R., Duenk, P., Topp, E., 2017. Impact of pre-application treatment on municipal sludge composition, soil dynamics of antibiotic resistance genes, and abundance of antibiotic-resistance genes on vegetables at harvest. Science of the Total Environment 587, 214–222. https://doi.org/10.1016/j.scitotenv.2017.02.123.

Layden, N.M., Mavinic, D.S., Kelly, H.G., Moles, R., Bartlett, J., 2007. Autothermal thermophilic aerobic digestion (ATAD)—Part I: review of origins, design, and process operation. Journal of Environmental Engineering and Science 6 (6), 665–678. https://doi.org/10.1139/s07-015.

Leiva, A.M., Piña, B., Vidal, G., 2021. Antibiotic resistance dissemination in wastewater treatment plants: a challenge for the reuse of treated wastewater in agriculture. Reviews in Environmental Science and Biotechnology 20 (4), 1043–1072. https://doi.org/10.1007/s11157-021-09588-8.

Liao, J., Chen, Y., 2018. Removal of intl1 and associated antibiotics resistant genes in water, sewage sludge and livestock manure treatments. Reviews in Environmental Science and Biotechnology 17 (3), 471–500. https://doi.org/10.1007/s11157-018-9469-y.

Liu, S., Song, F., Zhu, N., Yuan, H., Cheng, J., 2010. Chemical and microbial changes during autothermal thermophilic aerobic digestion (ATAD) of sewage sludge. Bioresource Technology 101 (24), 9438–9444. https://doi.org/10.1016/j.biortech.2010.07.064.

Liu, S., Zhu, N., Li, L.Y., 2011. The one-stage autothermal thermophilic aerobic digestion for sewage sludge treatment. Chemical Engineering Journal 174 (2–3), 564–570. https://doi.org/10.1016/j.biortech.2011.11.041.

Liu, X., Zhang, G., Liu, Y., Lu, S., Qin, P., Guo, X., Bi, B., Wang, L., Xi, B., Wu, F., Wang, W., Zhang, T., 2019. Occurrence and fate of antibiotics and antibiotic resistance genes in typical urban water of Beijing, China. Environmental Pollution 246, 163–173. https://doi.org/10.1016/j.envpol.2018.12.005.

Lloret, E., Pastor, L., Martínez-Medina, A., Blaya, J., Pascual, J.A., 2012. Evaluation of the removal of pathogens included in the Proposal for a European Directive on spreading of sludge on land during autothermal thermophilic aerobic digestion (ATAD). Chemical Engineering Journal 198, 171–179. https://doi.org/10.1016/j.cej.2012.05.06.

Manaia, C.M., 2017. Assessing the risk of antibiotic resistance transmission from the environment to humans: non-direct proportionality between abundance and risk. Trends in Microbiology 25 (3), 173–181. https://doi.org/10.1016/j.tim.2016.11.014.

Manaia, M., Rocha, J., Scaccia, N., 2018. Antibiotic resistance in wastewater treatment plants: tackling the black box. Environment International 115, 312–324. https://doi.org/10.1016/j.envint.2018.03.044.

Mansilla, S., Portugal, J., Bayona, J.M., Matamoros, V., Leiva, A.M., Vidal, G., Pina, B., 2021. Compounds of emerging concern as new plant stressors linked to water reuse and biosolid application in agriculture. Journal of Environmental Chemical Engineering 9 (3), 105198. https://doi.org/10.1016/j.jece.2021.105198.

Markowicz, A., Bondarczuk, K., Cycoń, M., Sułowicz, S., 2021. Land application of sewage sludge: response of soil microbial communities and potential spread of antibiotic resistance. Environmental Pollution 271, 116317. https://doi.org/10.1016/j.envpol.2020.116317.

Martín, J., Santos, J.L., Aparicio, I., Alonso, E., 2015. Pharmaceutically active compounds in sludge stabilization treatments: anaerobic and aerobic digestion, wastewater stabilization ponds and composting. Science of the Total Environment 503, 97–104. https://doi.org/10.1016/j.scitotenv.2014.05.089.

Martínez, J., Coque, T., Baquero, F., 2014. What is a resistance gene? Ranking risk in resistomes. Nature Reviews Microbiology 13 (2), 116–123. https://doi.org/10.1038/nrmicro3399.

Monsalves, N., Leiva, A.M., Gómez, G., Vidal, G., 2022. Antibiotic-resistant gene behavior in constructed wetlands treating sewage: a critical review. Sustainability 14 (14), 8524. https://doi.org/10.3390/su14148524.

Morseletto, P., 2020. Targets for a circular economy. Resources, Conservation and Recycling 153, 104553. https://doi.org/10.1016/j.resconrec.2019.104553.

Murray, R., Tien, Y.C., Scott, A., Topp, E., 2019. The impact of municipal sewage sludge stabilization processes on the abundance, field persistence, and transmission of antibiotic resistant bacteria and antibiotic resistance genes to vegetables at harvest. Science of the Total Environment 651, 1680–1687. https://doi.org/10.1016/j.scitotenv.2018.10.030.

Murray, C.J., Ikuta, K.S., Sharara, F., Swetschinski, L., Aguilar, G.R., Gray, A., et al., 2022. Global burden of bacterial antimicrobial resistance in 2019: a systematic analysis. The Lancet 399 (10325), 629–655. https://doi.org/10.1016/S0140-6736(21)02724-0.

Neumann, P., Pesante, S., Venegas, M., Vidal, G., 2016. Developments in pre-treatment methods to improve anaerobic digestion of sewage sludge. Reviews in Environmental Science and Biotechnology 15 (2), 173–211. https://doi.org/10.1007/s11157-016-9396-8.

Neumann, P., González, Z., Vidal, G., 2017. Sequential ultrasound and low-temperature thermal pretreatment: process optimization and influence on sewage sludge solubilization, enzyme activity and anaerobic digestion. Bioresource Technology 234, 178–187. https://doi.org/10.1016/j.biortech.2017.03.029.

Nguyen, A., Vu, H., Nguyen, L., Wang, Q., Djordjevic, S., Donner, E., Yin, H., Nghiem, L., 2021. Monitoring antibiotic resistance genes in wastewater treatment: current strategies and future challenges. Science of the Total Environment 783, 146–964. https://doi.org/10.1016/j.scitotenv.2021.146964.

Pärnänen, K.M., Narciso-da-Rocha, C., Kneis, D., Berendonk, T.U., Cacace, D., Do, T.T., Manaia, C.M., 2019. Antibiotic resistance in European wastewater treatment plants mirrors the pattern of clinical antibiotic resistance prevalence. Science Advances 5 (3), 9124. https://doi.org/10.1126/sciadv.aau9124.

Pazda, M., Kumirska, J., Stepnowski, P., Mulkiewicz, E., 2019. Antibiotic resistance genes identified in wastewater treatment plant systems—a review. Science of the Total Environment 697 (134023), 134023. https://doi.org/10.1016/j.scitotenv.2019.134023.

Pei, J., Yao, H., Wang, H., Ren, J., Yu, X., 2016. Comparison of ozone and thermal hydrolysis combined with anaerobic digestion for municipal and pharmaceutical waste sludge with tetracycline resistance genes. Water Research 99, 122–128. https://doi.org/10.1016/j.watres.2016.04.058.

Piña, B., Bayona, J.M., Christou, A., Fatta-Kassinos, D., Guillon, E., Lambropoulou, D., Michael, C., Polesel, F., Sayen, S., 2020. On the contribution of reclaimed wastewater irrigation to the potential exposure of humans to antibiotics, antibiotic resistant bacteria and antibiotic resistance genes–NEREUS COST Action ES1403 position paper. Journal of Environmental Chemical Engineering 8 (1), 102131. https://doi.org/10.1016/j.jece.2018.01.011.

Rahube, T.O., Marti, R., Scott, A., Tien, Y.C., Murray, R., Sabourin, L., Duenk, P., Lapen, D.R., Topp, E., 2016. Persistence of antibiotic resistance and plasmid-associated genes in soil following application of sewage sludge and abundance on vegetables at harvest. Canadian Journal of Microbiology 62 (7), 600–607. https://doi.org/10.1139/cjm-2016-0034.

Redhead, S., Nieuwland, J., Esteves, S., Lee, D.-H., Kim, D.-W., Mathias, J., et al., 2020. Fate of antibiotic resistant *E. coli* and antibiotic resistance genes during full scale conventional and advanced anaerobic digestion of sewage sludge. PLoS One 15 (12), e0237283. https://doi.org/10.1371/journal.pone.0237283.

Sabri, N., Schmitt, H., van der Zaan, B., Gerritsen, H., Rijnaarts, H., Langenhoff, A., 2021. Performance of full scale constructed wetlands in removing antibiotics and antibiotic resistance genes. Science of the Total Environment 786, 147–368. https://doi.org/10.1016/j.scitotenv.2021.147368.

Syafiuddin, A., Boopathy, R., 2021. Role of anaerobic sludge digestion in handling antibiotic resistant bacteria and antibiotic resistance genes—a review. Bioresource Technology 330, 124970. https://doi.org/10.1016/j.biortech.2021.124970.

Tang, Z., Zhang, Y., Zhang, S., Gao, Y., Duan, Y., Zeng, T., Zhou, S., 2022. Temporal dynamics of antibiotic resistant bacteria and antibiotic resistance genes in activated sludge upon exposure to starvation. Science of the Total Environment 840 (156594), 156594. https://doi.org/10.1016/j.scitotenv.2022.156594.

Tong, J., Fang, P., Zhang, J., Wei, Y., Su, Y., Zhang, Y., 2019. Microbial community evolution and fate of antibiotic resistance genes during sludge treatment in two full-scale anaerobic digestion plants with thermal hydrolysis pretreatment. Bioresource Technology 288, 121575. https://doi.org/10.1016/j.biortech.2019.121575.

Tong, J., Liu, J., Zheng, X., Zhang, J., Ni, X., Chen, M., Wei, Y., 2016. Fate of antibiotic resistance bacteria and genes during enhanced anaerobic digestion of sewage sludge by microwave pretreatment. Bioresource Technology 217, 37–43. https://doi.org/10.1016/j.biortech.2016.02.130.

Urra, J., Alkorta, I., Mijangos, I., Epelde, L., Garbisu, C., 2019. Application of sewage sludge to agricultural soil increases the abundance of antibiotic resistance genes without altering the composition of prokaryotic communities. Science of the Total Environment 647, 1410–1420. https://doi.org/10.1016/j.scitotenv.2018.08.092.

Venegas, M., Leiva, A.M., Reyes-Contreras, C., Neumann, P., Piña, B., Vidal, G., 2021. Presence and fate of micropollutants during anaerobic digestion of sewage and their implications for the circular economy: a short review. Journal of Environmental Chemical Engineering 9 (1), 104931. https://doi.org/10.1016/j.jece.2020.104931.

Wang, J., Chu, L., Wojnárovits, L., Takács, E., 2020. Occurrence and fate of antibiotics, antibiotic resistant genes (ARGs) and antibiotic resistant bacteria (ARB) in municipal wastewater treatment plant: an overview. Science of the Total Environment 744, 140–997. https://doi.org/10.1016/j.scitotenv.2020.140997.

Wang, F.H., Qiao, M., Chen, Z., Su, J.Q., Zhu, Y.G., 2015. Antibiotic resistance genes in manure-amended soil and vegetables at harvest. Journal of Hazardous Materials 299, 215–221. https://doi.org/10.1016/j.jhazmat.2015.05.028.

Wengenroth, L., Berglund, F., Blaak, H., Chifiriuc, M.C., Flach, C.F., Pircalabioru, G.G., Larsson, D.G.J., Marutescu, L., van Passel, M.W.J., Popa, M., Radon, K., de Roda Husman, A.M., Rodríguez-Molina, D., Weinmann, T., Wieser, A., Schmitt, H., 2021. Antibiotic resistance in wastewater treatment plants and transmission risks for employees and residents: the concept of the AWARE study. Antibiotics 10 (5), 478. https://doi.org/10.3390/antibiotics10050478.

WHO, 2017. Antibacterial Agents in Clinical Development: An Analysis of the Antibacterial Clinical Development Pipeline, Including Tuberculosis. World Health Organization, 45 pp.

WHO, 2022. Global Antimicrobial Resistance and Use Surveillance System (GLASS) Report 2022. World Health Organization, 82 pp.

Xu, R., Yang, Z.H., Wang, Q.P., Bai, Y., Liu, J.B., Zheng, Y., Zhang, Y.R., Xiong, W.P., Ahmad, K., Fan, C.Z., 2018. Rapid startup of thermophilic anaerobic digester to remove tetracycline and sulfonamides resistance genes from sewage sludge. Science of the Total Environment 612, 788–798. https://doi.org/10.1016/j.scitotenv.2017.08.295.

Yang, L., Liu, W., Zhu, D., Hou, J., Ma, T., Wu, L., Zhu, Y., Christie, P., 2018. Application of biosolids drives the diversity of antibiotic resistance genes in soil and lettuce at harvest. Soil Biology and Biochemistry 122, 131–140. https://doi.org/10.1016/j.soilbio.2018.04.017.

Zhang, J., Sui, Q., Tong, J., Zhong, H., Wang, Y., Chen, M., Wei, Y., 2018. Soil types influence the fate of antibiotic-resistant bacteria and antibiotic resistance genes following the land application of sludge composts. Environment International 118, 34–43. https://doi.org/10.1016/j.envint.2018.05.029.

Zhang, J., Liu, J., Lu, T., Shen, P., Zhong, H., Tong, J., Wei, Y., 2019. Fate of antibiotic resistance genes during anaerobic digestion of sewage sludge: role of solids retention times in different configurations. Bioresource Technology 274, 488–495. https://doi.org/10.1016/j.biortech.2018.12.008.

Zhang, Y., Mao, Q., Su, Y.-A., Zhang, H., Liu, H., Fu, B., Su, Z., Wen, D., 2021. Thermophilic rather than mesophilic sludge anaerobic digesters possess lower antibiotic resistant genes abundance. Bioresource Technology 329, 124924. https://doi.org/10.1016/j.biortech.2021.124924.

Zheng, G., Lu, Y., Wang, D., Zhou, L., 2019. Importance of sludge conditioning in attenuating antibiotic resistance: removal of antibiotic resistance genes by bioleaching and chemical conditioning with Fe [III]/CaO. Water Research 152, 61–73. https://doi.org/10.1016/j.watres.2018.12.053.

2

Fate and behavior of microplastics in biosolids

Sartaj Ahmad Bhat[1], Zaw Min Han[2], Shiamita Kusuma Dewi[3], Guangyu Cui[4], Yongfen Wei[1] and Fusheng Li[1]

[1]RIVER BASIN RESEARCH CENTER, GIFU UNIVERSITY, GIFU, JAPAN; [2]GRADUATE SCHOOL OF ENGINEERING, GIFU UNIVERSITY, GIFU, JAPAN; [3]UNITED GRADUATED SCHOOL OF AGRICULTURAL SCIENCE, GIFU UNIVERSITY, GIFU, JAPAN; [4]SCHOOL OF ENVIRONMENT AND ENERGY, PEKING UNIVERSITY SHENZHEN GRADUATE SCHOOL, SHENZHEN, CHINA

1. Introduction

Microplastics (MPs) have surfaced as a cause of concern because of their small size (<5 mm) and serious pollutant interactions (Gigault et al., 2018). MPs are present all over, in the oceans, soils, wetlands, and in various organisms (Woodall et al., 2014; Rozman et al., 2021; Yu et al., 2021). Recent studies of MPs have focused on their impact on the marine ecosystem (Jambeck et al., 2015; Gola et al., 2021). However, researchers suggest that the release of plastic annually in the soil is around 4–23 times higher than in the ocean (Horton et al., 2017; Huang et al., 2020). Major sources of MPs in soil are plastic mulching, littering, atmospheric deposition, and the use of biosolids as fertilizer (Dris et al., 2016; Duis and Coors, 2016; Huang et al., 2020; Vithanage et al., 2021; Yang et al., 2021). More than 90% of MPs removed in wastewater treatment plants (WWTPs) are mostly retained in the sludge/biosolids (Sol et al., 2020; Reddy and Nair, 2022). It is a common trend to use this biosolid as a fertilizer on agricultural soils (Bhat et al., 2022). According to Zhang et al. (2020), the physical properties of soil are affected by MP incorporation. Similarly, MPs are transferred into roots via agricultural soil (Li et al., 2021). In addition, these MPs have been observed to hamper various physiological and gut microbiota attributes as well as bioaccumulate in soil fauna (Ju et al., 2019; Ya et al., 2021). MPs are persistent and could remain in the soil for a very long time, which can be a possible threat to ecological functions, soil biodiversity, global food production, and human health (Scheurer and Bigalke, 2018; Sajjad et al., 2022; Gudeta et al., 2023; Sharma et al., 2023). MPs are liable for several variations in the soil's physicochemical properties, together with porosity, plant growth, and microbial and enzymatic activities (Fei et al., 2020; Cheng et al., 2021; Sajjad et al., 2022). Qi et al. (2018) reported the increase of residual MPs in agricultural soils, which caused negative effects on the production of

Occurrence and Behavior of Emerging Contaminants in Organic Wastes and Their Control Strategies
https://doi.org/10.1016/B978-0-443-13585-9.00012-4

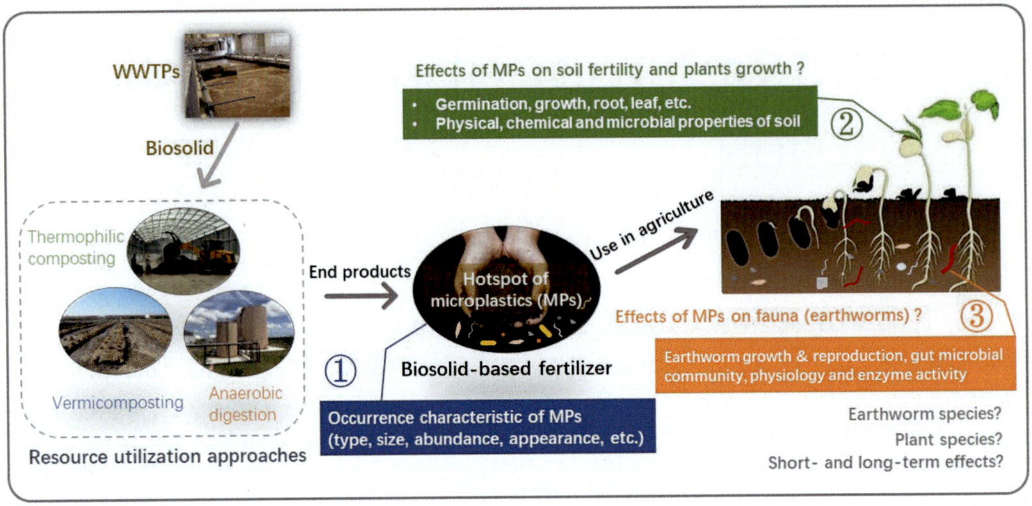

FIGURE 2.1 The occurrence characteristics of MPs and effect on soil fertility, plant growth and soil fauna.

wheat. The study also observed that the presence of earthworms had a positive effect on the growth of crops and mostly eased the damage done by plastic residues. According to Zhang et al. (2019), MPs have a large specific surface area; they can also absorb other organic contaminants in soil systems and shift them to terrestrial organisms, raising fears about the bioaccumulation of toxins through the food chain. Information about the MPs in biosolids and their impact on soil systems needs proper evaluation. The foremost sources of MPs in the soil are surface/wastewater and biosolids, but slightly is observed about their fate and their impact on soil, fauna, and plants. An investigation would help to evaluate the effects of microplastic on the soil properties, plant response, and physiology of earthworms as it is an emerging contaminant. Therefore, the present chapter aims to evaluate the microplastic concentration in biosolids and its effects on agricultural soils, plants, and earthworms. Through this chapter, we will help to understand the fate and behavior of MPs from biosolid to soil and how they affect plants and earthworms in the soil system. The occurrence characteristics of MPs and their effect on soil fertility, plant growth, and soil fauna are presented in Fig. 2.1.

2. Occurrence and characteristics of microplastics in biosolids

WWTPs can efficiently eliminate the MPs from the wastewater; however, these eliminated MPs can be accumulated in activated sludge/biosolids (Li et al., 2018; Corradini et al., 2019). According to Mahon et al. (2017), the concentration of MPs in sewage sludge of WWTPs can be as great as 15,385 particles kg^{-1}. According to Carr et al. (2016), MPS end up in the sludge, with concentrations differing greatly between findings, from 1.00

Table 2.1 The range of microplastic in sludge samples obtained from WWTPs in different countries.

S. No	Location	Microplastics size range	Concentration (MPs/kg dw of sludge)	Study
1	USA	45–400 μm	1×10^3	Carr et al. (2016)
2	Scotland	1.2–2 mm	1.20×10^3	Murphy et al. (2016)
3	Norway	54 μm–4.99 mm	$(1.70–19.84) \times 10^3$	Lusher et al. (2017)
4	Germany	<500 μm	$(1.00–24.00) \times 10^3$	Mintenig et al. (2017)
5	Korea	>300 μm	14.90×10^3	Lee and Kim (2018)
6	Finland	250 μm–5 mm	$(23.00 \pm 4.20) \times 10^3$	Lares et al. (2018)
7	Italy	10 μm–5.0 mm	$(113.00 \pm 57.00) \times 10^3$	Magni et al. (2019)
8	China	60 μm–4.2 mm	$(240.30 \pm 31.40) \times 10^3$	Liu et al. (2019)
9	Spain	36 μm–4.72 mm	$(133.00 \pm 59.00) \times 10^3$	Edo et al. (2020)
10	France	<500 μm	$(40.50 \pm 11.90) \times 10^3$	El Hayany et al. (2020)
11	Australia	>25 μm	$(15.90–45.70) \times 10^3$	Ziajahromi et al. (2021)
12	Italy	300 μm–5.0 mm	1.67×10^3	Pittura et al. (2021)
13	UK	<50 μm	1.61×10^{10} and 1.02×10^{10}	Harley-Nyang et al. (2022)
14	Spain	20 μm and 5 mm	4.23 ± 1.26 MPs/L to 0.075 ± 0.019 MPs/L	Dronjak et al. (2022)

MPs/gram (MPs/g) to 169,000 MPs/g. Independent of the nation or region, there are wide variations in the MP concentrations in the sludge. In WWTPs, sludge is further processed in digesters, dewatered, thickened, and finally dried. The treated sludge/biosolid contains important elements; therefore, it follows that they can be used for a variety of useful applications, including agricultural land applications. Studies reveal that, compared to MPs entering the oceans and freshwater sediments from various sources, the annual amount of MPs moving into the terrestrial environment through sludge use on land is significantly higher nowadays (Gong and Xie, 2020; Zhang et al., 2020). Soil MP contamination will primarily originate from the application of sludge/biosolids to farmland, which will increase the amount of MP in the agricultural soil (Edo et al., 2020; van den Berg et al., 2020). Organic manure, produced from biowaste and composting, can also serve as a vehicle for MPs to infiltrate the soil (Weithmann et al., 2018). According to Bläsing and Amelung (2018), compost may contain up to 1200 mg/kg of MPs. Table 2.1 shows the range of MPs in sludge samples obtained from WWTPs in different countries.

3. Effect of microplastics on soil fertility and plant growth

MPs in soil and their detrimental impact on the health and fertility of the soil have drawn increased attention in recent years. MPs can alter the physical, chemical, and microbiological characteristics of soil, but the outcomes vary and rely on the kind, shape, dose, and size of the polymer. Plant performance, agricultural yield, and crop safety may be threatened by MP-induced changes in soil fertility and the availability of contaminants. MP buildup in soil will unavoidably have an impact on soil characteristics, both directly

Table 2.2 Effects of microplastics on soil properties.

S. No	Soil property	Microplastics type/ dose/size	Effects	Study
1	Soil structure	PS/2.0%/547 −555 μm	Decreased water stable aggregates	de Souza Machado et al. (2019)
2	Soil porosity	PES/1.0%/<2 mm	The volume of <30 μm pores in PES soils were larger than that the control	Guo et al. (2021)
3	Soil bulk density	PES/0.4%/5000 μm	Bulk density decreased	de Souza Machado et al. (2018)
4	Soil water	PE/1.0%/10 mm	Water evaporation increased	Wan et al. (2019)
5	Soil pH	PS/0.5%/10−100 μm	Soil pH decreased	Dong et al. (2021)
6	Soil organic matter	PE/5.0%/<150 μm	No significant effect	Ren et al. (2020)
7	Soil nutrients	PLA/2.0%/20−50 μm	Increased NO_2^- and NO_3^- content	Chen et al. (2020a)
8	Soil pollutants	PE/10%/8.68 −500 μm	Soil Cd availability increased	Wang et al. (2021)

and indirectly. MPs have a hydrophobic surface, which may alter the soil's ability to hold and transport water (de Souza Machado et al., 2019). Three key indices of soil health are its physical, chemical, and microbial components (Kibblewhite et al., 2008). In addition to influencing soil microbes, terrestrial plants, and soil faunal assemblages, soil physical features such as soil structure, porosity, soil moisture, soil air, and aeration also interact with soil chemical properties. Studies have revealed that MPs have an impact on the function, metabolic rate, and community structure of soil microbes (Huang et al., 2019; Liu et al., 2021). MPs have been found to have an impact on plant performance (Qi et al., 2020; Zang et al., 2020), change the microbial community and activity in soil (Fei et al., 2020; Qi et al., 2020), and alter soil microbial activity (de Souza Machado et al., 2019); however, their effects greatly vary with polymer dose, type, and shape. The application of biochar on agricultural land may enhance the quality of the soil and plant productivity. Biochar's highly fragrant and porous qualities could raise the pH, amount of carbon, and nutrient retention of soil (Yi et al., 2020). In the soil-plant system, MPs are an emerging contaminant. Farmland normally contains biochar and bits of degradable mulching film. It is currently unknown how the soil-plant system would react to the combined effects of MPs and other contaminants. The effect of MPs on soil physical and chemical properties is presented in Table 2.2.

4. Effect of microplastics on earthworms

Earthworms have been frequently employed in recent years as a bioindicator of soil pollution to assess the effects of MPs on soil ecosystems. Earthworms have a crucial role in fertility, metabolism, and the preservation of the structure and functionality of soil ecosystems. They are significant creatures in the soil food chain (OECD, 2004).

Additionally, earthworms are highly reproducible and adaptable, yet they are also susceptible to harmful and dangerous environmental contaminants. Therefore, it is common practice to test the toxicity of contaminants using earthworms, particularly *Eisenia fetida* species of earthworms. MPs are highly hydrophobic and have a large specific surface area, which allows them to adsorb a range of contaminants from the surrounding environment (Horton et al., 2017). Thus, the combined effects of MPs and the pollutants that have been adsorbed on earthworms have received a lot of attention (Shi et al., 2020; Huang et al., 2021). According to several studies, earthworms exposed to MPs experience oxidative stress, histopathological alterations, neurotoxicities, DNA damage, and oxidative damage (Wang et al., 2019; Rodriguez-Seijo et al., 2017; Prendergast-Miller et al., 2019; Kwak and An, 2021). According to studies, polyethylene plastics in soil dramatically slowed *Lumbricus terrestris* development rates and increased inflammation and congestion in *Eisenia andrei*'s gut walls (Huerta Lwanga et al., 2016; Rodriguez-Seijo et al., 2017). Additionally, a number of studies (Rillig et al., 2017; Maaβ et al., 2017) have demonstrated that earthworms are capable of moving MPs from the topsoil layers to deeper soil layers, increasing contamination. Body weight is a crucial growth indicator for earthworms. Presently, the impacts of MPs on earthworm growth mostly depend on the concentration/size of MPs and the joint consequences of MPs and other contaminants (Hodson et al., 2017). Cao et al. (2017) observed that low concentrations of polystyrene MPs ($\leq$0.5%, w/w) have no impact on earthworm growth, but when exposed to high concentrations (1% and 2%, w/w), earthworm growth was hindered, and greater mortalities were observed. The accumulation of MPs in the gut and stomach of earthworms may be the primary cause of these negative effects. The accumulation can harm the immune systems of earthworms and subsequently affect their feeding habits and development (Rist et al., 2017; Yin et al., 2019). According to Zhou et al. (2020), exposure to MPs and heavy metals together had more detrimental effects on earthworm growth than exposure to MPs alone. This is because MPs have the potential to increase the bioavailability of heavy metal ions, which can also prevent earthworms from growing (Huang et al., 2021). It's possible that earthworms' ability to feed and store energy is hampered by ingesting MPs (Wright et al., 2013). Earthworms' avoidance behavior is a defense mechanism against soil pollutants like MPs. High MP concentrations can negatively alter soil structure, making it less conducive to earthworm movement (de Souza Machado et al., 2018). MPs have been revealed to negatively affect the reproduction of earthworms. According to Ding et al. (2021), the quantity of cocoons and hatchlings dramatically dropped as the number of MPs increased. The reproductive systems of earthworms may be harmed when they ingest MPs (Kwak and An, 2021). Earthworms exposed to MPs may experience oxidative stress, which could alter their enzyme activity (Jiang et al., 2020). Jiang et al. (2020) examined that superoxide dismutase activities were reduced and glutathione levels were raised in *E. fetida* when subjected to 100 and 1300 nm polystyrene particles. The effects of MPs on the growth, behavior, and oxidative stress of earthworms are presented in Table 2.3.

Table 2.3 Effects of microplastics on the growth, behavior, and oxidative stress of earthworms.

S. No	Earthworm species	Microplastics type/dose/size	Toxic effects	Study
1	*Eisenia fetida*	PE/1000 mg/kg/250–300 μm	Male reproductive organ damaged	Kwak and An (2021)
2	*Eisenia fetida*	PE/0–500 g/kg/<150 μm	Avoidance behavior and mortality increased	Ding et al. (2021)
3	*Eisenia fetida*	HDPE/0.25%/28–145 μm	Altered the activities of superoxide dismutase, catalase and glutathione S-transferase	Li et al. (2021)
4	*Eisenia fetida*	PS/1000 μg/kg/1300 nm	Superoxide dismutase activity decreased	Jiang et al. (2020)
5	*Eisenia fetida*	LDPE/1.5 g/kg/<400 μm	Catalase activities increased	Chen et al. (2020b)
6	*Lumbricus terrestris*	Microfiber/1.0%/35–45 μm	No change in mortality and avoidance responses	Prendergast-Miller et al. (2019)
7	*Enchytraeus crypticus*	Nylon/2.0%/63–150 μm	Reproduction reduced	Lahive et al. (2019)
8	*Eisenia fetida*	LDPE/1000 mg/kg/250–1000 μm	Glutathione S-transferase activities increased; catalase activities reduced	Rodríguez-Seijo et al. (2018)
9	*Eisenia fetida*	PS/2.0%/58 μm	Growth inhibited; mortality increased	Cao et al. (2017)
10	*Lumbricus terrestris*	LDPE/60%/<400 μm	Loss of weight; impact on burrow characteristics	Huerta Lwanga et al. (2017)

5. Conclusions and perspectives

This chapter discussed the occurrence and characteristics of MPs and their effects on agricultural soils, plants, and earthworms. The application of biosolids/sewage sludge, flooding, littering, and atmospheric input are a few of the sources of soil MPs. Earthworms have been known to preferentially ingest MPs with smaller particle sizes. MPs with higher concentrations and smaller sizes have more harmful effects on earthworms. MPs can also enable the buildup of heavy metals and other contaminants by earthworms and cause harsher damage. To reduce MPs in soils, the use of biodegradable plastics, solid waste management, and wastewater treatment technologies are essential. Therefore, future research should concentrate on field investigations, particularly on real microplastic exposure levels and their cumulative long-term consequences on earthworms from MPs and other pollutants. For the protection of terrestrial ecosystems, it is still a significant issue to figure out how to remove MPs directly from the soil.

Acknowledgment

Sartaj Ahmad Bhat acknowledges the Japan Society for the Promotion of Science (JSPS) for the JSPS International Postdoctoral Fellowship.

References

Bhat, S.A., Cui, G., Yaseera, N., Lei, X., Ameen, F., Li, F., 2022. Removal potential of microplastics in organic solid wastes via biological treatment approaches. In: Chowdhary, P., Mani, S., Chaturvedi, P. (Eds.), Microbial Biotechnology: Role in Ecological Sustainability and Research. https://doi.org/10.1002/9781119834489.ch14.

Bläsing, M., Amelung, W., 2018. Plastics in soil: analytical methods and possible sources. Science of the Total Environment 612, 422–435.

Cao, D., Wang, X., Luo, X., Liu, G., Zheng, H., 2017. Effects of polystyrene microplastics on the fitness of earthworms in an agricultural soil. IOP Conference Series: Earth and Environmental Science 61, 012148.

Carr, S.A., Liu, J., Tesoro, A.G., 2016. Transport and fate of microplastic particles in wastewater treatment plants. Water Research 91, 174–182.

Chen, H., Wang, Y., Sun, X., Peng, Y., Xiao, L., 2020a. Mixing effect of polylactic acid microplastic and straw residue on soil property and ecological function. Chemosphere 243, 125271.

Chen, Y., Liu, X., Leng, Y., Wang, J., 2020b. Defense responses in earthworms (*Eisenia fetida*) exposed to low-density polyethylene microplastics in soils. Ecotoxicology and Environmental Safety 187, 109788.

Cheng, Y., Song, W., Tian, H., Zhang, K., Li, B., Du, Z., Zhang, W., Wang, J., Wang, J., Zhu, L., 2021. The effects of high-density polyethylene and polypropylene microplastics on the soil and earthworm Metaphire guillelmi gut microbiota. Chemosphere 267, 129219.

Corradini, F., Meza, P., Eguiluz, R., Casado, F., Huerta-Lwanga, E., Geissen, V., 2019. Evidence of microplastic accumulation in agricultural soils from sewage sludge disposal. Science of the Total Environment 671, 411–420.

de Souza Machado, A.A., Lau, C.W., Till, J., Kloas, W., Lehmann, A., Becker, R., Rillig, M.C., 2018. Impacts of microplastics on the soil biophysical environment. Environmental Science and Technology 52 (17), 9656–9665.

de Souza Machado, A.A., Lau, C.W., Kloas, W., Bergmann, J., Bachelier, J.B., Faltin, E., Becker, R., Görlich, A.S., Rillig, M.C., 2019. Microplastics can change soil properties and affect plant performance. Environmental Science and Technology 53 (10), 6044–6052.

Ding, W., Li, Z., Qi, R., Jones, D.L., Liu, Q., Liu, Q., Yan, C., 2021. Effect thresholds for the earthworm *Eisenia fetida*: toxicity comparison between conventional and biodegradable microplastics. Science of the Total Environment 781, 146884.

Dong, Y., Gao, M., Qiu, W., Song, Z., 2021. Effect of microplastics and arsenic on nutrients and microorganisms in rice rhizosphere soil. Ecotoxicology and Environmental Safety 211, 111899.

Dris, R., Gasperi, J., Saad, M., Mirande, C., Tassin, B., 2016. Synthetic fibers in atmospheric fallout: a source of microplastics in the environment? Marine Pollution Bulletin 104 (1–2), 290–293.

Dronjak, L., Exposito, N., Rovira, J., Florencio, K., Emiliano, P., Corzo, B., Schuhmacher, M., Valero, F., Sierra, J., 2022. Screening of microplastics in water and sludge lines of a drinking water treatment plant in Catalonia, Spain. Water Research, 119185.

Duis, K., Coors, A., 2016. Microplastics in the aquatic and terrestrial environment: sources (with a specific focus on personal care products), fate and effects. Environmental Sciences Europe 28 (1), 1–25.

Edo, C., González-Pleiter, M., Leganés, F., Fernández-Piñas, F., Rosal, R., 2020. Fate of microplastics in wastewater treatment plants and their environmental dispersion with effluent and sludge. Environmental Pollution 259, 113837.

El Hayany, B., El Fels, L., Quénéa, K., Dignac, M.F., Rumpel, C., Gupta, V.K., Hafidi, M., 2020. Microplastics from lagooning sludge to composts as revealed by fluorescent staining-image analysis, Raman spectroscopy and pyrolysis-GC/MS. Journal of Environmental Management 275, 111249.

Fei, Y., Huang, S., Zhang, H., Tong, Y., Wen, D., Xia, X., Wang, H., Luo, Y., Barceló, D., 2020. Response of soil enzyme activities and bacterial communities to the accumulation of microplastics in an acid cropped soil. Science of the Total Environment 707, 135634.

Gigault, J., Ter Halle, A., Baudrimont, M., Pascal, P.Y., Gauffre, F., Phi, T.L., El Hadri, H., Grassl, B., Reynaud, S., 2018. Current opinion: what is a nanoplastic? Environmental Pollution 235, 1030–1034.

Gola, D., Tyagi, P.K., Arya, A., Chauhan, N., Agarwal, M., Singh, S.K., Gola, S., 2021. The impact of microplastics on marine environment: a review. Environmental Nanotechnology, Monitoring and Management 16, 100552.

Gong, J., Xie, P., 2020. Research progress in sources, analytical methods, eco-environmental effects, and control measures of microplastics. Chemosphere 254, 126790.

Guo, Q.Q., Xiao, M.R., Zhang, G.S., 2021. The persistent impacts of polyester microfibers on soil bio-physical properties following thermal treatment. Journal of Hazardous Materials 420, 126671.

Gudeta, K., Kumar, V., Bhagat, A., Julka, J.M., Bhat, S.A., Ameen, F., Qadri, H., Singh, S., Amarowicz, R., 2023. Ecological adaptation of earthworms for coping with plant polyphenols, heavy metals, and microplastics in the soil: a review. Heliyon 9, e14572.

Harley-Nyang, D., Memon, F.A., Jones, N., Galloway, T., 2022. Investigation and analysis of microplastics in sewage sludge and biosolids: a case study from one wastewater treatment works in the UK. Science of the Total Environment 823, 153735.

Hodson, M.E., Duffus-Hodson, C.A., Clark, A., Prendergast-Miller, M.T., Thorpe, K.L., 2017. Plastic bag derived-microplastics as a vector for metal exposure in terrestrial invertebrates. Environmental Science and Technology 51 (8), 4714–4721.

Horton, A.A., Walton, A., Spurgeon, D.J., Lahive, E., Svendsen, C., 2017. Microplastics in freshwater and terrestrial environments: evaluating the current understanding to identify the knowledge gaps and future research priorities. Science of the Total Environment 586, 127–141.

Huang, Y., Zhao, Y., Wang, J., Zhang, M., Jia, W., Qin, X., 2019. LDPE microplastic films alter microbial community composition and enzymatic activities in soil. Environmental Pollution 254, 112983.

Huang, Y., Liu, Q., Jia, W., Yan, C., Wang, J., 2020. Agricultural plastic mulching as a source of microplastics in the terrestrial environment. Environmental Pollution 260, 114096.

Huang, C., Ge, Y., Yue, S., Zhao, L., Qiao, Y., 2021. Microplastics aggravate the joint toxicity to earthworm *Eisenia fetida* with cadmium by altering its availability. Science of the Total Environment 753, 142042.

Huerta Lwanga, E., Gertsen, H., Gooren, H., Peters, P., Salánki, T., Van Der Ploeg, M., Besseling, E., Koelmans, A.A., Geissen, V., 2016. Microplastics in the terrestrial ecosystem: implications for Lumbricus terrestris (Oligochaeta, Lumbricidae). Environmental Science and Technology 50 (5), 2685–2691.

Huerta Lwanga, E., Gertsen, H., Gooren, H., Peters, P., Salanki, T., van der Ploeg, M., Besseling, E., Koelmans, A.A., Geissen, V., 2017. Incorporation of microplastics from litter into burrows of Lumbricus terrestris. Environmental Pollution 220 (Pt A), 523–531.

Jambeck, J.R., Geyer, R., Wilcox, C., Siegler, T.R., Perryman, M., Andrady, A., Narayan, R., Law, K.L., 2015. Plastic waste inputs from land into the ocean. Science 347 (6223), 768–771.

Jiang, X., Chang, Y., Zhang, T., Qiao, Y., Klobucar, G., Li, M., 2020. Toxicological effects of polystyrene microplastics on earthworm (*Eisenia fetida*). Environmental Pollution 259, 113896.

Ju, H., Zhu, D., Qiao, M., 2019. Effects of polyethylene microplastics on the gut microbial community, reproduction and avoidance behaviors of the soil springtail, Folsomia candida. Environmental Pollution 247, 890–897.

Kibblewhite, M.G., Ritz, K., Swift, M.J., 2008. Soil health in agricultural systems. Philosophical Transactions of the Royal Society B: Biological Sciences 363 (1492), 685–701.

Kwak, J.I., An, Y.J., 2021. Microplastic digestion generates fragmented nanoplastics in soils and damages earthworm spermatogenesis and coelomocyte viability. Journal of Hazardous Materials 402, 124034.

Lahive, E., Walton, A., Horton, A.A., Spurgeon, D.J., Svendsen, C., 2019. Microplastic particles reduce reproduction in the terrestrial worm *Enchytraeus crypticus* in a soil exposure. Environmental Pollution 255, 113174.

Lares, M., Ncibi, M.C., Sillanpää, M., Sillanpää, M., 2018. Occurrence, identification and removal of microplastic particles and fibers in conventional activated sludge process and advanced MBR technology. Water Research 133, 236–246.

Lee, H., Kim, Y., 2018. Treatment characteristics of microplastics at biological sewage treatment facilities in Korea. Marine Pollution Bulletin 137, 1–8.

Li, X., Chen, L., Mei, Q., Dong, B., Dai, X., Ding, G., Zeng, E.Y., 2018. Microplastics in sewage sludge from the wastewater treatment plants in China. Water Research 142, 75–85.

Li, H.Z., Zhu, D., Lindhardt, J.H., Lin, S.M., Ke, X., Cui, L., 2021. Long-term fertilization history alters effects of microplastics on soil properties, microbial communities, and functions in diverse farmland ecosystem. Environmental Science and Technology 55 (8), 4658–4668.

Liu, X., Yuan, W., Di, M., Li, Z., Wang, J., 2019. Transfer and fate of microplastics during the conventional activated sludge process in one wastewater treatment plant of China. Chemical Engineering Journal 362, 176–182.

Liu, Y., Huang, Q., Hu, W., Qin, J., Zheng, Y., Wang, J., Wang, Q., Xu, Y., Guo, G., Hu, S., Xu, L., 2021. Effects of plastic mulch film residues on soil-microbe-plant systems under different soil pH conditions. Chemosphere 267, 128901.

Lusher, A.L., Hurley, R., Vogelsang, C., Nizzetto, L., Olsen, M., 2017. Mapping Microplastics in Sludge. Norsk Institutt for Vannforskning. NIVA-Rapport, p. 7215.

Maaß, S., Daphi, D., Lehmann, A., Rillig, M.C., 2017. Transport of microplastics by two collembolan species. Environmental Pollution 225, 456–459.

Magni, S., Binelli, A., Pittura, L., Avio, C.G., Della Torre, C., Parenti, C.C., Gorbi, S., Regoli, F., 2019. The fate of microplastics in an Italian wastewater treatment plant. Science of the Total Environment 652, 602–610.

Mahon, A.M., O'Connell, B., Healy, M.G., O'Connor, I., Officer, R., Nash, R., Morrison, L., 2017. Microplastics in sewage sludge: effects of treatment. Environmental Science and Technology 51 (2), 810–818.

Mintenig, S.M., Int-Veen, I., Löder, M.G., Primpke, S., Gerdts, G., 2017. Identification of microplastic in effluents of waste water treatment plants using focal plane array-based micro-Fourier-transform infrared imaging. Water Research 108, 365–372.

Murphy, F., Ewins, C., Carbonnier, F., Quinn, B., 2016. Wastewater treatment works (WwTW) as a source of microplastics in the aquatic environment. Environmental Science and Technology 50 (11), 5800–5808.

OECD, 2004. Earthworm Reproduction Test. Guideline for Testing Chemicals. No. 222. OECD, Paris, France.

Pittura, L., Foglia, A., Akyol, Ç., Cipolletta, G., Benedetti, M., Regoli, F., Eusebi, A.L., Sabbatini, S., Tseng, L.Y., Katsou, E., Gorbi, S., 2021. Microplastics in real wastewater treatment schemes: comparative assessment and relevant inhibition effects on anaerobic processes. Chemosphere 262, 128415.

Prendergast-Miller, M.T., Katsiamides, A., Abbass, M., Sturzenbaum, S.R., Thorpe, K.L., Hodson, M.E., 2019. Polyester-derived microfibre impacts on the soil-dwelling earthworm Lumbricus terrestris. Environmental Pollution 251, 453–459.

Qi, Y., Ossowicki, A., Yang, X., Lwanga, E.H., Dini-Andreote, F., Geissen, V., Garbeva, P., 2020. Effects of plastic mulch film residues on wheat rhizosphere and soil properties. Journal of Hazardous Materials 387, 121711.

Qi, Y., Yang, X., Pelaez, A.M., Lwanga, E.H., Beriot, N., Gertsen, H., Garbeva, P., Geissen, V., 2018. Macro- and micro-plastics in soil-plant system: effects of plastic mulch film residues on wheat (*Triticum aestivum*) growth. Science of the Total Environment 645, 1048–1056.

Reddy, A.S., Nair, A.T., 2022. The fate of microplastics in wastewater treatment plants: an overview of source and remediation technologies. Environmental Technology and Innovation, 102815.

Ren, X., Tang, J., Liu, X., Liu, Q., 2020. Effects of microplastics on greenhouse gas emissions and the microbial community in fertilized soil. Environmental Pollution 256, 113347.

Rillig, M.C., Ziersch, L., Hempel, S., 2017. Microplastic transport in soil by earthworms. Scientific Reports 7 (1), 1–6.

Rist, S., Baun, A., Hartmann, N.B., 2017. Ingestion of micro- and nanoplastics in Daphnia magna—quantification of body burdens and assessment of feeding rates and reproduction. Environmental Pollution 228, 398–407.

Rodriguez-Seijo, A., Lourenço, J., Rocha-Santos, T.A.P., Da Costa, J., Duarte, A.C., Vala, H., Pereira, R., 2017. Histopathological and molecular effects of microplastics in *Eisenia andrei* Bouché. Environmental Pollution 220, 495–503.

Rodríguez-Seijo, A., da Costa, J.P., Rocha-Santos, T., Duarte, A.C., Pereira, R., 2018. Oxidative stress, energy metabolism and molecular responses of earthworms (*Eisenia fetida*) exposed to low-density polyethylene microplastics. Environmental Science and Pollution Research 25 (33), 33599–33610.

Rozman, U., Turk, T., Skalar, T., Zupančič, M., Korošin, N.Č., Marinšek, M., Olivero-Verbel, J., Kalčíková, G., 2021. An extensive characterization of various environmentally relevant microplastics—material properties, leaching and ecotoxicity testing. Science of the Total Environment 773, 145576.

Sajjad, M., Huang, Q., Khan, S., Khan, M.A., Yin, L., Wang, J., Lian, F., Wang, Q., Guo, G., 2022. Microplastics in the soil environment: a critical review. Environmental Technology and Innovation 27, 102408.

Scheurer, M., Bigalke, M., 2018. Microplastics in Swiss floodplain soils. Environmental Science & Technology 52 (6), 3591–3598.

Sharma, U., Sharma, S., Rana, V.S., Rana, N., Kumar, V., Sharma, S., Qadri, H., Kumar, V., Bhat, S.A., 2023. Assessment of microplastics pollution on soil health and eco-toxicological risk in horticulture. Soil Systems 7 (1), 7. https://doi.org/10.3390/soilsystems7010007.

Shi, Z., Wen, M., Zhang, J., Tang, Z., Wang, C., 2020. Effect of phenanthrene on the biological characteristics of earthworm casts and their relationships with digestive and anti-oxidative systems. Ecotoxicology and Environmental Safety 193, 110359.

Sol, D., Laca, A., Laca, A., Díaz, M., 2020. Approaching the environmental problem of microplastics: importance of WWTP treatments. Science of the Total Environment 740, 140016.

van den Berg, P., Huerta-Lwanga, E., Corradini, F., Geissen, V., 2020. Sewage sludge application as a vehicle for microplastics in eastern Spanish agricultural soils. Environmental Pollution 261, 114198.

Vithanage, M., Ramanayaka, S., Hasinthara, S., Navaratne, A., 2021. Compost as a carrier for micro-plastics and plastic-bound toxic metals into agroecosystems. Current Opinion in Environmental Science and Health 24, 100297.

Wan, Y., Wu, C., Xue, Q., Hui, X., 2019. Effects of plastic contamination on water evaporation and desiccation cracking in soil. Science of the Total Environment 654, 576–582.

Wang, J., Coffin, S., Sun, C., Schlenk, D., Gan, J., 2019. Negligible effects of microplastics on animal fitness and HOC bioaccumulation in earthworm *Eisenia fetida* in soil. Environmental Pollution 249, 776–784.

Wang, F., Wang, X., Song, N., 2021. Polyethylene microplastics increase cadmium uptake in lettuce (Lactuca sativa L.) by altering the soil microenvironment. Science of the Total Environment 784, 147133.

Weithmann, N., Möller, J.N., Löder, M.G., Piehl, S., Laforsch, C., Freitag, R., 2018. Organic fertilizer as a vehicle for the entry of microplastic into the environment. Science Advances 4 (4) eaap8060.

Woodall, L.C., Sanchez-Vidal, A., Canals, M., Paterson, G.L., Coppock, R., Sleight, V., Calafat, A., Rogers, A.D., Narayanaswamy, B.E., Thompson, R.C., 2014. The deep sea is a major sink for microplastic debris. Royal Society Open Science 1 (4), 140317.

Wright, S.L., Rowe, D., Thompson, R.C., Galloway, T.S., 2013. Microplastic ingestion decreases energy reserves in marine worms. Current Biology 23 (23), R1031–R1033.

Ya, H., Jiang, B., Xing, Y., Zhang, T., Lv, M., Wang, X., 2021. Recent advances on ecological effects of microplastics on soil environment. Science of the Total Environment 798, 149338.

Yang, L., Zhang, Y., Kang, S., Wang, Z., Wu, C., 2021. Microplastics in soil: a review on methods, occurrence, sources, and potential risk. Science of the Total Environment 780, 146546.

Yi, Q., Liang, B., Nan, Q., Wang, H., Zhang, W., Wu, W., 2020. Temporal physicochemical changes and transformation of biochar in a rice paddy: insights from a 9-year field experiment. Science of the Total Environment 721, 137670.

Yin, L., Liu, H., Cui, H., Chen, B., Li, L., Wu, F., 2019. Impacts of polystyrene microplastics on the behavior and metabolism in a marine demersal teleost, black rockfish (Sebastes schlegelii). Journal of Hazardous Materials 380, 120861.

Yu, H., Qi, W., Cao, X., Hu, J., Li, Y., Peng, J., Hu, C., Qu, J., 2021. Microplastic residues in wetland ecosystems: do they truly threaten the plant-microbe-soil system? Environment International 156, 106708.

Zang, H., Zhou, J., Marshall, M.R., Chadwick, D.R., Wen, Y., Jones, D.L., 2020. Microplastics in the agroecosystem: are they an emerging threat to the plant-soil system? Soil Biology and Biochemistry 148, 107926.

Zhang, B., Yang, X., Chen, L., Chao, J., Teng, J., Wang, Q., 2020. Microplastics in soils: a review of possible sources, analytical methods and ecological impacts. Journal of Chemical Technology and Biotechnology 95 (8), 2052–2068.

Zhang, S., Wang, J., Liu, X., Qu, F., Wang, X., Wang, X., Li, Y., Sun, Y., 2019. Microplastics in the environment: a review of analytical methods, distribution, and biological effects. TrAC, Trends in Analytical Chemistry 111, 62–72.

Ziajahromi, S., Neale, P.A., Silveira, I.T., Chua, A., Leusch, F.D., 2021. An audit of microplastic abundance throughout three Australian wastewater treatment plants. Chemosphere 263, 128294.

Zhou, Y., Liu, X., Wang, J., 2020. Ecotoxicological effects of microplastics and cadmium on the earthworm *Eisenia foetida*. Journal of Hazardous Materials 392, 122273.

3

Occurrence, fate, detection, ecological impact and mitigation of antimicrobial resistance genes derived from animal waste

Muhammad Adil[1] and Pragya Tiwari[2]

[1]PHARMACOLOGY & TOXICOLOGY SECTION, UNIVERSITY OF VETERINARY & ANIMAL SCIENCES, LAHORE, JHANG CAMPUS, JHANG, PAKISTAN; [2]DEPARTMENT OF BIOTECHNOLOGY, YEUNGNAM UNIVERSITY, GYEONGSAN, GYEONGBUK, REPUBLIC OF KOREA

1. Introduction

Antimicrobial resistance is a global issue of environmental and public health significance. Antimicrobial drugs are utilized for prophylactic, therapeutic and/or growth promotion purposes in veterinary medicine. The worldwide consumption of antimicrobial agents in livestock sector was greater than 63,000 tons in 2010 (Checcucci et al., 2020). Accordingly, the global utilization of antimicrobials in animal production is expected to reach 150,000 tons until the year 2030 on account of progressively growing global population and consequent rise in the demand for animal-derived food items (Tasho and Cho, 2016). Around 150 antimicrobial agents are recently consumed for livestock production and healthcare, and their residual concentrations are transferred to agricultural land owing to the utilization of animal manure as fertilizer (Checcucci et al., 2020). Antimicrobial growth promoters are usually administered at sub-therapeutic concentrations for longer periods of time to enhance the feed conversion ratio of poultry and young animals, including broilers and beef calves. However, the consumption of antimicrobial growth promoters is no longer permitted in Australia and the European Union since 2006 (Checcucci et al., 2020). Conversely, antimicrobial growth promoters are still utilized in several other countries of the world. In contrast to cattle and fish farming, the prophylactic or therapeutic usage of antimicrobials and the subsequent plenitude of antimicrobial resistance genes are comparatively higher in case of swine and poultry production units (Kim et al., 2011). The excretion rates of various antimicrobial drugs used in veterinary practice have been listed in Table 3.1. The

Occurrence and Behavior of Emerging Contaminants in Organic Wastes and Their Control Strategies
https://doi.org/10.1016/B978-0-443-13585-9.00006-9

Table 3.1 Excretion rates of various antimicrobial drugs used in veterinary practice.

Antimicrobial agent	Excretion rate	References
Erythromycin	5–10	McArdell et al. (2003)
Sulphamethoxine	15	Jjemba (2002)
Oxytetracycline	21	Montforts (1999)
Virginiamycin	0–31	Jjemba (2002)
Ampicillin	60	Hirsch et al. (1999)
Chlortetracycline	65	Arikan et al. (2009)
Streptomycin	66	Jjemba (2002)
Amoxicillin	90	Park and Choi (2008)
Difloxacin	90	Sukul et al. (2009)

majority of the antimicrobial drugs are either excreted unchanged from the animal bodies as parent compounds or as bioactive metabolic products in animal manure (Sarmah et al., 2006).

Bacteria may either intrinsically possess the antimicrobial resistance by undergoing spontaneous mutation or alternatively acquire it through the reception of specific antimicrobial resistance genes from other bacteria via horizontal gene transfer (Woodford and Ellington, 2007). Target modification, efflux-mediated drug extrusion and enzymatic inactivation of antimicrobials constitute the three common mechanisms of antimicrobial drug resistance (Manaia, 2017). Mutations are essentially required for the evolution of antimicrobial resistance genes and the production of multiple types of variants that exhibit remarkable similarity and enormous ecotoxic potential (Woodford and Ellington, 2007). Transformation, conjugation and transduction are the three basic mechanisms of horizontal gene transfer for antimicrobial resistance genes from animal waste. Conjugation represents the most prevalent pathway, whereas transduction is the least occurring mechanism of horizontal gene transfer (Wang et al., 2018). Conjugative plasmids, usually mediating the total quantity of resistance genes in an ecosystem, are referred to as resistomes, which enhance the survival of microorganisms in hostile environments through conferring resistance against heavy metals and antimicrobials (Bennett, 2008).

Mobile genetic elements usually transport the antimicrobial resistance genes. The persistence and spread of several types of human commensals as well as pathogenic bacteria typically occur following the horizontal transfer of mobile genetic elements (Checcucci et al., 2020). Animal waste usually contains the mobile genetic elements, for instance, transposons, integrons and plasmids, that are critically required for conjugative horizontal gene transfer (Pärnänen et al., 2016). Integrons consisting of an integration domain for external gene cassettes and an integrase-encoding gene can be readily acquired by microorganisms (particularly bacteria) and subjected to functional expression (Mazel, 2006). The correlation of integrons with transposons or plasmids enables their horizontal dispersion across the bacterial populations (Rowe-Magnus and Mazel, 2002). The transmission of antimicrobial resistance genes through horizontal gene transfer has been frequently documented in several types of animal production units (Hammerum et al., 2006). The co-occurrence of antimicrobial residues and increased density of microorganisms in animal manure enhances the prospects of horizontal gene transfer (Blau et al., 2018). Moreover, bacteriophages and extracellular DNA from animal waste have been linked with the transmission of antimicrobial resistance genes via transduction and transformation mechanisms (Zeng and Lin, 2017).

Around 60 forms of antimicrobial resistance genes have been recorded in different types of animal wastes, such as manure, sediments, lagoon slurry and wastewater (Yuan et al., 2019). Table 3.2 indicates the several types of antimicrobial resistance genes documented in animal waste. Frequently recorded antimicrobial resistance genes in animal waste include *bla, fca, erm, sul* and *tet* (He et al., 2020). Mobile genetic elements encompassing transposons, integrons and plasmids are extensively found in animal waste (Johnson et al., 2016). Resistance against the major classes of antimicrobial agents has been associated with nine specific types of antimicrobial resistance genes, including *aaa* for aminoglycosides, *bla* for β-lactams, *fca* for fluoroquinolones, quinolones, florfenicol, and chloramphenicol, *erm* for macrolides, lincosamides, and streptogramins B, *mcr* for colistin, *mdr* for multidrugs, *sul* for sulphonamides, *tet* for tetracyclines, and *van* for vancomycin, respectively (He et al., 2020). Despite the extensive research on various aspects of antimicrobial resistance, the causal role of animal waste has been insufficiently addressed. Keeping in view the global significance of antimicrobial resistance, the current chapter outlines the occurrence, fate, detection methods, ecological impact and mitigation of antimicrobial resistance genes derived from animal waste.

Table 3.2 Various types of antimicrobial resistance genes recorded in animal waste.

Antimicrobial agent(s)	Antimicrobial resistance genes	References
Aminoglycosides	*acc*	Tien et al. (2017)
β-lactams	*bla*	Gou et al. (2018)
Chloramphenicol	*cfr, cmlA, fexA, floR*	Cui et al. (2016)
Colistin	*mcr*	Yuan et al. (2019)
Macrolide-lincosamide-streptogramin B (MLSB)	*erm*	Tien et al. (2017)
Quinolones, Amphenicols	*fca*	Gou et al. (2018)
Streptogramin	*satA/satG*	Werner et al. (2000)
Sulphonamides	*sul*	Gou et al. (2018)
Multi drugs	*mdr*	Qian et al. (2018)
Tetracyclines	*tet*	Tien et al. (2017)
Vancomycin	*van*	Gou et al. (2018)

2. Occurrence and fate of antimicrobial resistance genes in animal waste

Intensive poultry and livestock farming is regarded as the leading cause of enhanced environmental transmission and persistence of antimicrobial residues. The environmental contamination of antimicrobial residues and propagation of antimicrobial resistance genes have been attributed to the enhanced incidence of resistant microorganisms, including commensal as well as pathogenic bacteria. Fig. 3.1 illustrates the environmental fate of antimicrobial resistance genes derived from animal waste (manure). The propagation of antimicrobial resistance genes, antimicrobial residues and antimicrobial-resistant bacteria occurs following the consumption of animal manure as fertilizer on agricultural land. Antimicrobial resistance genes and antimicrobial residues can be transferred from animal manure to the human microbiota as a result of horizontal gene transfer induced by mobile genetic elements (Jechalke et al., 2014). The contamination of soil with manure-containing antimicrobial residues substantially enhances the retention and horizontal dispersion of antimicrobial resistance genes by means of plasmids and integrons, thus promoting the environmental propagation of antimicrobial-resistant bacteria (Sengeløv et al., 2003). Despite the failure of particular bacteria to undergo environmental adaptation, antimicrobials promote the dissemination of certain bacterial taxa in soil by diminishing the growth of other microorganisms and exerting positive selection pressure (Ding et al., 2014). The adsorption of antimicrobial residues with soil components leads to their environmental accumulation and prolonged retention in agricultural land (Du and Liu, 2012).

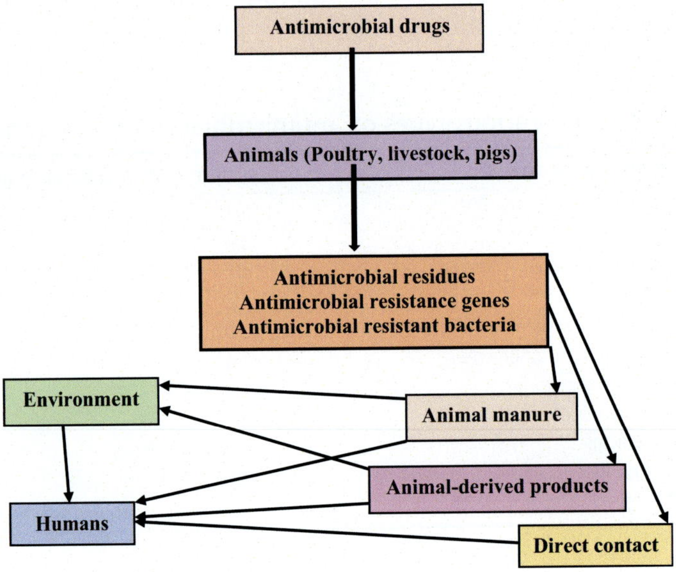

FIGURE 3.1 Environmental fate of antimicrobial resistance genes derived from animal waste.

From upstream water to background soil, the concentration of antimicrobial resistance genes was 28,000 times greater in various types of animal wastes, such as sediments, lagoon slurry, wastewater and solids used for manure synthesis (Zhu et al., 2013). Wastewater derived from chicken and swine farms exhibited a significantly higher level of antimicrobial resistance genes in comparison to municipal and hospital wastewaters, whereas, the prevalence of antimicrobial resistance genes in wastewater obtained from fish and cattle farms was equivalent to that of municipal and hospital wastewaters (He et al., 2020). Moreover, livestock waste contained significantly higher levels of antimicrobial resistance genes than human waste (Ekpeghere et al., 2017). Varying dose rates and consumption of antimicrobials have been postulated as the underlying causes of considerable discrepancies in the levels of antimicrobial resistance genes among the several types of animal wastes. Generally, the waste material of poultry and swine origins exhibits a relatively higher concentration of antimicrobial resistance genes than that collected from fish and cattle farms (Huang et al., 2019). Furthermore, the greatest diversity in antimicrobial resistance genes was recorded in the excretory material derived from poultry and swine farms (Qian et al., 2018).

The introduction of manure-based antimicrobial resistance genes is accredited for enhancing their abundance and diversity in the soil by almost 10^5-fold (Han et al., 2018). The single-time manure application has been correlated with the persistence of antimicrobial resistance genes for more than 120 days in the soil, which may eventually require at least 3–6 months for reduction to a safer limit. Nevertheless, these values may differ with the application techniques and types of manure. Free and intracellular antimicrobial resistance genes located in the air, water and soil can be transmitted to indigenous bacteria through the phenomenon of horizontal gene transfer (Fang et al., 2018; Sancheza et al., 2016). Water-borne antimicrobial resistance genes can be acquired by the intestinal microflora of fish and other aquatic organisms (Marti et al., 2018). The propagation of antimicrobial-resistant bacteria and their resistance genes is more likely in areas with poor sewage conditions and inadequate water treatment (Collignon and McEwen, 2019). Antimicrobial residues, antimicrobial resistance genes and resistant bacteria may also be transmitted to humans through the utilization of contaminated animal-derived food products, including eggs, meat and milk (Marshall and Levy, 2011).

3. Detection and quantification of antimicrobial resistance

Antimicrobial sensitivity testing is routinely carried out in the research as well as clinical microbiology laboratories for examining the susceptibility pattern of resistant pathogens against commonly used antimicrobial agents (Tang et al., 2014). Antimicrobial resistance can be determined through a variety of simple as well as highly sophisticated techniques (Fig. 3.2). Frequently used methods of antimicrobial sensitivity assessment encompass the disk diffusion test (Kirby-Baur technique), agar well-diffusion test, broth macrodilution test (tube dilution test), broth microdilution test, E-test strips and automated

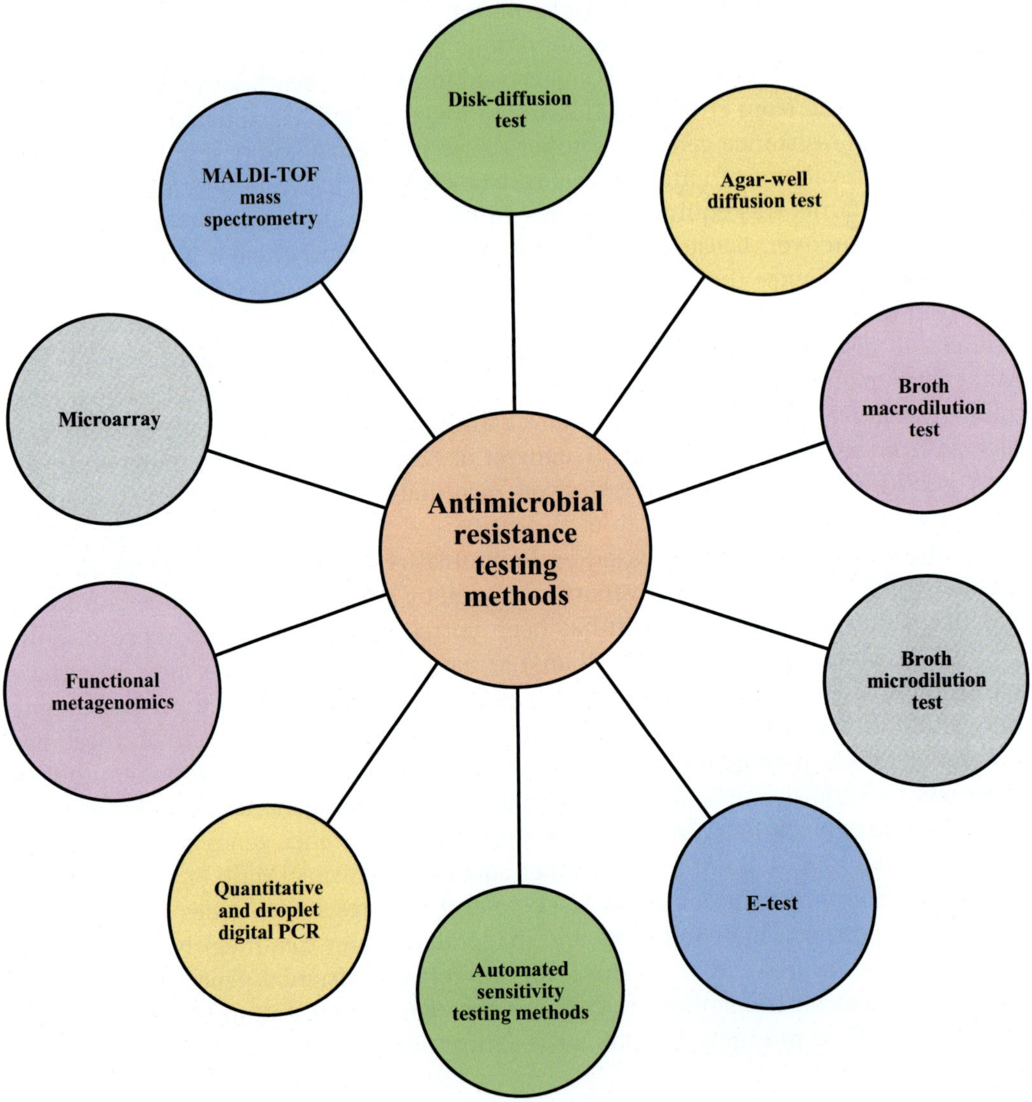

FIGURE 3.2 Different methods for the assessment of antimicrobial resistance.

systems (March-Rosselló, 2017). For this purpose, a particular amount of test microorganisms is grown on a culture medium containing sequentially varying concentrations of antimicrobial drug(s), and the endpoint is determined either in terms of the zone(s) of inhibition or the presence/absence of turbidity. Nevertheless, the precise mechanisms of antimicrobial resistance cannot be described using phenotypic-based methods of susceptibility assessment. Moreover, the susceptibility patterns of non-culturable environmental bacteria cannot be determined by means of phenotypic techniques (Franklin et al., 2016). Quantitative polymerase chain reaction (qPCR) is a culture-independent technique for analyzing antimicrobial resistance genes in environmental DNA samples. However, DNA extraction from a large number of diverse DNA samples is essential for qPCR-based comparison (Franklin et al., 2016). Alternatively, the samples can be subjected to droplet digital PCR for the accurate and independent estimation of nucleic acids without using a standard curve (Martinez-Hernandez et al., 2019).

Currently, more accurate and cost-effective methods like whole-genome sequencing are preferred for detecting the antimicrobial resistance (Holden et al., 2013). Metagenomics can be applied for detecting the antimicrobial resistance genes and determining the entire resistome in an environment with the help of known sequences of resistance genes. Metagenomic data can be used for reconstructing complete or partial genomes and estimating the environmental occurrence of antimicrobial resistance genes (Hultman et al., 2015). Functional metagenomics involves the cloning of environment-derived DNA in suitable laboratory organisms (for example, *E. coli*) and subsequently determining the sensitivity of the host to different antimicrobial drugs. Sub-cloning, mutagenesis or silicon analysis can be used for further assessment of the clones, demonstrating the phenotypic resistance (Iwu et al., 2020). Functional metagenomics is superior to PCR and metagenomics as it does not need the former knowledge regarding antimicrobial resistance genes. Microarrays can be applied to determine several antimicrobial resistance genes, including carbapenemases, ampicillinase C and extended-spectrum β-lactamases (Iwu et al., 2020). Based on the detection of bacterial hydrolytic enzymes, the immunochromatographic technique offers an economical, rapid and easily available system for antimicrobial sensitivity analysis (March-Rosselló, 2017). Nevertheless, bacterial hydrolytic enzymes cannot be properly characterized by means of this method. Besides, colorimetric methods are also available for detecting the bacterial hydrolytic enzymes. Clindamycin and chloramphenicol resistance can be projected through MALDI-TOF mass spectrometry method, which is based upon the detection of 16S ribosomal RNA methylation using the methyltransferase enzymes (Iwu et al., 2020).

4. Ecological impact of animal waste-derived antimicrobial resistance genes

The currently practiced misuse and overuse of antimicrobial agents in animal production, health care and human medicine represent serious public health concerns. Consequently, animal waste constitutes an important hotspot for the environmental contamination of antimicrobial residues, heavy metals and antimicrobial resistance genes. The sub-inhibitory concentrations of antimicrobials in manure can influence the regulatory and expression levels of genes as well as the complex interactions among different bacterial strains (Brüssow, 2015). Moreover, soil microbial diversity can be altered or disrupted through the environmental contamination of antimicrobial residues (Han et al., 2018; Kemper, 2008). Manure-derived antimicrobial-resistant bacteria and antimicrobial residues may deleteriously influence the growth and development of crops as well as the quality of vegetables, fruits and other agricultural products (Muhammad et al., 2020; Zhou et al., 2020). Vegetables and fruits grown on manure-fertilized soil usually contain antimicrobial-resistant bacteria and antimicrobial residues. The utilization of contaminated fruits and vegetables in raw form leads to the ingestion of antimicrobial-resistant bacteria that may invade the human gastrointestinal tract or alternatively get excreted in the stool and constitute a serious public health hazard (Blaak et al., 2014). The growing incidence of severe and difficult-to-treat microbial infections has been attributed to the issue of antimicrobial resistance.

The extensive application of antimicrobials in poultry and livestock production has been linked with the globally recorded several types of antimicrobial resistance genes in animal waste (He et al., 2020). The partial elimination of antimicrobial resistance genes by conventional waste processing techniques results in their introduction into water and soil. Following their transmission to the human body through several routes of exposure, including ingestion and inhalation, the antimicrobial-resistant bacteria can cause clinical infections that may remain refractory to conventional antimicrobial therapy. The global mortality ascribed to antimicrobial-resistant infections was documented as 700,000 per annum in 2014 and is projected to reach 10 million until the year 2050, whereas, productivity losses and cumulative healthcare costs were estimated as 103 trillion US dollars (He et al., 2020). The concurrent transmission of genes conferring resistance to heavy metals and antimicrobials through mobile genetic elements has been recorded in the human gut, sediments and soil, and consequently associated with the environmental maintenance of antimicrobial resistance genes (Rosewarne et al., 2010). Although the ingestion and inhalation of contaminated material can give rise to the transmission of antimicrobial-resistance genes and antimicrobial-resistant bacteria into the human body, the precise underlying mechanism has not been elucidated (Marti et al., 2013; Sanganyado and Gwenzi, 2019). Normally, the gut microbiota inhibits the colonization of the human body by foreign, pathogenic microorganisms. However, the use of antimicrobial drugs even at residual concentrations may lead to the destruction of dominant, beneficial microflora with the subsequent predisposition of the patient to the invasion of the gastrointestinal tract by opportunistic pathogens and the replication of antimicrobial resistance genes (Fig. 3.3). In some cases, the antimicrobial-induced changes in the gut microbiome may likely persist for several years (Dethlefsen and Relman, 2011; Jakobsson et al., 2010). Prolonged colonization of the human gut by antimicrobial-resistant bacteria may facilitate horizontal gene transfer with the subsequent emergence of multidrug-resistant pathogens (Krezalek et al., 2018). Opportunistic pathogens inhabiting the human gastrointestinal tract may receive antimicrobial resistance genes from anaerobic commensal bacteria

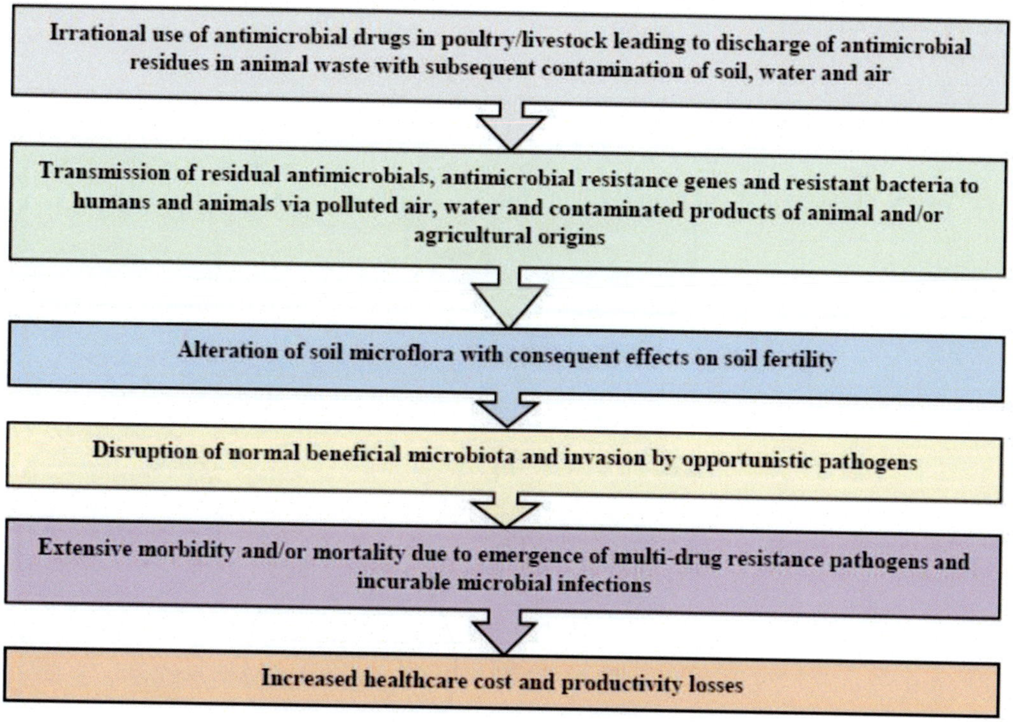

FIGURE 3.3 Ecological impact of animal waste-derived antimicrobial resistance genes.

(Van Schaik, 2015). Immune response, infection barrier and maladaptation may preclude some livestock-derived antimicrobial-resistant bacteria from invading humans (Groeger and Meyle, 2015; Iwasaki and Medzhitov, 2015).

5. Mitigation strategies

Several strategies are currently available for the mitigation of antimicrobial resistance genes derived from animal waste (Fig. 3.4). Intensive animal farms are currently practicing biological treatments and anaerobic digestion for the processing of animal waste (Van Epps and Blaney, 2016). Vegetated aquatic systems meant for the processing of agricultural drainage water and wastewater known as constructed wetlands, diminished the level of the antimicrobial resistance genes associated with swine wastewater by 0.183 log (Lavrnić et al., 2018). The removal capacities of biological techniques differ among the various kinds of antimicrobial resistance genes on account of variations in the operating conditions (Guo et al., 2015b). Constructed wetlands are usually superior to bioreactors on account of their various removal mechanisms, including biodegradation, chemical and physical adsorption, and filtration (Huang et al., 2017). The densities of pathogenic bacteria and antimicrobial residues are slightly diminished when wastewater

FIGURE 3.4 Various strategies for mitigation of antimicrobial resistance genes in animal waste.

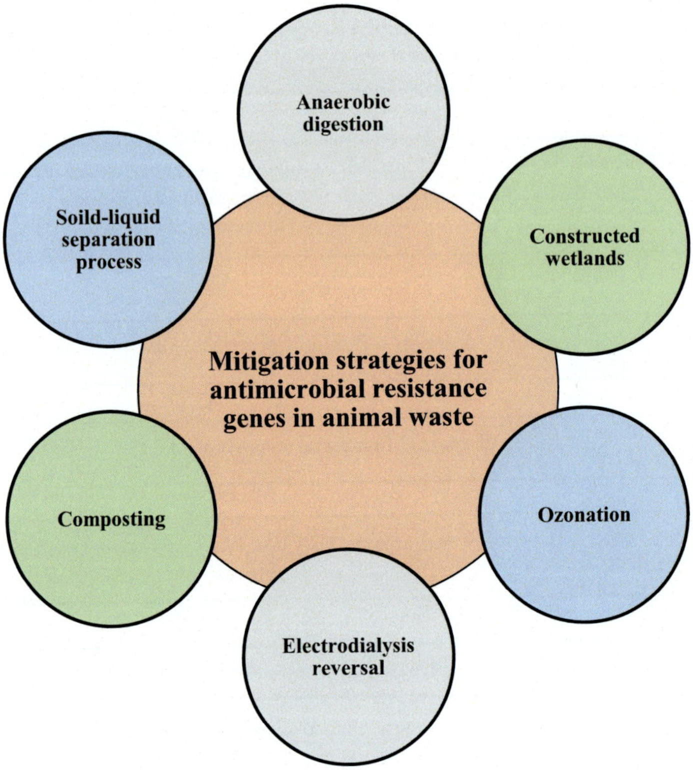

and animal manure are subjected to biological treatments to reduce the environmental contents of nitrates (Van den Meersche et al., 2019). The application of bacterial strains for the pre-treatment of manure enhanced the degradation of antimicrobials and biogas production but had no impact on the level of antimicrobial-resistant genes or antimicrobial-resistant bacteria (Liu et al., 2019). The efficient excretion of antimicrobial resistance genes through constructed wetlands can be ensured by reducing the hydraulic loading rate, providing appropriate vegetation and fillings, and optimizing operating conditions (Huang et al., 2017).

Anaerobic digestion is the microbial breakdown of organic compounds in an oxygen-free environment. Although this process can be potentially employed for the elimination of antimicrobial resistance genes, it may give rise to a gene-specific reduction effect (Pu et al., 2018). Even, the enrichment of some antimicrobial resistance genes has been associated with anaerobic digestion. Despite being an expensive process, thermophilic anaerobic digestion is better than mesophilic digestion in terms of methane production, pathogen elimination and waste stabilization (Youngquist et al., 2016). Mesophilic antimicrobial-resistant bacteria can be effectively eliminated by exposure to a temperature that is higher than 55°C in an aerobic digester (Diehl and LaPara, 2010). Although anaerobic digestion has been coupled with the effective removal of antimicrobial resistance genes, it may enhance the fecundity of some other types of antimicrobial resistance genes, such as *fca* and *sul* (Song et al., 2017). Thermophilic anaerobic digestion was more efficient than ambient temperature or mesophilic anaerobic digestion in the removal of some antimicrobials (sulphonamides and tetracyclines) and antimicrobial-resistant bacteria such as proteobacteria and bacteroidetes (Sun et al., 2016). Besides, enhanced concentrations of total volatile solids resulted in better excretion of antimicrobial-resistant bacteria and associated resistance genes through anaerobic digestion (Sun et al., 2019). The elimination of heavy metals, antimicrobials and antimicrobial resistance genes through the digestion process can be improved through the inclusion of wheat straw and biochar (Zhang et al., 2017).

Composting is usually regarded as more efficacious than stockpiling (Youngquist et al., 2016). Besides converting the animal waste into a humus-rich material, composting also leads to heat generation and the destruction of pathogenic microorganisms. Compared to mesophilic composting, the antimicrobial resistance genes were efficiently removed through a thermophilic composting technique (Qian et al., 2016). The addition of superphosphate, calcium zeolite, superabsorbent polymers, biochar and clay enhanced the effective elimination of antimicrobial resistance genes by means of composting (Awasthi et al., 2019; Guo et al., 2017). Moreover, the antimicrobial resistance genes contained in swine manure were considerably reduced as a result of anaerobic composting (Li et al., 2021). Bulking agents, including wheat straw and cauliflower substantially diminished the abundance of antimicrobial resistance genes in the composted swine manure (Li et al., 2020). The multiplicity and abundance of antimicrobial resistance genes in animal manure can be considerably reduced through composting. Mesophilic composting is inferior to thermophilic composting in the elimination of antimicrobial resistance genes (Sun et al., 2016). Despite considerably diminishing the antimicrobial load, particularly in the thermophilic stage, composting may lead to the aggregation of recalcitrant antimicrobials in the amended soils and compost products (Zhang et al., 2019). Based on the operational conditions and type of manure, a 0.72–2 log reduction in the level of antimicrobial resistance genes can be achieved following the thermophilic composting of cattle, swine and poultry manure (He et al., 2020).

The accumulation of fibrous solids during long-term storage can be prevented by removing them from the manure slurries through the solid-liquid separation process (Oliver et al., 2020). The resultant solids can be subjected to the digestion process in terms of feedstock or alternatively utilized as a renewable bedding material (Garcia et al., 2009). However, some types of antimicrobial resistance genes contained in the isolated solids can be concentrated by the solid-liquid separation and consequent processing is not carried out. By virtue of reactive oxygen species, oxidizing post-treatments such as Fenton conditions or ozonation can lead to effective degradation of antimicrobials in the treated wastewater or animal manure (Uslu and Balcıoğlu, 2009). More advanced and relatively expensive oxidation techniques can be employed for the destruction of microbial DNA using ionizing radiation. Regardless of their high environmental and biological risks, affordable combinations of oxidation techniques and ionizing radiation can be used for the destruction of antimicrobial resistance genes and antimicrobial residues in organic matrices. Tetracycline and sulphadiazine were successfully removed from pig manure by means of electrodialysis reversal (Shi et al., 2020). Iron nanoparticles facilitated methane production and anaerobic digestion while diminishing the abundance of antimicrobial resistance genes (Zhang et al., 2020). Zerovalent iron nanoparticles augmented the anaerobic digestion of cattle manure and the resultant decomposition of antimicrobial resistance genes (Wang et al., 2020).

Animal waste previously stored in lagoons is subjected to composing or anaerobic digestion for the separation of liquid and solid material. The solid animal waste forms manure, which is typically used for enhancing the fertility of agricultural lands. Whereas, the wastewater can be either left untreated and discharged into the water bodies or alternatively treated by means of a constructed wetland or a bioreactor through sequential aerobic and anaerobic processes (Liu et al., 2013; Tao et al., 2014). The utilization of land application methods like injection or incorporation and stabilization of manure with lime prior to its utilization have been recommended for mitigating the diffusion of antimicrobial resistance genes in the soil (Joy et al., 2013). Additives are known for enhancing the elimination of antimicrobial resistance genes, heavy metals and resistant bacteria from animal waste (Peng et al., 2018). Moreover, the wastewater obtained from animal and poultry production units can be conveniently disinfected using ultraviolet radiation or the chlorination technique (Shi et al., 2013). Discouraging the consumption of animal manure for fertilizing agricultural soil may diminish the exposure of human pathogens to antimicrobial resistance genes. Apart from the availability of host microorganisms, the proliferation and attenuation of antimicrobial resistance genes also depend upon environmental conditions, including temperature, light/UV light intensity, accessibility of oxygen and substrate, as well as the concentrations of heavy metals and residual antimicrobials (Guo et al., 2015a; Rysz et al., 2013; Yuan et al., 2014). UV light can cause the destruction of antimicrobial resistance genes due to the synthesis of cytotoxic reactive oxygen species (Maclean et al., 2009). Therefore, temporal and spatial variations in the afore-mentioned conditions may potentially give rise to fluctuations in the concentrations of antimicrobial resistance genes.

6. Conclusions and future perspectives

Antimicrobial resistance constitutes one of the well-established one-health challenges that necessitates appropriate control strategies at the animal-human-environment interface. Being a multifactorial process the environmental propagation of antimicrobial resistance genes is dependent upon several extrinsic and intrinsic factors, including the decomposition of the microbiome, environmental conditions, horizontal gene transfer dynamics and replication mechanisms of antimicrobial-resistant bacteria. Antimicrobial resistance is a global issue of a highly complex nature and growing rates of clinical infections caused by resistant microorganisms are still documented in countries that have considerably reduced their antimicrobial consumption. Several pathways can be used for the propagation of antimicrobial-resistant bacteria to humans, which may eventually lead to prolonged silent colonization or acute infections. The mechanisms of propagation and disruption of antimicrobial resistance genes in the human gut, air, soil and water require thorough consideration. Currently, mathematical modeling techniques have been exploited for predicting and quantifying the spread of antimicrobial resistance genes in agricultural waste and the gut microbiota of animals and birds.

The global consumption of antimicrobials in poultry and livestock production significantly contributes to the risk of antimicrobial resistance. Prophylactic administration of antimicrobials directly enhances the selective pressure and thereby facilitates the replication of antimicrobial-resistant bacteria. Accordingly, the consumption of antimicrobials should be restricted in livestock and poultry production through the adoption of advanced farming strategies and improved waste management. Alternative microbial control strategies include the utilization of prebiotics, probiotics, antimicrobial peptides, organic acids, plant-derived essential oils and nanoparticle-based drug formulations. Besides, the therapeutic efficacy of existing antimicrobial agents can be enhanced through the designing of more efficient drug delivery systems. Biological and chemical sanitizing techniques can help in reducing the density of antimicrobial-resistant bacteria. The environmental dissemination of antimicrobial resistance genes can be minimized by improving the efficiency of recently used animal waste treatments. For instance, optimizing vegetation selection, concentration of dissolved oxygen and contact time may augment the removal of antimicrobial resistance genes using the constructed wetlands. Good agricultural practices should be developed and implemented to minimize the propagation of antimicrobial resistance genes, antimicrobial residues and resistant bacteria via humans-animals-environment interactions.

Environmental risk assessment studies primarily rely on culture-independent methods like PCR and metagenomic sequencing for quantifying the ecological levels of antimicrobial residues, antimicrobial resistance genes and antimicrobial-resistant bacteria. Whereas, culture-based techniques are typically used during clinical investigation for characterizing the particular strains of antimicrobial-resistant bacteria within a specified region. Accordingly, the correlation and comparison of data collected from clinical and environmental risk assessment studies require collaborative efforts for designing a more holistic approach of data collection, sharing and analysis. Effective collaboration between clinical microbiologists and environmental experts is critically needed for modeling the propagation of antimicrobial resistance genes from animal waste and quantifying their influence on human health. Improved farming practices can lead to a significant reduction in the plenitude and multiplicity of animal waste-derived antimicrobial resistance genes. Appropriate biosecurity measures can minimize the prophylactic use of antimicrobials by suppressing the multiplication of antimicrobial resistance genes in the premises and waste products of livestock and poultry farms. Although minimising the consumption of antimicrobials may reduce the levels of certain antimicrobial-resistant bacteria and multidrug-resistant pathogens, the abundance of antimicrobial-resistance genes usually remains unchanged. Accordingly, mechanistic insight into the propagation of antimicrobial-resistant bacteria in animal manure and wastewater should be targeted by future research using the predictive modeling techniques.

References

Arikan, O.A., Mulbry, W., Rice, C., 2009. Management of antibiotic residues from agricultural sources: use of composting to reduce chlortetracycline residues in beef manure from treated animals. Journal of Hazardous Materials 164 (2–3), 483–489.

Awasthi, M.K., Chen, H., Awasthi, S.K., Duan, Y., Liu, T., Pandey, A., et al., 2019. Application of metagenomic analysis for detection of the reduction in the antibiotic resistance genes (ARGs) by the addition of clay during poultry manure composting. Chemosphere 220, 137–145.

Bennett, P.M., 2008. Plasmid encoded antibiotic resistance: acquisition and transfer of antibiotic resistance genes in bacteria. British Journal of Pharmacology 153 (S1), S347–S357.

Blaak, H., van Hoek, A.H., Veenman, C., van Leeuwen, A.E.D., Lynch, G., van Overbeek, W.M., de Roda Husman, A.M., 2014. Extended spectrum ß-lactamase-and constitutively AmpC-producing Enterobacteriaceae on fresh produce and in the agricultural environment. International Journal of Food Microbiology 168, 8–16.

Blau, K., Bettermann, A., Jechalke, S., Fornefeld, E., Vanrobaeys, Y., Stalder, T., et al., 2018. The transferable resistome of produce. mBio 9 (6), e01300–e01800.

Brüssow, H., 2015. Microbiota and the human nature: know thyself. Environmental Microbiology 17 (1), 10–15.

Checcucci, A., Trevisi, P., Luise, D., Modesto, M., Blasioli, S., Braschi, I., Mattarelli, P., 2020. Exploring the animal waste resistome: the spread of antimicrobial resistance genes through the use of livestock manure. Frontiers in Microbiology 11, 1416.

Collignon, P.J., McEwen, S.A., 2019. One health—its importance in helping to better control antimicrobial resistance. Tropical Medicine and Infectious Disease 4 (1), 22.

Cui, E., Wu, Y., Zuo, Y., Chen, H., 2016. Effect of different biochars on antibiotic resistance genes and bacterial community during chicken manure composting. Bioresource Technology 203, 11–17.

Dethlefsen, L., Relman, D.A., 2011. Incomplete recovery and individualized responses of the human distal gut microbiota to repeated antibiotic perturbation. Proceedings of the National Academy of Sciences 108 (supplement_1), 4554–4561.

Diehl, D.L., LaPara, T.M., 2010. Effect of temperature on the fate of genes encoding tetracycline resistance and the integrase of class 1 integrons within anaerobic and aerobic digesters treating municipal wastewater solids. Environmental Science and Technology 44 (23), 9128–9133.

Ding, G.-C., Radl, V., Schloter-Hai, B., Jechalke, S., Heuer, H., Smalla, K., Schloter, M., 2014. Dynamics of soil bacterial communities in response to repeated application of manure containing sulfadiazine. PLoS One 9 (3), e92958.

Du, L., Liu, W., 2012. Occurrence, fate, and ecotoxicity of antibiotics in agro-ecosystems. A review. Agronomy for Sustainable Development 32, 309–327.

Ekpeghere, K.I., Lee, J.-W., Kim, H.-Y., Shin, S.-K., Oh, J.-E., 2017. Determination and characterization of pharmaceuticals in sludge from municipal and livestock wastewater treatment plants. Chemosphere 168, 1211–1221.

Fang, H., Han, L., Zhang, H., Long, Z., Cai, L., Yu, Y., 2018. Dissemination of antibiotic resistance genes and human pathogenic bacteria from a pig feedlot to the surrounding stream and agricultural soils. Journal of Hazardous Materials 357, 53–62.

Franklin, A.M., Aga, D.S., Cytryn, E., Durso, L.M., McLain, J.E., Pruden, A., et al., 2016. Antibiotics in agroecosystems: introduction to the special section. Journal of Environmental Quality 45 (2), 377–393.

Garcia, M., Szogi, A., Vanotti, M., Chastain, J., Millner, P., 2009. Enhanced solid–liquid separation of dairy manure with natural flocculants. Bioresource Technology 100 (22), 5417–5423.

Gou, M., Hu, H.-W., Zhang, Y.-J., Wang, J.-T., Hayden, H., Tang, Y.-Q., He, J.-Z., 2018. Aerobic composting reduces antibiotic resistance genes in cattle manure and the resistome dissemination in agricultural soils. Science of the Total Environment 612, 1300–1310.

Groeger, S.E., Meyle, J., 2015. Epithelial barrier and oral bacterial infection. Periodontology 2000 69 (1), 46–67.

Guo, A., Gu, J., Wang, X., Zhang, R., Yin, Y., Sun, W., et al., 2017. Effects of superabsorbent polymers on the abundances of antibiotic resistance genes, mobile genetic elements, and the bacterial community during swine manure composting. Bioresource Technology 244, 658–663.

Guo, M.-T., Yuan, Q.-B., Yang, J., 2015a. Distinguishing effects of ultraviolet exposure and chlorination on the horizontal transfer of antibiotic resistance genes in municipal wastewater. Environmental Science and Technology 49 (9), 5771–5778.

Guo, M.-T., Yuan, Q.-B., Yang, J., 2015b. Insights into the amplification of bacterial resistance to erythromycin in activated sludge. Chemosphere 136, 79–85.

Hammerum, A.M., Sandvang, D., Andersen, S.R., Seyfarth, A.M., Porsbo, L.J., Frimodt-Møller, N., Heuer, O.E., 2006. Detection of sul1, sul2 and sul3 in sulphonamide resistant Escherichia coli isolates obtained from healthy humans, pork and pigs in Denmark. International Journal of Food Microbiology 106 (2), 235–237.

Han, X.-M., Hu, H.-W., Chen, Q.-L., Yang, L.-Y., Li, H.-L., Zhu, Y.-G., et al., 2018. Antibiotic resistance genes and associated bacterial communities in agricultural soils amended with different sources of animal manures. Soil Biology and Biochemistry 126, 91–102.

He, Y., Yuan, Q., Mathieu, J., Stadler, L., Senehi, N., Sun, R., Alvarez, P.J., 2020. Antibiotic resistance genes from livestock waste: occurrence, dissemination, and treatment. NPJ Clean Water 3 (1), 1–11.

Hirsch, R., Ternes, T., Haberer, K., Kratz, K.-L., 1999. Occurrence of antibiotics in the aquatic environment. Science of the Total Environment 225 (1–2), 109–118.

Holden, M.T., Hsu, L.-Y., Kurt, K., Weinert, L.A., Mather, A.E., Harris, S.R., et al., 2013. A genomic portrait of the emergence, evolution, and global spread of a methicillin-resistant Staphylococcus aureus pandemic. Genome Research 23 (4), 653–664.

Huang, L., Xu, Y., Xu, J., Ling, J., Zheng, L., Zhou, X., Xie, G., 2019. Dissemination of antibiotic resistance genes (ARGs) by rainfall on a cyclic economic breeding livestock farm. International Biodeterioration and Biodegradation 138, 114–121.

Huang, X., Zheng, J., Liu, C., Liu, L., Liu, Y., Fan, H., 2017. Removal of antibiotics and resistance genes from swine wastewater using vertical flow constructed wetlands: effect of hydraulic flow direction and substrate type. Chemical Engineering Journal 308, 692–699.

Hultman, J., Waldrop, M.P., Mackelprang, R., David, M.M., McFarland, J., Blazewicz, S.J., et al., 2015. Multi-omics of permafrost, active layer and thermokarst bog soil microbiomes. Nature 521 (7551), 208–212.

Iwasaki, A., Medzhitov, R., 2015. Control of adaptive immunity by the innate immune system. Nature Immunology 16 (4), 343–353.

Iwu, C.D., Korsten, L., Okoh, A.I., 2020. The incidence of antibiotic resistance within and beyond the agricultural ecosystem: a concern for public health. Microbiologyopen 9 (9), e1035.

Jakobsson, H.E., Jernberg, C., Andersson, A.F., Sjölund-Karlsson, M., Jansson, J.K., Engstrand, L., 2010. Short-term antibiotic treatment has differing long-term impacts on the human throat and gut microbiome. PLoS One 5 (3), e9836.

Jechalke, S., Heuer, H., Siemens, J., Amelung, W., Smalla, K., 2014. Fate and effects of veterinary antibiotics in soil. Trends in Microbiology 22 (9), 536–545.

Jjemba, P.K., 2002. The potential impact of veterinary and human therapeutic agents in manure and biosolids on plants grown on arable land: a review. Agriculture, Ecosystems and Environment 93 (1–3), 267–278.

Johnson, T.A., Stedtfeld, R.D., Wang, Q., Cole, J.R., Hashsham, S.A., Looft, T., et al., 2016. Clusters of antibiotic resistance genes enriched together stay together in swine agriculture. mBio 7 (2), e02214–e02215.

Joy, S.R., Bartelt-Hunt, S.L., Snow, D.D., Gilley, J.E., Woodbury, B.L., Parker, D.B., et al., 2013. Fate and transport of antimicrobials and antimicrobial resistance genes in soil and runoff following land application of swine manure slurry. Environmental Science and Technology 47 (21), 12081–12088.

Kemper, N., 2008. Veterinary antibiotics in the aquatic and terrestrial environment. Ecological Indicators 8 (1), 1–13.

Kim, K.-R., Owens, G., Kwon, S.-I., So, K.-H., Lee, D.-B., Ok, Y.S., 2011. Occurrence and environmental fate of veterinary antibiotics in the terrestrial environment. Water, Air, and Soil Pollution 214, 163–174.

Krezalek, M.A., Hyoju, S., Zaborin, A., Okafor, E., Chandrasekar, L., Bindokas, V., et al., 2018. Can methicillin-resistant Staphylococcus aureus silently travel from the gut to the wound and cause postoperative infection? Modeling the "Trojan Horse Hypothesis". Annals of Surgery 267 (4), 749–758.

Lavrnić, S., Braschi, I., Anconelli, S., Blasioli, S., Solimando, D., Mannini, P., Toscano, A., 2018. Long-term monitoring of a surface flow constructed wetland treating agricultural drainage water in Northern Italy. Water 10 (5), 644.

Li, H., Zheng, X., Cao, H., Tan, L., Yang, B., Cheng, W., Xu, Y., 2021. Reduction of antibiotic resistance genes under different conditions during composting process of aerobic combined with anaerobic. Bioresource Technology 325, 124710.

Li, S., Liu, Y., Ge, R., Yang, S., Zhai, Y., Hua, T., et al., 2020. Microbial electro-Fenton: a promising system for antibiotics resistance genes degradation and energy generation. Science of the Total Environment 699, 134160.

Liu, L., Liu, C., Zheng, J., Huang, X., Wang, Z., Liu, Y., Zhu, G., 2013. Elimination of veterinary antibiotics and antibiotic resistance genes from swine wastewater in the vertical flow constructed wetlands. Chemosphere 91 (8), 1088–1093.

Liu, M., Ni, H., Yang, L., Chen, G., Yan, X., Leng, X., et al., 2019. Pretreatment of swine manure containing β-lactam antibiotics with whole-cell biocatalyst to improve biogas production. Journal of Cleaner Production 240, 118070.

Maclean, M., MacGregor, S.J., Anderson, J.G., Woolsey, G., 2009. Inactivation of bacterial pathogens following exposure to light from a 405-nanometer light-emitting diode array. Applied and Environmental Microbiology 75 (7), 1932–1937.

Manaia, C.M., 2017. Assessing the risk of antibiotic resistance transmission from the environment to humans: non-direct proportionality between abundance and risk. Trends in Microbiology 25 (3), 173–181.

March-Rosselló, G.A., 2017. Rapid methods for detection of bacterial resistance to antibiotics. Enfermedades Infecciosas y Microbiología Clínica 35 (3), 182–188.

Marshall, B.M., Levy, S.B., 2011. Food animals and antimicrobials: impacts on human health. Clinical Microbiology Reviews 24 (4), 718–733.

Marti, E., Huerta, B., Rodríguez-Mozaz, S., Barceló, D., Marcé, R., Balcázar, J.L., 2018. Abundance of antibiotic resistance genes and bacterial community composition in wild freshwater fish species. Chemosphere 196, 115–119.

Marti, R., Scott, A., Tien, Y.-C., Murray, R., Sabourin, L., Zhang, Y., Topp, E., 2013. Impact of manure fertilization on the abundance of antibiotic-resistant bacteria and frequency of detection of antibiotic resistance genes in soil and on vegetables at harvest. Applied and Environmental Microbiology 79 (18), 5701–5709.

Martinez-Hernandez, F., Garcia-Heredia, I., Lluesma Gomez, M., Maestre-Carballa, L., Martínez Martínez, J., Martinez-Garcia, M., 2019. Droplet digital PCR for estimating absolute abundances of widespread pelagibacter viruses. Frontiers in Microbiology 10, 1226.

Mazel, D., 2006. Integrons: agents of bacterial evolution. Nature Reviews Microbiology 4 (8), 608–620.

McArdell, C.S., Molnar, E., Suter, M.J.-F., Giger, W., 2003. Occurrence and fate of macrolide antibiotics in wastewater treatment plants and in the Glatt Valley Watershed, Switzerland. Environmental Science and Technology 37 (24), 5479–5486.

Montforts, M., 1999. Environmental Risk Assessment for Veterinary Medicinal Products. Part 1. Other than GMO-Containing and Immunological Products. First Update. National Institute of Public Health and the Environment (NIPHE), Bilthoven, The Netherland.

Muhammad, J., Khan, S., Su, J.Q., Hesham, A.E.-L., Ditta, A., Nawab, J., Ali, A., 2020. Antibiotics in poultry manure and their associated health issues: a systematic review. Journal of Soils and Sediments 20, 486–497.

Oliver, J.P., Gooch, C.A., Lansing, S., Schueler, J., Hurst, J.J., Sassoubre, L., et al., 2020. Invited review: fate of antibiotic residues, antibiotic-resistant bacteria, and antibiotic resistance genes in US dairy manure management systems. Journal of Dairy Science 103 (2), 1051–1071.

Park, S., Choi, K., 2008. Hazard assessment of commonly used agricultural antibiotics on aquatic ecosystems. Ecotoxicology 17, 526–538.

Pärnänen, K., Karkman, A., Tamminen, M., Lyra, C., Hultman, J., Paulin, L., Virta, M., 2016. Evaluating the mobility potential of antibiotic resistance genes in environmental resistomes without metagenomics. Scientific Reports 6 (1), 1–9.

Peng, S., Li, H., Song, D., Lin, X., Wang, Y., 2018. Influence of zeolite and superphosphate as additives on antibiotic resistance genes and bacterial communities during factory-scale chicken manure composting. Bioresource Technology 263, 393–401.

Pu, C., Liu, H., Ding, G., Sun, Y., Yu, X., Chen, J., et al., 2018. Impact of direct application of biogas slurry and residue in fields: in situ analysis of antibiotic resistance genes from pig manure to fields. Journal of Hazardous Materials 344, 441–449.

Qian, X., Gu, J., Sun, W., Wang, X.-J., Su, J.-Q., Stedfeld, R., 2018. Diversity, abundance, and persistence of antibiotic resistance genes in various types of animal manure following industrial composting. Journal of Hazardous Materials 344, 716–722.

Qian, X., Sun, W., Gu, J., Wang, X.-J., Zhang, Y.-J., Duan, M.-L., et al., 2016. Reducing antibiotic resistance genes, integrons, and pathogens in dairy manure by continuous thermophilic composting. Bioresource Technology 220, 425–432.

Rosewarne, C.P., Pettigrove, V., Stokes, H.W., Parsons, Y.M., 2010. Class 1 integrons in benthic bacterial communities: abundance, association with Tn 402-like transposition modules and evidence for coselection with heavy-metal resistance. FEMS Microbiology Ecology 72 (1), 35–46.

Rowe-Magnus, D.A., Mazel, D., 2002. The role of integrons in antibiotic resistance gene capture. International Journal of Medical Microbiology 292 (2), 115–125.

Rysz, M., Mansfield, W.R., Fortner, J.D., Alvarez, P.J., 2013. Tetracycline resistance gene maintenance under varying bacterial growth rate, substrate and oxygen availability, and tetracycline concentration. Environmental Science and Technology 47 (13), 6995–7001.

Sancheza, H.M., Echeverria, C., Thulsiraj, V., Zimmer-Faust, A., Flores, A., Laitz, M., et al., 2016. Antibiotic resistance in airborne bacteria near conventional and organic beef cattle farms in California, USA. Water, Air, and Soil Pollution 227, 1–12.

Sanganyado, E., Gwenzi, W., 2019. Antibiotic resistance in drinking water systems: occurrence, removal, and human health risks. Science of the Total Environment 669, 785–797.

Sarmah, A.K., Meyer, M.T., Boxall, A.B., 2006. A global perspective on the use, sales, exposure pathways, occurrence, fate and effects of veterinary antibiotics (VAs) in the environment. Chemosphere 65 (5), 725–759.

Sengeløv, G., Agersø, Y., Halling-Sørensen, B., Baloda, S.B., Andersen, J.S., Jensen, L.B., 2003. Bacterial antibiotic resistance levels in Danish farmland as a result of treatment with pig manure slurry. Environment International 28 (7), 587–595.

Shi, L., Hu, Z., Simplicio, W.S., Qiu, S., Xiao, L., Harhen, B., Zhan, X., 2020. Antibiotics in nutrient recovery from pig manure via electrodialysis reversal: Sorption and migration associated with membrane fouling. Journal of Membrane Science 597, 117633.

Shi, P., Jia, S., Zhang, X.-X., Zhang, T., Cheng, S., Li, A., 2013. Metagenomic insights into chlorination effects on microbial antibiotic resistance in drinking water. Water Research 47 (1), 111–120.

Song, W., Wang, X., Gu, J., Zhang, S., Yin, Y., Li, Y., et al., 2017. Effects of different swine manure to wheat straw ratios on antibiotic resistance genes and the microbial community structure during anaerobic digestion. Bioresource Technology 231, 1–8.

Sukul, P., Lamshöft, M., Kusari, S., Zühlke, S., Spiteller, M., 2009. Metabolism and excretion kinetics of 14C-labeled and non-labeled difloxacin in pigs after oral administration, and antimicrobial activity of manure containing difloxacin and its metabolites. Environmental Research 109 (3), 225–231.

Sun, W., Gu, J., Wang, X., Qian, X., Peng, H., 2019. Solid-state anaerobic digestion facilitates the removal of antibiotic resistance genes and mobile genetic elements from cattle manure. Bioresource Technology 274, 287–295.

Sun, W., Qian, X., Gu, J., Wang, X.-J., Duan, M.-L., 2016. Mechanism and effect of temperature on variations in antibiotic resistance genes during anaerobic digestion of dairy manure. Scientific Reports 6 (1), 30237.

Tang, S.S., Apisarnthanarak, A., Hsu, L.Y., 2014. Mechanisms of β-lactam antimicrobial resistance and epidemiology of major community-and healthcare-associated multidrug-resistant bacteria. Advanced Drug Delivery Reviews 78, 3–13.

Tao, C.-W., Hsu, B.-M., Ji, W.-T., Hsu, T.-K., Kao, P.-M., Hsu, C.-P., et al., 2014. Evaluation of five antibiotic resistance genes in wastewater treatment systems of swine farms by real-time PCR. Science of the Total Environment 496, 116–121.

Tasho, R.P., Cho, J.Y., 2016. Veterinary antibiotics in animal waste, its distribution in soil and uptake by plants: a review. Science of the Total Environment 563, 366–376.

Tien, Y.-C., Li, B., Zhang, T., Scott, A., Murray, R., Sabourin, L., et al., 2017. Impact of dairy manure preapplication treatment on manure composition, soil dynamics of antibiotic resistance genes, and abundance of antibiotic-resistance genes on vegetables at harvest. Science of the Total Environment 581, 32–39.

Uslu, M.Ö., Balcıoğlu, I.A., 2009. Comparison of the ozonation and Fenton process performances for the treatment of antibiotic containing manure. Science of the Total Environment 407 (11), 3450–3458.

Van den Meersche, T., Rasschaert, G., Haesebrouck, F., Van Coillie, E., Herman, L., Van Weyenberg, S., et al., 2019. Presence and fate of antibiotic residues, antibiotic resistance genes and zoonotic bacteria during biological swine manure treatment. Ecotoxicology and Environmental Safety 175, 29–38.

Van Epps, A., Blaney, L., 2016. Antibiotic residues in animal waste: occurrence and degradation in conventional agricultural waste management practices. Current Pollution Reports 2, 135–155.

Van Schaik, W., 2015. The human gut resistome. Philosophical Transactions of the Royal Society B: Biological Sciences 370 (1670), 20140087.

Wang, Q., Gu, J., Wang, X., Ma, J., Hu, T., Peng, H., et al., 2020. Effects of nano-zerovalent iron on antibiotic resistance genes and mobile genetic elements during swine manure composting. Environmental Pollution 258, 113654.

Wang, X., Yang, F., Zhao, J., Xu, Y., Mao, D., Zhu, X., et al., 2018. Bacterial exposure to ZnO nanoparticles facilitates horizontal transfer of antibiotic resistance genes. NanoImpact 10, 61–67.

Werner, G., Klare, I., Heier, H., Hinz, K.-H., Böhme, G., Wendt, M., Witte, W., 2000. Quinupristin/dalfopristin-resistant enterococci of the satA (vatD) and satG (vatE) genotypes from different ecological origins in Germany. Microbial Drug Resistance 6 (1), 37–47.

Woodford, N., Ellington, M.J., 2007. The emergence of antibiotic resistance by mutation. Clinical Microbiology and Infection 13 (1), 5–18.

Youngquist, C.P., Mitchell, S.M., Cogger, C.G., 2016. Fate of antibiotics and antibiotic resistance during digestion and composting: a review. Journal of Environmental Quality 45 (2), 537–545.

Yuan, Q.-B., Guo, M.-T., Yang, J., 2014. Monitoring and assessing the impact of wastewater treatment on release of both antibiotic-resistant bacteria and their typical genes in a Chinese municipal wastewater treatment plant. Environmental Science: Processes and Impacts 16 (8), 1930–1937.

Yuan, Q.B., Zhai, Y.F., Mao, B.Y., Schwarz, C., Hu, N., 2019. Fates of antibiotic resistance genes in a distributed swine wastewater treatment plant. Water Environment Research 91 (12), 1565–1575.

Zeng, X., Lin, J., 2017. Factors influencing horizontal gene transfer in the intestine. Animal Health Research Reviews 18 (2), 153–159.

Zhang, J., Wang, Z., Wang, Y., Zhong, H., Sui, Q., Zhang, C., Wei, Y., 2017. Effects of graphene oxide on the performance, microbial community dynamics and antibiotic resistance genes reduction during anaerobic digestion of swine manure. Bioresource Technology 245, 850–859.

Zhang, M., He, L.-Y., Liu, Y.-S., Zhao, J.-L., Liu, W.-R., Zhang, J.-N., et al., 2019. Fate of veterinary antibiotics during animal manure composting. Science of the Total Environment 650, 1363–1370.

Zhang, Y., Yang, Z., Xiang, Y., Xu, R., Zheng, Y., Lu, Y., et al., 2020. Evolutions of antibiotic resistance genes (ARGs), class 1 integron-integrase (intI1) and potential hosts of ARGs during sludge anaerobic digestion with the iron nanoparticles addition. Science of the Total Environment 724, 138248.

Zhou, X., Wang, J., Lu, C., Liao, Q., Gudda, F.O., Ling, W., 2020. Antibiotics in animal manure and manure-based fertilizers: occurrence and ecological risk assessment. Chemosphere 255, 127006.

Zhu, Y.-G., Johnson, T.A., Su, J.-Q., Qiao, M., Guo, G.-X., Stedtfeld, R.D., et al., 2013. Diverse and abundant antibiotic resistance genes in Chinese swine farms. Proceedings of the National Academy of Sciences 110 (9), 3435–3440.

4

Electrical and electronic waste: An emerging global contaminant

Gratien Twagirayezu[1], Kui Huang[2], Hongguang Cheng[1], Christian Sekomo Birame[3], Abias Uwimana[4] and Olivier Irumva[5]

[1]STATE KEY LABORATORY OF ENVIRONMENTAL GEOCHEMISTRY, INSTITUTE OF GEOCHEMISTRY, CHINESE ACADEMY OF SCIENCES, GUIYANG, GUIZHOU, CHINA; [2]SCHOOL OF ENVIRONMENTAL AND MUNICIPAL ENGINEERING, LANZHOU JIAOTONG UNIVERSITY, LANZHOU, CHINA; [3]NATIONAL INDUSTRIAL RESEARCH AND DEVELOPMENT AGENCY, KIGALI, RWANDA; [4]COLLEGE OF SCIENCE AND TECHNOLOGY, UNIVERSITY OF RWANDA, KIGALI, RWANDA; [5]SCHOOL OF SCIENCE AND ENGINEERING, TONGJI UNIVERSITY, SHANGHAI, CHINA

1. Introduction

The demand for EEE is growing all across the globe due to the growth of the information, communication, and technology sectors and the quest for the discovery of new equipment (Shittu et al., 2021; Twagirayezu et al., 2021). For instance, developing nations hastily integrate networks and technology that offer unlimited access to government, education, business, and healthcare service data (Baldé et al., 2017). This results in an extraordinary amount of e-waste being discarded. In 2019, e-waste generated worldwide was approximately 53.6 million metric tons (Mt) and was estimated to increase to an enormous 74.7 Mt by 2030 (Forti et al., 2020). About 83.0% of the entire amount of e-waste created in 2019 was not recorded and either burned publicly or discarded illegally, while only 17% was collected and recycled appropriately (Shahabuddin et al., 2022). E-waste is currently an emerging contaminant worldwide due to several harmful substances that can contaminate our environment and human health (Borthakur and Singh, 2021; Twagirayezu et al., 2022a). The leachate that follows from an area where e-waste has been dumped is full of poisonous heavy metals and organic compounds that are terrible for human and animal health as well as the environment (Roy et al., 2022a; Twagirayezu et al., 2022a).

Typically, e-waste contains a variety of precious metals, such as copper, gold, silver, platinum, and other high-value components. Tragically, several amounts of e-waste are being either burned or disposed of in landfills rather than collected and recycled (Shahabuddin et al., 2022; Siddiqi et al., 2020). When it comes to the production of e-waste, there is often a gaping chasm between rates in developed and underdeveloped

Occurrence and Behavior of Emerging Contaminants in Organic Wastes and Their Control Strategies
https://doi.org/10.1016/B978-0-443-13585-9.00008-2

nations (Srivastava and Pathak, 2020; Twagirayezu et al., 2022a). Remarkably, under the regulatory umbrella, developed countries export or donate their e-waste to developing and underdeveloped countries while they have advanced management and technology systems for e-waste processing (Mohamed, 2019; Turaga et al., 2019). The largest quantities of e-waste are typically disposed of in nations located in Africa and Asia. Approximately 75%–80% of the 20–50 Mt of e-waste produced globally every year is transferred to poor countries, particularly in Asia and Africa, to be recycled and disposed of (Golev et al., 2016), where the top two importers are India and China. Nigeria, Ghana, Bangladesh, Pakistan, and Kenya are major countries that import e-waste for recycling. For instance, in 2015 and 2016, Nigeria imported between 60,000 and 71,000 tons of used EEE via two main ports in Lagos, with the majority (77%) produced by the European Union (Maes and Preston-Whyte, 2022). In addition, Ghana registered 0.215 Mt of e-waste brought into the country in 2019, of which 30%, 14%, and 56% were new products or second-hand e-waste that required processing (Maes and Preston-Whyte, 2022). However, these nations lack comprehensive, suitable handling and recycling methods with little regard for worker safety and minimal concern for environmental preservation (Garlapati, 2016), which are restricted by the 1992 Basel Convention and any other relevant national environmental legislation (Shamim et al., 2015). In such nations, vast quantities of e-waste are thrown directly into landfill sites or offered to local vendors who engage in unscientific recycling procedures. These practices include the unsafe combustion of wires, micro-plastics, and printed circuit boards (PCBs) to sort metal and polymer substances (Chen et al., 2021; Pathak et al., 2019). The inappropriate recycled methods are inadequate for preserving the environment and ensuring public health safety because they involve an inefficient recovery of metals from e-waste (Hoang et al., 2022; Li and Achal, 2020). Contrary to popular belief, developed countries protect themselves from the negative effects of e-waste, while developing and underdeveloped countries struggle to find effective methods of dealing with this issue (Twagirayezu et al., 2023). However, there has yet to be any recent work reviewing e-waste as an emerging global contaminant for protecting human beings when dealing with e-waste.

This chapter was aimed at outlining the commonly used EEE. In addition, the effects of e-waste on the environment and human health were highlighted. It was also intended to present green and sustainable approaches for e-waste management, perspectives, and mitigation measures.

2. E-waste classification worldwide

As depicted in Fig. 4.1 (Baldé et al., 2017), e-waste encompasses six distinct types worldwide. However, the most common EEEs that are shipped in developing and underdeveloped countries are listed as follows: (1) Televisions and their accessories (including television, decoder, DVD player, receiver cables, satellite dish); (2) computers and their accessories (including computer speakers, desktop computers, laptops, CD-R, CD-RW & DVD, computers, notebooks, keyboards & mice, hard drives, CDMA sticks,

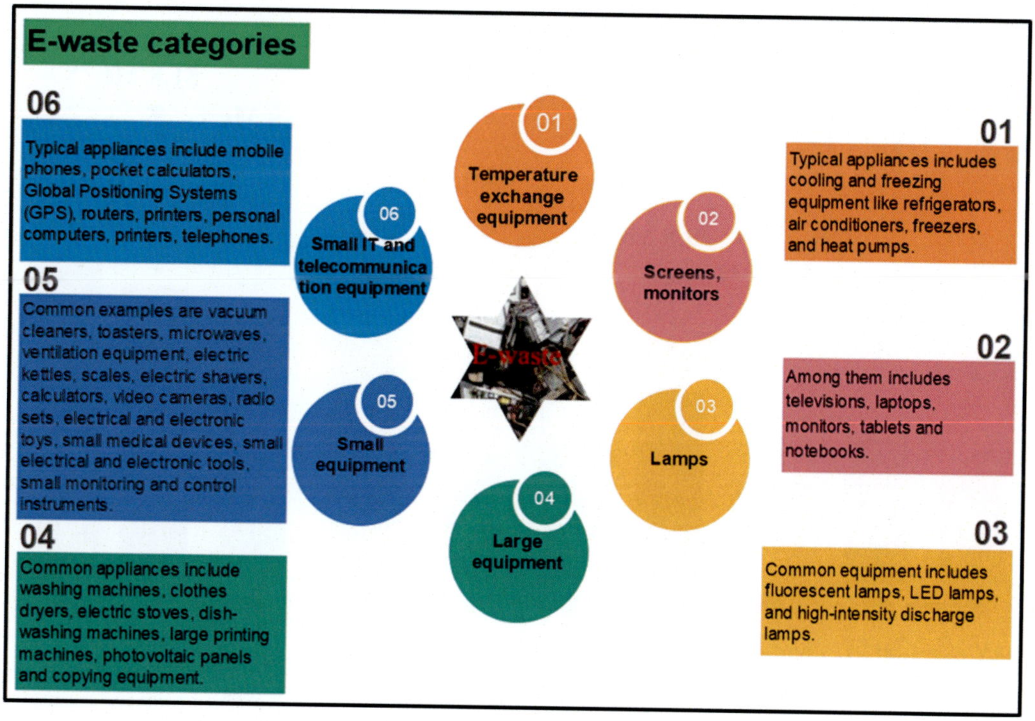

FIGURE 4.1 E-waste categories.

printers, USB sticks); (3) mobile devices and their accessories (including headsets, mobile chargers (separate), mobile phones, mobile phone batteries); and (4) other objects (such as power adapters, power cables, juice makers, radios, refrigerators, stoves, washing machines, tape recorders, air conditioners, ironing machines, men and women beauty appliances, vacuum cleaners, power dividers, varieties of lamps, rechargeable batteries, dry cell batteries, coffee grinders, and kettles).

3. Global e-waste production

E-waste production globally is at a rate of around 57.4 Mt/year (Roy et al., 2022b). This is because technology keeps getting better, which means that technology companies keep upgrading their equipment, which is part of their business model. E-waste overflow has led to a 5% rise in municipal solid waste generated in developed nations (Bhat et al., 2012; Shamim et al., 2015). As depicted in Fig. 4.2, there has been a tremendous increase in the quantity of e-waste produced globally from 2010 to 2019 (Andeobu et al., 2021; Forti et al., 2020). The global e-waste quantity is predicted to increase to 74.7 Mt by 2030 (Forti et al., 2020), representing high EEE consumption. According to the United States Environmental Protection Agency data from 2009, the total number of new electronic items sold was 438 million tons in the United States (Ilankoon et al., 2018). EEE in

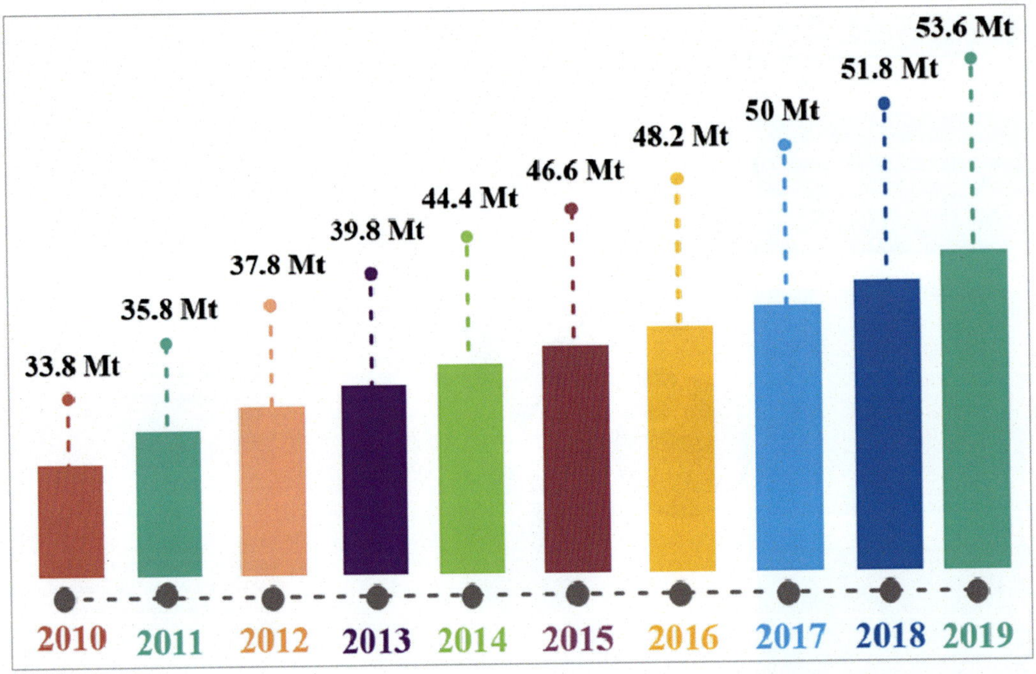

FIGURE 4.2 Global e-waste generation rate in a million metric tons.

storage accounted for approximately 5 million tons, with ready-to-EoL electronic products accounting for 2.37 million tons, with only 25% of e-waste collected for recycling (Awasthi and Li, 2017).

However, one billion laptops and other accessories were thrown away in 2008, and another one billion devices were thrown away 5 years later (Thavalingam and Karunasena, 2016). Presently, e-waste is one of the most concerning environmental problems on a global scale. For instance, e-waste produced by countries in 2012 is represented in Fig. 4.3 (Live Science, 2013), where there was 10.933 Mt in Europe, 10.3 Mt in the USA, 7.995 Mt in China, 3.033 Mt in India, 3.022 Mt in Japan, 1.556 Mt in Russia, 1.530 Mt in Brazil, and 1.138 Mt in Mexico. However, as the country's economic development improves, the amount of e-waste produced also fluctuates. According to the amount of e-waste generated in 2018 (Baldé et al., 2017), about 50 Mt of e-waste was developed around the world, with Asia at the forefront in terms of overall e-waste creation with 18.2 Mt. The fact that China ranks as the most technologically advanced nation in the whole world and therefore generates the highest quantity of e-waste. In addition, e-waste is imported into South Asia for recycling purposes. Asia is followed by Europe, America, Africa, and Oceania, which produce 12.3, 11.3, 2.2, and 0.7 Mt, respectively. However, Oceania, the world's smallest population, comes out on top when one counts the per capita e-waste output since it generates 17.3 kg/inhabitant. The shares of the other regions are distributed as follows: Europe with 16.6, America with 11.6, Asia with

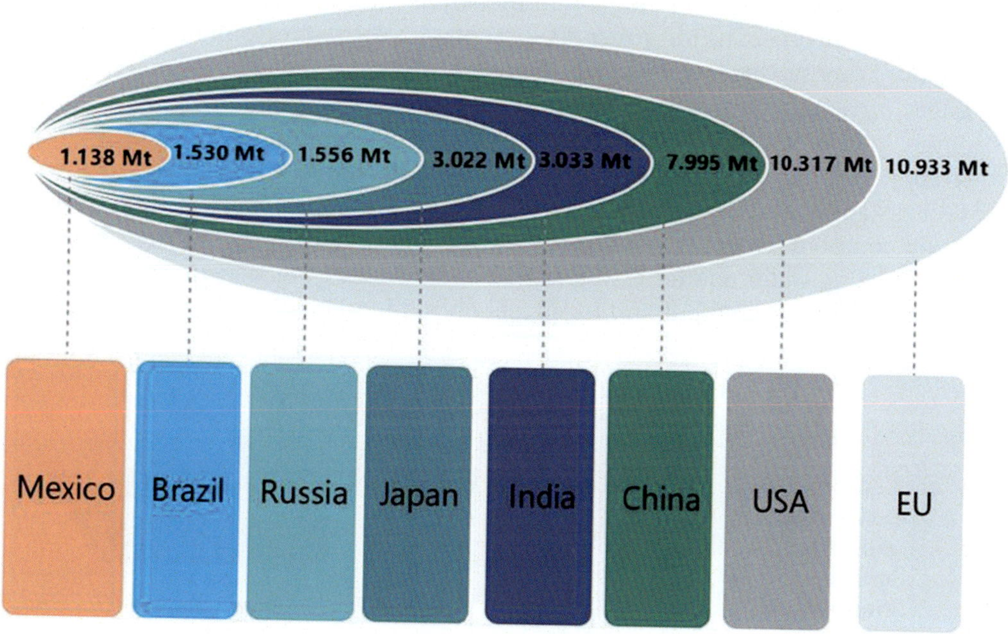

FIGURE 4.3 Amount of e-waste produced in different countries.

4.4, and Africa with 2.2 kg/person. However, Europe ranks first in the world in terms of e-waste collection, with a rate of 35%. The United States comes in second with 17%, whereas 15%, 6%, and 0% come from Asia, Oceania, and Africa, respectively.

4. Effects of e-waste on environmental and human health

E-waste components are processed to induce environmental issues such as air pollution, land contamination, water pollution, climate change, etc. (Zhu et al., 2022). Typically, the environment is involved in a wide variety of things on the entire planet, and it will undoubtedly remain an important factor in determining the health and happiness of future generations. The soil in the environment is a critical component of food production, biological productivity, plant and animal health promotion, water conservation, and environmental quality maintenance (Twagirayezu et al., 2022b). However, high concentrations of heavy metals (such as lead, arsenic, and cadmium) in soil represent a greater danger to the environment, food safety, human and animal health, etc. (Ren et al., 2019; Sonone et al., 2020) due to improper management of e-waste.

Waste management techniques, particularly in developing and underdeveloped nations, entail negative consequences for soil and human health (Ferronato and Torretta, 2019; Vaccari et al., 2018). Several studies have shown that pollutants are often everywhere, from cultivated fields to hot spots at e-waste sites and open burning areas, making them difficult to control (Chai et al., 2020; Wu et al., 2019). If we do nothing, pollution may advance to the point where the whole world will be depleted and cause serious health concerns. The improper disposal of e-waste might result in environmental degradation. Disposal sites and illegally dumped e-waste threaten the contents of soil and affect agricultural output (Chen et al., 2022; Twagirayezu et al., 2022a). Environmental difficulties caused by e-waste around the world are mostly caused by a lack of knowledge about the disposal of e-waste, ineffective enforcement of e-waste disposal legislation, and a lack of a system or organizations to oversee the disposal of EEE. Toxic compounds from abandoned e-waste items may seep into the earth and harm the drinking water aquifer (Twagirayezu et al., 2022a). Fig. 4.4 summarizes the e-waste cycle in the environment and its effects on human health.

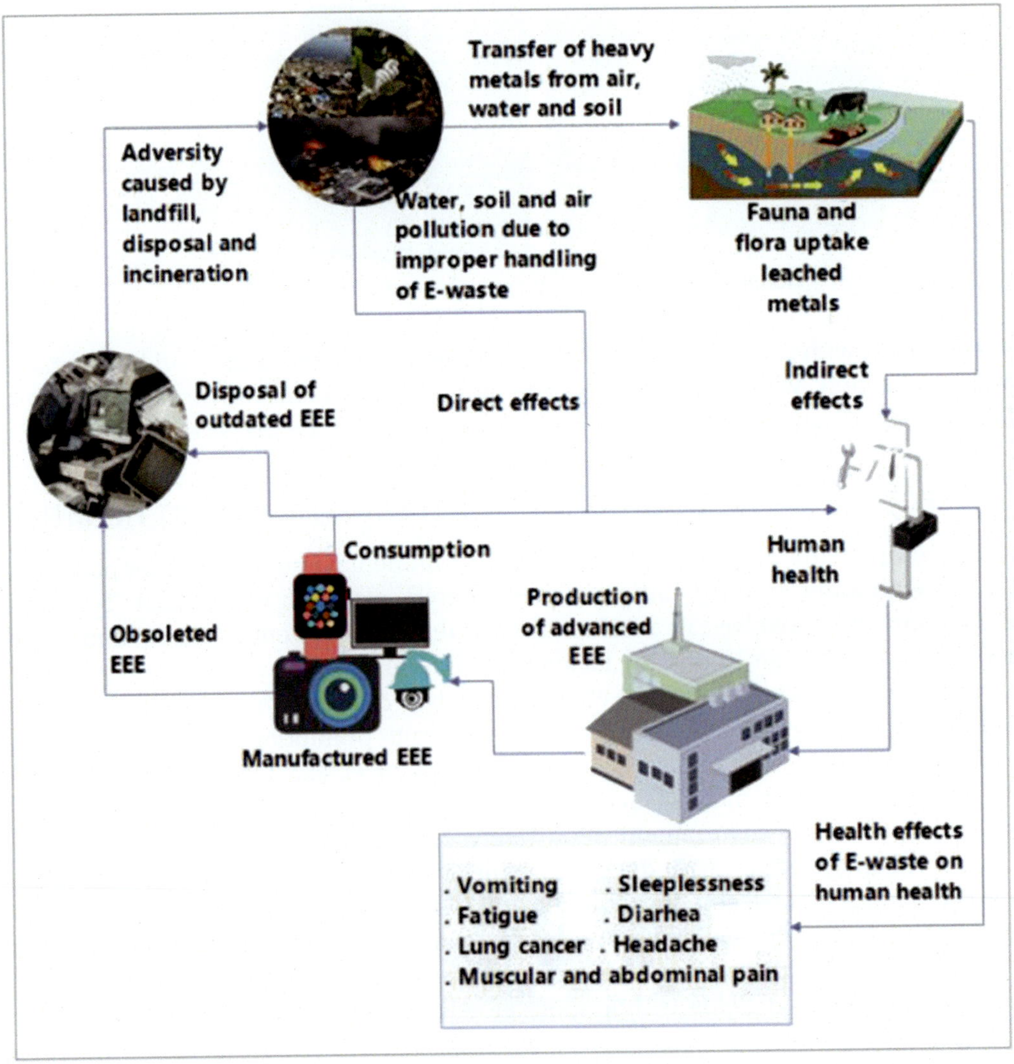

FIGURE 4.4 E-waste cycle in the environment and its effects on human health.

On the planet, there are 26 typical kinds of e-waste based on the contamination factor of more than 1000 hazardous substances, such as organic compounds and heavy metals, that threaten human health and the environment when improperly disposed of (Garlapati, 2016). The main components of e-waste are ferrous and non-ferrous metals, plastic, glass, and others (e.g., wood, ceramic, and rubber) (Arshadi et al., 2020; Tran and Salhofer, 2018). E-waste disposal is a problem worldwide since most of the population has a limited understanding of what constitutes e-waste. E-waste incineration and improper disposal emit toxic elements detrimental to human health. E-waste collectors, especially youths, are adversely impacted since they do not wear protective clothing when collecting e-waste, resulting in serious health consequences (Rautela et al., 2021). As depicted in Fig. 4.4, exposure to e-waste might result in headaches, nausea, irritability, eye pain, vomiting, etc. Recyclers are at risk of developing liver problems, renal problems, and neurological illnesses (Li and Achal, 2020). In addition, due to their lack of awareness, they should endanger their own health and the health of others in their immediate vicinity (Shaikh et al., 2020).

Environmentally dangerous effects of e-waste have been introduced into the environment mistakenly by careless social citizens on agricultural land, in open places, and in bodies of water. Many studies have shown that high amounts of harmful metals such as lead (Pb), copper (Cu), and organic pollutants such as polybrominated dibenzo-p-dioxins/dibenzofurans (PCBB/Fs), polycyclic aromatic hydrocarbons, mercury (Hg), and polychlorinated biphenyls (PCBs) can be found in soil, water, air, and sediments that are near to the location of the specific activities (Twagirayezu et al., 2023). Therefore, workers and surrounding people may be exposed to these harmful contaminants through skin exposure, direct inhalation, and oral ingestion of water and food. E-waste includes a broad range of items, including hazardous compounds; however, if not correctly managed, it might pollute the environment and affect human health. Incineration and landfilling are two options for e-waste disposal, both of which pose major environmental risks due to contamination (Ahirwar and Tripathi, 2021; Ikhlayel, 2018). Toxic materials may be transferred into groundwater through landfill leachate, and poisonous gases can be discharged into the atmosphere during combustion. E-waste recycling has the potential to release hazardous compounds into the atmosphere, especially if the recovery industry is only marginally lucrative and often unable to take the necessary precautions to safeguard the environment and the health of workers. E-waste disposal is a significant challenge to both the principal components and the ecological health of the domain (Guo et al., 2018; Twagirayezu et al., 2022a).

E-waste is not separated from other waste in different countries and is kept in inappropriate containers such as pre-packaged rice, sugar, and flavor. While these packages hold wet or liquid substances, they may be destroyed and discarded again in homes. If containers are fully packed, they are transported and leaned along the roadway, and unloaded waste is removed and returned to its owner for reuse. If waste trucks are loaded or unloaded manually by unskilled workers without the proper equipment, the waste workers and the environment are kept secure. Local governments and waste collection companies should establish sound health and safety procedures for loading and unloading these waste vehicles to safeguard the safety of their staff. Another thing to consider is educational campaigns regarding the dangers of heavy metals in e-waste and their hazards to human health (Karel Houessionon et al., 2021; Uhunamure et al., 2021).

Table 4.1 lists the dangerous compounds, their locations, and their impacts on humans and the environment (Pathak et al., 2017). Heavy metals such as mercury, cadmium, lead, chromium, and other elements, as well as halogenated components such as CFCs and PCBs and brominated flame retardants, may also be discovered in those chemicals as potentially harmful components (Ghimire and Ariya, 2020). These elements react with one another as catalysts, producing toxic constituents for both human and environmental health (Hsu et al., 2019). Many different kinds of metals may be found in e-waste that people throw away, and if these metals are left untreated and exposed to the outside world, they could be very dangerous to humans. Countries should ensure that e-waste is properly segregated from household waste to avoid harming or compromising human health or the environment. Arsenic (As), nickel, mercury (Hg), and selenium, among others, have reached a point where they should be fought owing to their chronic and progressive illnesses, as well as their adverse impacts on human health and the environment (Beula and Sureshkumar, 2021).

Table 4.1 Components of e-waste that are hazardous, their incidence, and consequences. The different letters at superscripts a, b, c, and d behind the toxic metals represent critical, hazardous, and toxic, radioactive waste, and limit in serum/blood, respectively.

Number	Substance	Hazardous metals	Limit (ppm)	A disease caused by exposure to a quantity greater than the allowable limit.
1	Ceramic capacitors, switches, batteries	Ag^a	5.0	The kidney, brain, lungs, and liver may all be damaged by excessive levels of blue pigments in the body.
2	Gallium arsenide is utilized in light-emitting	As^b	5.0	Long exposure has a detrimental impact on the skin, nerve signaling, and lung cancer.
3	Electron tube, fluorescent lamp, CRT gun, lubricant	Ba^b	Less than 100	Induce swelling of the brain, muscular weakness, and heart damage.
4	Motherboard, power supply boxes	Be^b	0.75	This leads to berylliosis, lung cancer, carcinogens, and skin disease.
5	PCBs, PVC Cables, casing	Br^b	0.1	Thyroid gland damage, skin disorders, hearing loss, hormonal issues, DNA damages
6	PCBs, CRTs, battery, infrared detectors, semiconductors, toners, printer ink	Cd^b	1.0	They pose an existential threat to human health, particularly the kidney.
7	Printed circuit boards (PCBs)	CN^b	Less than 0.5	Cyanide poisoning at a concentration greater than 2.5 parts per million (ppm) may result in coma and death.

Table 4.1 Components of e-waste that are hazardous, their incidence, and consequences. The different letters at superscripts a, b, c, and d behind the toxic metals represent critical, hazardous, and toxic, radioactive waste, and limit in serum/blood, respectively.—cont'd

Number	Substance	Hazardous metals	Limit (ppm)	A disease caused by exposure to a quantity greater than the allowable limit.
8	Plastic computer hosing, hard discs, cabling, as a colorant in pigments	Cr (VI)[b]	5.0	The toxic substances in the environment induce DNA damage and irreversible visual impairment.
9	Batteries, switches, LCD, backlight bulbs, or lamps	Hg[b]	0.2	It causes damage to the brain, kidneys, and fetuses.
10	Mobile, batteries, telephone	Li[a]	Less than 10d	They should induce diarrhea, drowsiness, vomiting, and muscular weakness.
11	PCB, semiconductor, CRT, batteries	Ni[a]	20.0	They cause bronchitis, allergic reactions, reduce lung function, and lung cancers.
12	Transistor, PCBs, LED lead-acid battery, CRT, florescent tubes, solder	Pb[c]	5.0	They damage the brain, neurological system, kidneys, and reproductive system and induce acute and long-term health impacts for humans.
13	Plastic computer housing, CRT glass, and a solder alloy	Sb[b]	Less than 0.5	Carcinogen, vomiting, diarrhea, leading pain, and stomach ulcer.
14	Photoelectric cells, fax machine	Se[b]	1.0	A high level of concentration results in selenosis.
15	Batteries, CRT	Sr[c]	1.5	Somatic and genetic alterations result from this bone, skin, and lung cancer.
16	Luminous substances and batteries	Zn[b]	250.0	Nausea, cramps, vomiting, diarrhea, and pain.
17	Insulation foam and cooling units	CFCs[b]	Less than 1.0 for 8 h/day	It causes ozone layer impacts that might contribute to an increase in the prevalence of skin cancer.
18	Transformer, condensers, and capacitor	PCBs[b]	5.0	PCB is known to cause cancer in animals and may cause liver harm in people.
19	Monitors, cabling, plastic computer housing and keyboard	PVC[b]	0.03	Contaminants in the air that are hazardous and harmful, and the release of HCl induces respiratory problems.

5. Global e-waste legislative framework

It is known that improper handling of e-waste significantly harms human health and the environment, as well as causing the loss of valuable and essential metals. Therefore, a regulatory framework for e-waste is vital to reducing the negative effects of e-waste contaminants. In developing countries, unsatisfactory legislation associated with the rising financial worth of the recovered elements from discarded e-waste items fosters the emerging market for illegal recycling by unofficial recyclers (Kazancoglu et al., 2020; Ngo et al., 2021). Countries like China, Hong Kong, India, and Nigeria process e-waste, while other countries utilize unscientific facilities and approaches (Pathak et al., 2019). In this regard, legislation, including transboundary movement regulations and extended producer responsibility for the transportation and processing of e-waste, has been adopted in many areas in many countries for the last decade (Pathak et al., 2017). Concurrently, e-waste processing activities have been initiated on a wide scale in developing nations such as Sri Lanka, Pakistan, Bangladesh, and other Southeast Asian countries (Pradhan and Kumar, 2014; Sthiannopkao and Wong, 2013). For this, it is essential to analyze the current circumstances surrounding recycling e-waste in developing nations. In light of the actual situation, developing countries and regions must set appropriate environmental standards and regulations for managing e-waste. The danger of harmful contaminants being discharged into the environment may be controlled or reduced by taking these measures. Table 4.2 describes e-waste management legislative frameworks in different countries (Pathak et al., 2017).

Table 4.2 Legislative framework for the management of e-waste management in different countries.

Number	Country/ region	Legislation/regulation	Remark
1	Japan	Regulations for the control of import, export, and other commerce of specified hazardous and other wastes	Prohibited for export in the absence of prior approval from the importing country.
2	Singapore	Exports and import of e-waste and used EEE, 2008	Case-by-case approval is required for the transportation of hazardous e-waste.
3	Nigeria	Guidelines for importers of UEEE into Nigeria, 2011	Imports of e-waste banned with a compulsory registration of importers.
4	United States	HR 2284: Responsible for recycling EEE recycling act, 2011	Banned the export of WEEE items such as PCs, xerox, TVs, printers, batteries, phones, CRTs, and other materials containing Pb, Cd, Hg, Cr, Be, and organic solvents
5	United Kingdom	Under EU directives in 2007	Adopted the EU directives.
6	South Korea	Act of 1994 relating to the control of transboundary movement of hazardous waste and its disposal	Restrictions on exporting without approval from the nation doing the importing.
7	Finland	Government decree on WEEE, 2004	Exporting products outside of the EU is prohibited unless the exporter can provide evidence that reuse and/or recycling will be carried out under the requirements of this decree.
8	China	Catalog of restricted imports of solid wastes, 2008	Restriction on unwanted electrical and electrochemical goods and wires, primarily to recycle copper.
9	Belgium	Directive 2002/96/EC on WEEE, 2002	The public waste agency of flanders manages waste and ensures that producers follow up on their responsibilities.

Table 4.2 Legislative framework for the management of e-waste management in different countries.—cont'd

Number	Country/ region	Legislation/regulation	Remark
10	France, Germany and The Netherlands	Under EU directives in 2005	Producers must use less hazardous materials; distributors and municipalities are responsible for collecting and processing discarded electronic equipment; and France has instituted an "eco-cost" for treating WEEE.
11	Vietnam	Law on environmental protection, 2005	Restricting the importation of hazardous waste and establishing producer accountability for waste generators.
12	Norway	The revised EU directives, 2006	Every manufacturer and importer of e-waste must register with a certified take-back company.
13	Thailand	The criterion for import of used EEE (UEEE), 2007	Control of the classified UEEE
14	Pakistan	Import policy order, 2009	Banned the import of air conditioners and refrigerators, Cathode Ray Tubes (CRTs) can be imported only with used computers.
15	Hong Kong	Advice on the movement of UEEE, 2011	Legislative control on UEEE.

6. Green and sustainable e-waste management

E-waste management is among the most important issues facing today's lives (Islam and Huda, 2019). E-waste may begin its journey in numerous ways within the circular economy, like repairing, reusing, recycling, and remanufacturing by end users or customers (Islam et al., 2021). The need to transform EEE via the circular economy has been underlined to reduce resource consumption and extend product lifecycles (Ottoni et al., 2020; Reike et al., 2018). It is interested in recovering secondary raw resources from e-waste (Xavier et al., 2021). Thus, innovation for e-waste management can be done in collaboration with manufacturers, retailers, and labor investors to ascertain extensive control of the informal e-waste sector works under environmental quality and human health protection while employing workers whose livelihood is dependent on the flow of discarded e-waste (Awasthi et al., 2019). Fig. 4.5 shows a proposed green and sustainable approach to e-waste management.

With environmental protection initiatives, counties have the potential to move from a linear system to a circular economy, highlighted by minimum industrial waste and resource optimization. All countries should look into ways to promote a circular economy, also termed a zero-waste economy, extracting the greatest amount of value possible from all types of resources and waste. For instance, switching from disposable e-waste (e.g., water dispensers, phones, and TVs) to reusable ones is healthy for the environment and more cost-effective. Efforts have been made in countries to coordinate

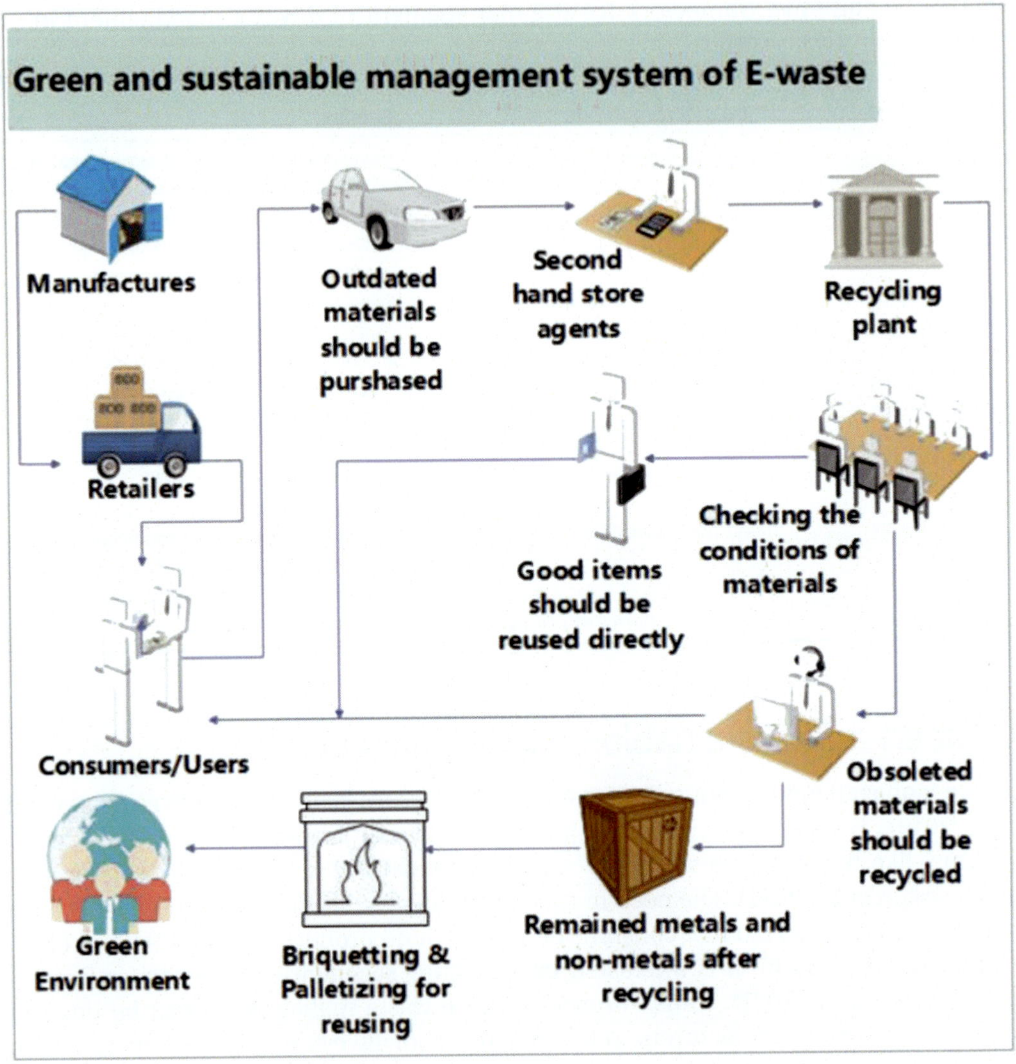

FIGURE 4.5 A proposed green and sustainable management system of e-waste.

the country's economic development with environmental preservation. Achieving this aim will only be possible with the support of local businesses that create in a manner that benefits both people and the earth. The countries' governments have underlined the necessity for industries and enterprises to incorporate the circular economy into their activities rather than continue with the present 'take, make, dispose of' paradigm of consumption. As represented in Fig. 4.6, in the linear economy, individuals take resources, convert them into items, and then dump them into waste at the end of their existence. In contrast, all commodities are reused in a circular economy by having long-

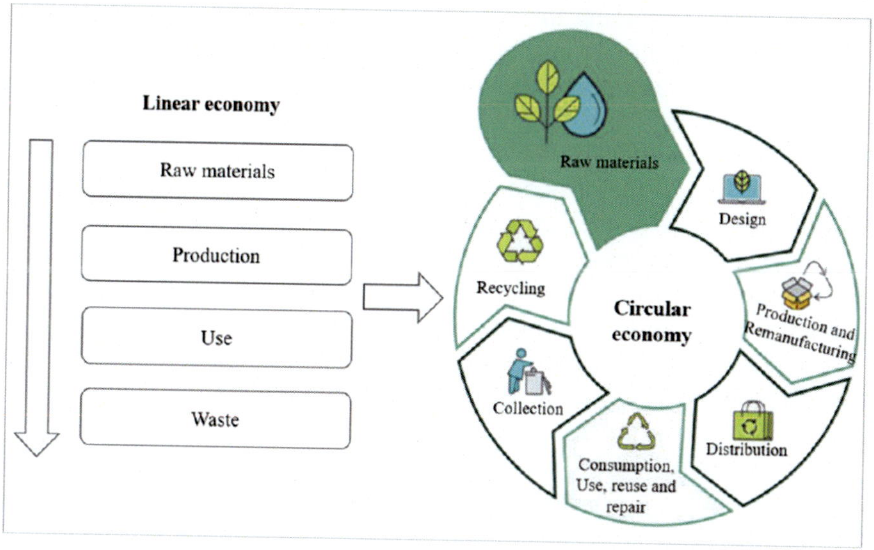

FIGURE 4.6 The circular economy approach of e-waste management.

lasting products, repairing and exchanging things, or recycling materials. A circular economy efficiently generates employment, enhances economic development, and raises living standards.

Herein, an economic analysis of e-waste should be considered. Typically, there is a considerable cost involved with e-waste, a significant portion of which may be recycled and reused. Therefore, the quantity of secondary material recovered from e-waste is calculated by multiplying the amount of waste material by the recycling yield of the relevant item. The recycling yield of items may be found in various types of literature (Islam et al., 2019; Santos and Alonso-García, 2018). In addition, although precious and vital metals account for a lesser amount of trash, their proportion to waste is remarkable, given their capacity to provide value and competitive advantage. Thus, determining the extent to which each metal contributes to the economy is critical. Various literature reviews, websites dealing with money management, and other sources provide evidence of the monetary worth of particular metals (Islam et al., 2019; Islam and Huda, 2019).

7. Recommendations and future prospects

Given the current status of e-waste deposits, the existing regulatory framework, and trading situations for EEE across the globe, ensuring the law is enforced is the most important thing that can be done. Countries should not be able to use charity as an excuse to dump waste, and NGOs should follow strict criteria to prevent receiving EEE as donations. Authorities should respond swiftly and decisively against those who break the rules if they find a breach of these criteria has occurred. It is especially important to teach young people the significance of doing their share in efficiently managing the e-waste they generate. Educating the public on proper e-waste collection, storage, and disposal is the most difficult part of the equation. This may be done via several means, including mass media such as television, newspapers, social media platforms, and awareness programs.

By raising people's levels of consciousness and making room for gathering spaces, rural education might be a major boon to the process. As soon as e-waste is collected, it may be sent to a reputable facility approved by the state or the Central Pollution Control Board for processing. A successful EPR system will include a prepayment for collection, transportation, recycling, and disposal. For this, a surcharge can be added to the sale price of goods and services to encourage recycling. These charges facilitate paying for the expense of recycling e-waste that would otherwise be dumped in a landfill. Some kind of government subsidy is required for the recycling of each EEE, whether it be to the original manufacturer or a private recycling company.

Manufacturer responsibilities can only be enforced by monitoring the secondary market and collecting data on used equipment sales. To effectively manage e-waste, basic services, including collection, transportation, separation, processing, storage, recycling, and disposal, must be implemented on a regional and state level, etc. The private sector should be encouraged to join government initiatives if it is actively involved. Incentives for designing and manufacturing more ecologically friendly items should be provided so that the social effect of the take-back legislation may be assessed. Countries may provide additional incentives and tax benefits to enterprises with effective waste management or recycling systems. Recyclers make a lot of money off the metals salvaged from discarded EEE, which means more jobs for people in the recycling industry if the practice is promoted. It is critical to consider the idea of a more unified approach that includes a refundable deposit in light of the significance of both the official and informal recycling procedures. Engineers might benefit from the input of environmentalists to better understand the ecological, social, and health implications of e-waste in the future and to incorporate this understanding into the design and development of goods moving forward.

8. Conclusion

This review article discusses the current state of e-waste and potential methods for managing e-waste around the globe. In addition, it offers recommendations for resolving a variety of e-waste issues that are prevalent all around the world. It is highly useful in reestablishing national policies for effective management and improvement, as well as in increasing public knowledge about the management of e-waste, which is quite advantageous. The quantitative classification of e-waste has been investigated, as have the potentially hazardous impacts of heavy and toxic metals as a consequence of inappropriate management and unlawful recycling. The predicted alarming yearly rise rate of imported EEE contributes to the considerable generation of e-waste in developing and underdeveloped countries, which is also likely to continue in the foreseeable future. This demonstrates that e-waste must be handled appropriately to prevent adverse environmental and human health impacts. E-waste management in some countries is restricted by the limited development of infrastructure and facilities, associated with a lack of appropriate funding. Consequently, adopted techniques in some countries are not as effective as they might be. Several effective strategies for the management of e-waste have been suggested in this piece of writing. The recommended procedures are essential to efficiently handling e-waste on a global scale. People all around the globe will be able to tackle the problem of managing e-waste by putting these alternative techniques into the system that's already in place to manage waste.

References

Ahirwar, R., Tripathi, A.K., 2021. E-waste management: a review of recycling process, environmental and occupational health hazards, and potential solutions. Environmental Nanotechnology, Monitoring & Management 15, 100409.

Andeobu, L., Wibowo, S., Grandhi, S., 2021. An assessment of e-waste generation and environmental management of selected countries in Africa, Europe and North America: a systematic review. Science of the Total Environment 792, 148078.

Arshadi, M., Yaghmaei, S., Esmaeili, A., 2020. Evaluating the optimal digestion method and value distribution of precious metals from different waste printed circuit boards. Journal of Material Cycles and Waste Management 22, 1690–1698.

Awasthi, A.K., Li, J., 2017. Management of electrical and electronic waste: a comparative evaluation of China and India. Renewable and Sustainable Energy Reviews 76, 434–447.

Awasthi, A.K., Li, J., Koh, L., Ogunseitan, O.A., 2019. Circular economy and electronic waste. Nature Electronics 2, 86–89. https://doi.org/10.1038/s41928-019-0225-2.

Baldé, P.,C., Forti, V., Gray, V., Kuehr, R., Stegmann, P., 2017. The Global e-Waste Monitor 2017: Quantities, Flows and Resources. United Nations University, International Telecommunication Union.

Beula, D., Sureshkumar, M., 2021. A review on the toxic E-waste killing health and environment–Today's global scenario. Materials Today Proceedings 47, 2168–2174.

Bhat, V., Rao, P., Patil, Y., 2012. Development of an integrated model to recover precious metals from electronic scrap-A novel strategy for e-waste management. Procedia - Social and Behavioral Sciences 37, 397–406.

Borthakur, A., Singh, P., 2021. The journey from products to waste: a pilot study on perception and discarding of electronic waste in contemporary urban India. Environmental Science & Pollution Research 28, 24511–24520.

Chai, B., Wei, Q., She, Y., Lu, G., Dang, Z., Yin, H., 2020. Soil microplastic pollution in an e-waste dismantling zone of China. Waste Management 118, 291–301.

Chen, H., Wang, L., Hu, B., Xu, J., Liu, X., 2022. Potential driving forces and probabilistic health risks of heavy metal accumulation in the soils from an e-waste area, southeast China. Chemosphere 289, 133182.

Chen, Z., Luo, X., Zeng, Y., Tan, S., Guo, J., Xu, Z., 2021. Polybrominated diphenyl ethers in indoor air from two typical E-waste recycling workshops in Southern China: emission, size-distribution, gas-particle partitioning, and exposure assessment. Journal of Hazardous Materials 402, 123667.

Ferronato, N., Torretta, V., 2019. Waste mismanagement in developing countries: a review of global issues. International Journal of Environmental Research and Public Health 16.

Forti, V., Balde, C.P., Kuehr, R., Bel, G., 2020. The Global E-Waste Monitor 2020: Quantities, Flows and the Circular Economy Potential.

Garlapati, V.K., 2016. E-waste in India and developed countries: management, recycling, business and biotechnological initiatives. Renewable and Sustainable Energy Reviews 54, 874–881.

Ghimire, H., Ariya, P.A., 2020. E-wastes: bridging the knowledge gaps in global production budgets, composition, recycling and sustainability implications. Sustainable Chemistry 1, 154–182.

Golev, A., Schmeda-Lopez, D.R., Smart, S.K., Corder, G.D., McFarland, E.W., 2016. Where next on e-waste in Australia? Waste Management 58, 348–358.

Guo, Q., Wang, E., Nie, Y., Shen, J., 2018. Profit or environment? A system dynamic model analysis of waste electrical and electronic equipment management system in China. Journal of Cleaner Production 194, 34–42.

Hoang, A.Q., Karyu, R., Tue, N.M., Goto, A., Matsukami, H., Suzuki, G., Takahashi, S., Viet, P.H., Kunisue, T., et al., 2022. Comprehensive characterization of halogenated flame retardants and organophosphate esters in settled dust from informal e-waste and end-of-life vehicle processing sites in Vietnam: occurrence, source estimation, and risk assessment. Environmental Pollution 310, 119809.

Hsu, E., Barmak, K., West, A.C., Park, A.-H.A., 2019. Advancements in the treatment and processing of electronic waste with sustainability: a review of metal extraction and recovery technologies. Green Chemistry 21, 919–936.

Ikhlayel, M., 2018. An integrated approach to establish e-waste management systems for developing countries. Journal of Cleaner Production 170, 119–130.

Ilankoon, I.M.S.K., Ghorbani, Y., Chong, M.N., Herath, G., Moyo, T., Petersen, J., 2018. E-waste in the international context – a review of trade flows, regulations, hazards, waste management strategies and technologies for value recovery. Waste Management 82, 258–275.

Islam, M.T., Huda, N., 2019. E-waste in Australia: generation estimation and untapped material recovery and revenue potential. Journal of Cleaner Production 237, 117787.

Islam, M.T., Huda, N., Baumber, A., Shumon, R., Zaman, A., Ali, F., Hossain, R., Sahajwalla, V., 2021. A global review of consumer behavior towards e-waste and implications for the circular economy. Journal of Cleaner Production 316, 128297.

Islam, M.T., Huda, N., Domínguez, A., Geyer, R., 2019. Photovoltaic waste assessment of major photovoltaic installations in the United States of America. Renewable Energy 237, 117787.

Karel Houessionon, M.G., Ouendo, E.M.D., Bouland, C., Takyi, S.A., Kedote, N.M., Fayomi, B., Fobil, J.N., Basu, N., 2021. Environmental heavy metal contamination from electronic waste (E-waste) recycling activities worldwide: a systematic review from 2005 to 2017. International Journal of Environmental Research and Public Health 18.

Kazancoglu, Y., Ozbiltekin, M., Ozen, Y.D.O., Sagnak, M., 2020. A proposed sustainable and digital collection and classification center model to manage e-waste in emerging economies. Journal of Enterprise Information Management 34.

Li, W., Achal, V., 2020. Environmental and health impacts due to e-waste disposal in China – a review. Science of the Total Environment 737, 139745. https://doi.org/10.1016/j.scitotenv.2020.139745.

Live Science, 2013. Tracking the World's E-Waste (Infographic).

Maes, T., Preston-Whyte, F., 2022. E-waste it wisely: lessons from Africa. SN Applied Sciences 4.

Mohamed, A.T., 2019. Sustainability of e-Waste Management: Egypt Case Study.

Ngo, H.T.T., Watchalayann, P., Nguyen, D.B., Doan, H.N., Liang, L., 2021. Environmental health risk assessment of heavy metal exposure among children living in an informal e-waste processing village in Vietnam. Science of the Total Environment 763, 142982.

Ottoni, M., Dias, P., Xavier, L.H., 2020. A circular approach to the e-waste valorization through urban mining in Rio de Janeiro, Brazil. Journal of Cleaner Production 261, 120990.

Pathak, P., Srivastava, R.R., et al., 2019. Environmental management of e-waste. In: Electronic Waste Management and Treatment Technology. Elsevier, pp. 103–132.

Pathak, P., Srivastava, R.R., et al., 2017. Assessment of legislation and practices for the sustainable management of waste electrical and electronic equipment in India. Renewable and Sustainable Energy Reviews 78, 220–232.

Pradhan, J.K., Kumar, S., 2014. Informal e-waste recycling: environmental risk assessment of heavy metal contamination in Mandoli industrial area, Delhi, India. Environmental Science & Pollution Research 21, 7913–7928. https://doi.org/10.1007/s11356-014-2713-2.

Rautela, R., Arya, S., Vishwakarma, S., Lee, J., Kim, K.H., Kumar, S., 2021. E-waste management and its effects on the environment and human health. Science of the Total Environment 773, 145623.

Reike, D., Vermeulen, W.J.V., Witjes, S., 2018. The circular economy: new or refurbished as CE 3.0?—exploring controversies in the conceptualization of the circular economy through a focus on history and resource value retention options. Resources, Conservation and Recycling 135, 246–264.

Ren, Z., Xiao, R., Zhang, Z., Lv, X., Fei, X., 2019. Risk assessment and source identification of heavy metals in agricultural soil: a case study in the coastal city of Zhejiang Province, China. Stochastic Environmental Research and Risk Assessment 33, 2109–2118.

Roy, H., Islam, M.S., Haque, S., Riyad, M.H., 2022a. Electronic waste management scenario in Bangladesh: policies, recommendations, and case study at Dhaka and Chittagong for a sustainable solution. Sustainable Technology and Entrepreneurship 1, 100025.

Roy, H., Rahman, T.U., Suhan, M.B.K., Al-Mamun, M.R., Haque, S., Islam, M.S., 2022b. A comprehensive review on hazardous aspects and management strategies of electronic waste: Bangladesh perspectives. Heliyon 8, e09802.

Santos, J.D., Alonso-García, M.C., 2018. Projection of the photovoltaic waste in Spain until 2050. Journal of Cleaner Production 196, 1613–1628.

Shahabuddin, M., Uddin, M.N., Chowdhury, J.I., Ahmed, S.F., Uddin, M.N., Mofijur, M., Uddin, M.A., 2022. A review of the recent development, challenges, and opportunities of electronic waste (e-waste). International Journal of Environmental Science and Technology 20.

Shaikh, S., Thomas, K., Zuhair, S., 2020. An exploratory study of e-waste creation and disposal: upstream considerations. Resources, Conservation and Recycling 155, 104662.

Shamim, A., Mursheda, A.K., Rafiq, I., 2015. E-waste trading impact on public health and ecosystem services in developing countries. Journal of Waste Resources 5, 2.

Shittu, O.S., Williams, I.D., Shaw, P.J., 2021. Global E-waste management: can WEEE make a difference? A review of e-waste trends, legislation, contemporary issues and future challenges. Waste Management 120, 549–563.

Siddiqi, M.M., Naseer, M.N., Abdul Wahab, Y., Hamizi, N.A., Badruddin, I.A., Hasan, M.A., Zaman Chowdhury, Z., Akbarzadeh, O., Johan, M.R., Kamangar, S., 2020. Exploring e-waste resources recovery in household solid waste recycling. Processes 8, 1047.

Sonone, S.S., Jadhav, S.V., Sankhla, M.S., Kumar, R., 2020. Water contamination by heavy metals and their toxic effect on aquaculture and human health through food chain. Letters in Applied NanoBioScience 10, 2148–2166.

Srivastava, R.R., Pathak, P., 2020. Policy issues for efficient management of E-waste in developing countries. In: Handbook of Electronic Waste Management. Elsevier, pp. 81–99.

Sthiannopkao, S., Wong, M.H., 2013. Handling e-waste in developed and developing countries: initiatives, practices, and consequences. Science of the Total Environment 463, 1147–1153.

Thavalingam, V., Karunasena, G., 2016. Mobile phone waste management in developing countries: a case of Sri Lanka. Resources, Conservation and Recycling 109, 34–43.

Tran, C.D., Salhofer, S.P., 2018. Processes in informal end-processing of e-waste generated from personal computers in Vietnam. Journal of Material Cycles and Waste Management 20, 1154–1178.

Turaga, R.M.R., Bhaskar, K., Sinha, S., Hinchliffe, D., Hemkhaus, M., Arora, R., Chatterjee, S., Khetriwal, D.S., Radulovic, V., Singhal, P., et al., 2019. E-waste management in India: issues and strategies. Vikalpa 44, 127–162.

Twagirayezu, G., Irumva, O., Uwimana, A., Nizeyimana, J.C., Nkundabose, J.P., 2021. Current status of E-waste and future perspective in developing countries: benchmark Rwanda. Energy and Environmental Engineering 8, 1–12. https://doi.org/10.13189/eee.2021.080101.

Twagirayezu, G., Irumva, O., Huang, K., Xia, H., Uwimana, A., Nizeyimana, J.C., Manzi, H.P., Nambajemariya, F., Itangishaka, A.C., 2022a. Environmental effects of electrical and electronic waste on water and soil: a review. Polish Journal of Environmental Studies 31.

Twagirayezu, G., Huang, K., Xia, H., 2022b. Effects of bio-contaminants in organic waste products on the soil environment. Fate of Biological Contaminants During Recyling of Organic Wastes. Elsevier, pp. 187–212.

Twagirayezu, G., Uwimana, A., Kui, H., Birame S., C., Irumva, O., Nizeyimana C., J., Cheng, H., 2023. Towards a sustainable and green approach of electrical and electronicwaste management in Rwanda: a critical review. Environmental Science and Pollution Research 30, 77959–77980. https://doi.org/10.1007/s11356-023-27910-5.

Uhunamure, S.E., Nethengwe, N.S., Shale, K., Mudau, V., Mokgoebo, M., 2021. Appraisal of households' knowledge and perception towards e-waste management in Limpopo province, South Africa. Recycling 6.

Vaccari, M., Vinti, G., Tudor, T., 2018. An analysis of the risk posed by leachate from dumpsites in developing countries. Environments 5, 99.

Wu, Q., Du, Y., Huang, Z., Gu, J., Leung, J.Y.S., Mai, B., Xiao, T., Liu, W., Fu, J., 2019. Vertical profile of soil/sediment pollution and microbial community change by e-waste recycling operation. Science of the Total Environment 669, 1001–1010.

Xavier, L.H., Giese, E.C., Ribeiro-Duthie, A.C., Lins, F.A.F., 2021. Sustainability and the circular economy: a theoretical approach focused on e-waste urban mining. Resources Policy 74, 101467.

Zhu, M., Li, X., Ma, J., Xu, T., Zhu, L., 2022. Study on complex dynamics for the waste electrical and electronic equipment recycling activities oligarchs closed-loop supply chain. Environmental Science & Pollution Research 29, 4519–4539.

5

Occurrence, detection, and classification of microplastics in excess sludge

Chengchen Wei[1], Zhiquan Yan[1], Jin Yang[1] and Kui Huang[1,2]

[1]SCHOOL OF ENVIRONMENTAL AND MUNICIPAL ENGINEERING, LANZHOU JIAOTONG UNIVERSITY, LANZHOU, CHINA; [2]KEY LABORATORY OF YELLOW RIVER WATER ENVIRONMENT IN GANSU PROVINCE, LANZHOU, CHINA

1. Introduction

Microplastics are synthetic polymer particles with a diameter of less than 5 mm. They are highly stable, have a long residence time, and easily adsorb pollutants (Li et al., 2018). Due to their small size, microplastics can easily be ingested, causing nutritional intake disorders, inflammatory responses, stress responses, reproductive disorders, and even death (Alomar et al., 2017). Sewage treatment plants are the main recipients of microplastics in water (Kay et al., 2018), and most plastic fragments are removed, retained, and accumulated in excess sludge as a by-product (Alvim et al., 2020). Studies have shown that 79%–97% of microplastics were retained in excess sludge in seven sewage treatment plants in coastal cities in China (Wei et al., 2021). In 38 sewage treatment plants in 11 countries worldwide, the average amount of microplastics in excess sludge ranged from 4.4×10^3/kg to 2.4×10^5/kg (Liu et al., 2021). When the particle size range was 20–300 μm, 79%–97% of microplastics were retained in excess sludge; when the particle size was ≥300 μm, the retention rate exceeded 99% (Gouin et al., 2015). Excess sludge is an important carrier, allowing microplastics to enter the environment. Microplastics accumulated in excess sludge eventually enter the soil ecosystem due to the utilization of sludge as a resource, causing potential harm. Microplastics in the soil can change the physical properties of the soil, affect soil enzyme activity, affect the growth of soil biota, and accumulate in environmental microorganisms, forming biofilms on the soil surface, leading to compound pollution effects (Ding et al., 2022). They can also aggregate and adsorb on the surface of plant roots, damaging the structure of root cell walls and cytoplasmic membranes, inhibiting water and nutrient absorption, and affecting plant seedling height and fruit yield (Li et al., 2022). Because of these many issues, urgent

Occurrence and Behavior of Emerging Contaminants in Organic Wastes and Their Control Strategies
https://doi.org/10.1016/B978-0-443-13585-9.00002-1

attention is needed on microplastic pollution from excess sludge and its potential adverse effects on human health to develop effective control measures.

Although most existing research has focused on microplastics in aquatic environments, there have been relatively few systematic reviews on microplastics in excess sludge. This article aims to comprehensively review current research progress on microplastics in excess sludge from sewage treatment plants, examining the research status, detection methods, and classification of microplastics. It also analyzes future research directions and provides a theoretical basis for reducing microplastics in excess sludge. It is hoped that this will provide some theoretical support for the operation and management of sewage treatment plants and the standardized treatment of sludge.

2. Sources, content, and potential hazards of microplastics

2.1 Microplastic sources

Microplastic sources can be divided into primary and secondary sources (Andrady, 2017); primary sources refer to micron-sized plastics produced by industrial activities, while secondary sources are fragments and decomposition products resulting from the physical, chemical, and weathering interactions with plastic waste in the environment. In sewage treatment, secondary sources of microplastics are almost non-existent, as microplastics in sewage treatment plants mainly come from domestic sewage and industrial wastewater. Microplastics in domestic sewage mainly come from personal care products and cosmetics or the washing of textile clothing (Waller et al., 2017). Industrial production processes produce microplastics primarily from the wear of automobile tires as well as microfibers generated during the manufacturing, printing, and dyeing of textiles, which enter sewage treatment plants with industrial wastewater (Fig. 5.1).

In excess sludge, analysis of the shape of microplastics indicates that sources can be commonly used consumer plastic products as well as resin-type plastics used in industrial production activities, including foam boards, insulation boards, and adhesives. These plastics gradually split into fragments when subjected to physical, chemical, and biological processes (Antunes et al., 2018). The possible sources of fibrous microplastics are primarily synthetic clothing and industrial products; the main materials used in synthetic clothing are polyester and nylon fibers. Industrial products such as plastic films and coil skeletons can also produce fibrous microplastics (Browne et al., 2011). Particulate microplastics are usually related to microspheres found in personal care products such as facial cleansers, shower gels, and toothpaste (Carr et al., 2016). In addition, they may also be present in certain industries, such as electronics and automobile manufacturing (Aliabad et al., 2019). Film-like microplastics usually come from packaging products such as anti-corrosion films, drinking water bottles,and fast food boxes,or from industrial-grade plastic films such as photographic films, X-ray, and tapes (Zhou et al., 2018).

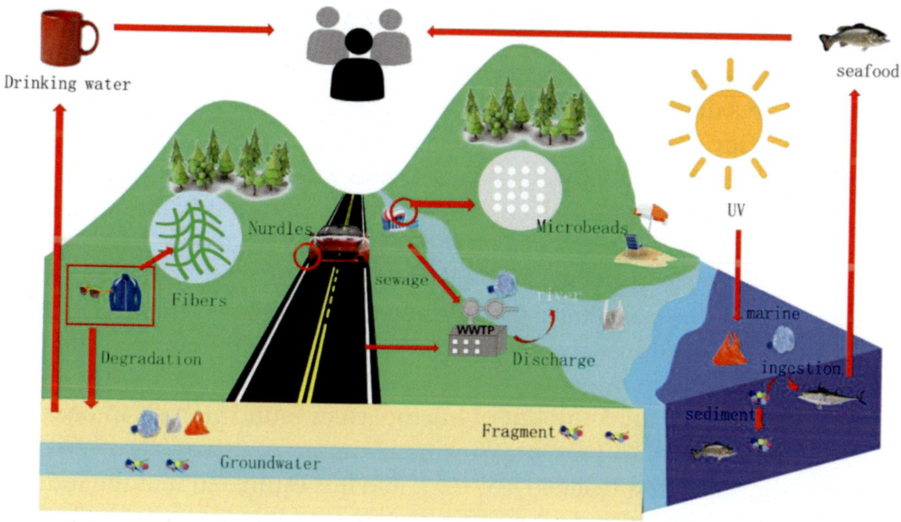

FIGURE 5.1 Microplastic distribution pathways.

2.2 Microplastic content in excess sludge

Sludge microplastic content is usually expressed as the number of microplastics per unit mass of sludge. Most of the microplastics in sewage treatment plants enter the sludge of different treatment units, with 20% μL (including microplastics) returning to the sewage pool through dewatering liquid; the remaining 80% eventually ends up in excess sludge for treatment. Microplastics in the returned activated sludge can reach 4.95×10^5 particles/kg dry sludge. The content of microplastics in sludge from sewage treatment plants in China was found to have both spatial and temporal distribution, which may be related to factors such as population density, economic development level, afforestation area, temperature, and rainfall. Additionally, process parameters such as the proportion of industrial wastewater in the influent, biochemical treatment processes, and sludge dewatering method also affect the microplastic content in sludge (Table 5.1).

Table 5.1 Classification and number of microplastics in excess sludge (Li et al., 2019).

Location	Particle size (diameter or length) (μm)	Concentration (piece)
Sweden	333	16.7 ± 1.96/g
USA	300	1.5–5.0/g
Netherlands	300	4.1–1.538/g
Germany	500	1–2.4/g
Norway	300	6.077/g
China	112	1.6–56.4/g

2.3 Fate and potential hazards of microplastics in excess sludge

Most microplastics (over 90%) in wastewater received by sewage treatment plants are retained in sludge residues and sludge cakes for subsequent treatment, including sanitary landfills, land use, incineration, and building material preparation. These plastics can then be released into environments such as soil and water bodies during land use and sanitary landfill processes. Studies have shown that a sewage treatment plant with a treatment capacity of $1 \times 10^4\ m^3/d$ in Finland discharged up to 4.6×10^8 microplastic particles through sludge. Norway has introduced up to 5×10^{11} microplastic particles per year into the soil with the agricultural use of sludge. It is estimated that the total amount of microplastics introduced into the soil due to land or agricultural use of sludge in the European Union and North America is 6.3×10^4–4.3×10^5 t and 4.4×10^4–3×10^5 t, respectively, exceeding the total amount of microplastics found globally in ocean surface water (Nizzetto et al., 2016). At present, the annual output of excess sludge from sewage treatment plants in China has reached 4×10^7 t, and it is estimated that the total amount of microplastics introduced into the soil ecosystem due to improper disposal or land use of sludge could reach $15 \times 10^{12} \sim 51 \times 10^{12}$ per year. Because of the large amount of potential pollution caused, the introduction of microplastics from the land application of sludge and their potential risks deserve attention.

The long-term accumulation of microplastics in the environment has adverse effects on human health and ecosystems. Excess sludge is the primary pathway by which microplastics enter terrestrial ecosystems, affecting the adaptability of soil organisms and reducing crop yields (Gao et al., 2019); they can adhere to vegetables and plants, entering the food chain and causing human health risks (Ziccardi et al., 2016). Many marine organisms, including plankton and invertebrates, ingest microplastics, causing health issues through blockages in the esophagus or the production of a false sense of satiety, resulting in reduced feeding efficiency, energy deficiency, injury, and even death. In addition, due to their hydrophobic surfaces and large specific surface areas, microplastics easily adsorb toxic substances (Zhang and Chen, 2020). Microplastics can act as carriers for heavy metals and pathogens, affecting the migration and fate of pollutants. Untreated plastic fragments contain chemical additives, including polybrominated diphenyl ethers and plasticizers such as phthalates and bisphenol A; these can cause toxic effects when ingested by living organisms, threatening human health and causing environmental damage (Fu et al., 2020). Microplastics may also adsorb heavy metal ions that can interact directly with DNA, causing cell apoptosis and endocrine system disruption (Kedzierski et al., 2018).

3. Detection methods for microplastics in excess sludge

3.1 Sampling

Microplastic sampling has mainly been done through grab sampling, which represents the sludge condition at the time of sampling; grab samples have an average weight range of 30–500 g (Liu et al., 2019). Sludge samples from sewage treatment plants are typically collected in sealable bags of around 250 g or 1.0 L glass sampling bottles (Van-den-Berg et al., 2020). According to standard sampling methods, a small amount (about 5–20 g) of sludge samples can be taken directly from the sludge treatment tank (Sun et al., 2019).

The sampling method and the quality of collected samples may be influenced by the microplastic extraction method and the characteristics of the water source. Large variations in the number of collected samples may also result in a lack of standardization in microplastic analysis (Li et al., 2018). The appropriate quality range for samples is considered to be 5–500 g. In addition, after collecting sludge samples, they are usually frozen in the dark at −20°C or refrigerated at 4°C to preserve and control important sample conditions before microplastic extraction is performed (Harley-Nyang et al., 2022).

3.2 Sample preparation

Sample preparation is used to separate microplastics from organic and inorganic substances in excess sludge (Nguyen et al., 2022). Currently, there is no standardized pre-treatment method for separating microplastics from environmental matrices (Koyuncuoğlu and Erden, 2021). To extract microplastics from sludge samples, pre-treatment methods mainly include sieving, oxidation, density separation, and filtration, which are usually combined (Rolsky et al., 2020). Sludge and biosolids contain organic and inorganic substances, such as clay particles, that must be removed before microplastic analysis. Pre-treating sludge with H_2O_2, Fenton's reagent, and HNO_3 has led to a higher microplastic extraction efficiency. Among oxidants, H_2O_2 has commonly been used to decompose natural organic matter as it does not affect microplastics and has been transferred to carboxylic acids, aldehydes, CO_2, and H_2O by generating hydroxyl radicals. Treatment has been performed at different solution strengths (10%–35%), temperatures (20–100°C), and periods (1 h–1 w) (Zarfl et al., 2019). Fenton's reagent can effectively destroy organic matter without degrading polymers and has an extraction efficiency of 79%–100%. Peroxide at 30% can be used for pre-digestion at 70°C, then floated (i.e., saturated with NaCl, $ZnCl_2$, and NaI), filtered with a nylon membrane, and further oxidized with a 30% $H_2O_2 + H_2SO_4$ solution at 70°C (Li et al., 2019).

3.2.1 Density separation

Density separation has been used to separate light component microplastics from heavy component impurities by utilizing the density difference between the target components and impurities in sludge samples. Specifically, saturated brine (the solute is generally NaCl or NaI) is added to the sample, which is shaken or stirred thoroughly to mix evenly, then settled until the heavy components separate from the water phase and re-settle while the microplastics continue to remain suspended or float on the surface of the solution; finally, microplastics are collected from the upper layer of the solution.

3.2.2 Digestion

Microplastics separated from sample media may contain organic particles that interfere with subsequent identification and need to be removed. In addition, organic materials in samples can be confused with microplastics, resulting in an overestimation of polymers (Table 5.2).

Table 5.2 Different digestion methods for removing organic residues from the surface of microplastics (Tirkey and Upadhyay, 2021).

Digestion	Treatment	Polymer degradation	Organic matter degradation	Recovery rate
	HNO_3 (55%) RT for 1 month	Whitening of polyvinyl chloride, degradation of polyamide	–	–
	100% HNO_3 80°C for 30 min	Some fraction of polymer	Oil residue	–
	HNO_3 (35%), 60°C for 1 h	Fusion of polyethylene terephthalate and high-density, polyethylene; destruction of polyamide	100%	–
	1:1 v/v of HCl and HNO_3, 80°C for 30 min	–	Smaller fragments	–
	HNO_3 (65%) + $HClO_4$ (65%) 4:1 overnight, boiled 10 min, diluted in 80°C distilled water	Polyamide degradation, yellowing	–	–
	37% HCl at 25°C, 96 h	Polyethylene terephthalate and polyvinyl chloride melted, forming clumps	>97%	–
Alkali	10% KOH 24 h at 60°C	Cellulose acetate		
	10% KOH at 40°C for 48–72 h	No	>95% reduced	–
	KOH (1 M), RT for 2 days	Degradation of low-density polyethylene, cellulose acetate, carbonyl, and polyamide	Most organic matter dissolved	–
	10 M NaOH 60°C, 24 h	Yellowing of polyvinyl chloride, polyethylene terephthalate degraded	–	–
	10 M NaOH 60°C	Discoloration of nylon, polyvinyl chloride, polyethylene, and damage induced	91%	–
	NaOH (20 g/L) + NaClO (14%) at 60°C for 24 h	–	Some tissues left	–
	1 M NaOH + SDS 0.5%, 50°C for 48 h, filtration,	Polyethylene terephthalate	Chitin partially digested	84 ± 15%

Table 5.2 Different digestion methods for removing organic residues from the surface of microplastics (Tirkey and Upadhyay, 2021).—cont'd

Digestion	Treatment	Polymer degradation	Organic matter degradation	Recovery rate
	if calciferous material remains at the filter, 2 M HCl is added followed by 30% H_2O_2	showed slight changes		
Oxidative	30% H_2O_2 for 7 days	Transparent, thinner, or smaller	50%	–
	H_2O_2 (35%), RT, 40°C for 96 h	Decrease in Raman peaks of polyvinyl chloride and polyamide	–	–
	H_2O_2 (30%), 60°C for 1 h, 100°C for 7 h	–	–	–
	H_2O_2 (6%) 70°C for 24 h	–	–	78% (Polyethylene)
Enzymatic	Proteinase K at 50 °C for 2 h with $NaClO_4$ and ultrasonication	No	>97%	–
	Pepsin (0.5%) and HCL (0.063 M), 35°C for 2 h	No	Partial	–
	Proteinase K with $CaCl_2$ h, shaken 20 min, Incubated at 60°C for 2 h, 30 mL H_2O_2 (30%) overnight	Formation of calcium layer	No	97%
	Trypsin incubated at 38–42°C for 30 min	No	88%	–
	Collagenase incubated at 38–42°C for 30 min	No	76%	–
	Papain incubate at 38–42°C for 30 min	No	72%	–
	SDS 50°C incubated for 24 h, Protease incubated 50°C for 24 h Cellulose TXL incubated at 50°C for 4 days, 35% H_2O_2 incubated at 50°C for 24 h, Chitinase incubated at 37°C for 5 days, H_2O_2 for 24 h at 37°C	–	98%	83%
	SDS, 24 h at 70°C + mixture of enzymes (1:1 Bio-enzyme F: bio-enzyme SE) for 48 h at 37.5°C	No	–	–

3.3 Identification methods for microplastics in excess sludge

After sampling and preparation, particles must then be detected and identified to confirm whether the separated particles are indeed microplastics. There have been various identification and chemical composition detection techniques developed, including SEM-EDS, FTIR, NIR, Raman, and NMR spectroscopy. SEM-EDS uses a high-intensity electron beam to irradiate the sample for imaging. The signals generated by the interaction between the sample and the electron beam emit secondary electrons that are used to generate information about the morphology and topography of the sample, providing information about the microplastic surface and the additives present on it. FTIR and Raman spectroscopy have been the best techniques for chemical characterization, as they have high accuracy, do not damage samples, and use spectra generated during molecular interaction with light. FTIR analysis has been found to be suitable for particles up to 20 μm, while Raman analysis has been suitable for particles up to 1 μm. Recently, NIR and NMR spectroscopy have also been used for microplastic identification. In near-infrared light, no pre-treatment is required for these techniques, and environmental samples up to 1 mm in size can be detected. Nuclear magnetic resonance analysis can also provide information on the number of microplastics present in a sample (Table 5.3).

Table 5.3 Identification methods for microplastics in excess sludge (Tirkey and Upadhyay, 2021).

Technique	Finding
SEM-EDS	SEM-EDS can quickly screen between non-plastic and plastic particles and can also detect small particles missed in visual inspection. Plastic particles show strong carbon peaks, accompanied by some weak element peaks; non-plastic particles do not have strong carbon peaks. SEM provides high-resolution topographic images of microplastics and has been used to confirm that plastic particles become brittle due to chemical and physical weathering.
ESEM	Environmental scanning electron microscopy (ESEM) is a type of SEM used for environmental sample analysis. Wet samples are kept in a low-pressure nitrogen environment, which makes it possible to image the samples; without this, the samples would degrade in a high vacuum. In ESEM-EDX, the sample is not coated with metal or carbon, making the sub-sample free of artifacts so that it may be used for further FTIR and Raman analysis.
FTIR	Direct, reliable, and non-destructive measurement is performed through specific infrared spectra to produce separate band diagrams for different types of plastics. When a sample is irradiated with infrared light, according to the molecular structure of microplastics, radiation is absorbed and measured in reflection or transmission mode. FTIR can not only accurately identify microplastic polymer types but can also provide more information about physical and chemical weathering by analyzing oxidation intensity. The limitation is that FTIR can only detect microplastics up to 20 μm.
Micro-FTIR	Micro-FTIR can overcome the shortcomings of FTIR and provide chemical imaging to detect microplastics as small as 10 μm. Characteristic bands can distinguish natural fibers from synthetic fibers. The scanning speed of filter paper is 20 min-19 s/mm^2.

Table 5.3 Identification methods for microplastics in excess sludge (Tirkey and Upadhyay, 2021).—cont'd

Technique	Finding
ATR-FTIR	ATR-FTIR is another reliable and fast detection tool suitable for analyzing large microplastic products, thick or opaque samples, and irregularly shaped microplastics and confirming polymer types without sample preparation. ATR-FTIR measurement is a form of surface contact analysis where plastic samples need to be in contact with the crystal. The disadvantage is that due to the high pressure generated by the probe, this method will damage fragile or highly weathered microplastics, and tiny microplastic particles may stick to the probe tip due to electrostatic forces.
FPA-FTIR	By scanning the separated microplastic residues on filter paper at a high lateral resolution without pre-classifying the filtered area, microplastics with particle sizes <20 μm can be analyzed and identified. Using automatic mapping software can provide information about particle number, size, and polymer type in a short time. FPA-FTIR filter images provide better information about small-sized microplastics at a faster speed with high resolution, thus providing reliable results with minimal analytical bias.
NIR	Near-infrared spectroscopy has advantages over FTIR in that it can penetrate deeper into plastic materials without sample preparation and can batch-check samples with a size detection limit of >1 mm to predict the concentration of microplastics in soil samples.
Raman spectroscopy	Raman spectroscopy is a vibrational spectroscopy technique that conveys information in the form of vibrational spectra based on the inelastic scattering of light. The obtained spectrum can identify particles present in the sample. It shows better spatial resolution for small samples and is highly sensitive to non-polar functional groups with narrowed spectral bands. Raman microspectroscopy can detect microplastics as small as 1 μm and provide their chemical and structural characteristics, which other spectroscopic techniques cannot.
NMR	Independent of size and is fast for microplastic analysis. This method quantifies the number of microplastics in environmental samples by reducing organic matrix signals through peak fitting. This allows it to qualitatively and quantitatively measure microplastics in samples.
Py-GC-MS	Py-GC-MS is a thermal analysis method that uses pyrolysis to simultaneously identify microplastics and their additives by directly introducing the sample with minimal pre-treatment. This method can identify individual particles and is suitable for a measurement with a small amount of sample (0.1 −0.5 mg). The disadvantage of this method is that it is destructive, the sample must be manually placed in the instrument, and high-molecular-weight compounds can condense from pyrolysis into the capillaries of the GC-MS system.

4. Classification of microplastics in excess sludge

4.1 Classification by source

Microplastics can be divided into primary and secondary types according to their sources. Primary microplastics refer to industrial products such as the particles contained in cosmetics or plastic and resin particles used as industrial raw materials that are discharged into the environment after passing through wastewater and sewage treatment plants. Secondary microplastics are plastic particles formed by the splitting and reduction in volume of large plastic waste through physical, chemical, and biological processes. Primary microplastics have mainly been identified as being sourced from microscopic plastic particles produced by factories, are present in many cosmetics, and can also be generated during the washing of clothing. Secondary microplastics have mainly been identified as being sourced from larger plastic fragments that decompose under exposure to environmental factors such as sunlight, wind, water flow, and bacteria.

4.2 Classification by morphology

Common shapes of microplastics include granular, fibrous, film-like, and fragmentary. In many cases, the shape of microplastic particles in the environment has been found to be irregular, with large differences in the shapes of microplastics in different ecosystems; fibers have been the most common shape found in aquatic and atmospheric environments, while the main shapes in soil have been fragmentary, film-like, and granular (Fig. 5.2).

4.3 Classification by composition

Based on their chemical composition, microplastics can be divided into low-density polyethylene, high-density polyethylene, polypropylene (PP), polystyrene (PS), polyethylene

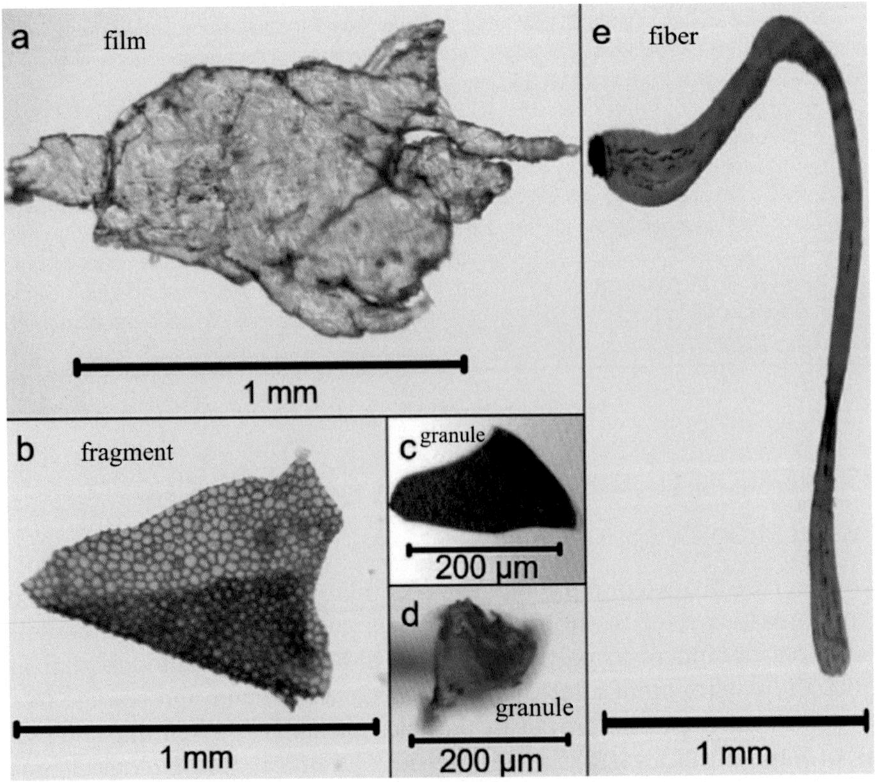

FIGURE 5.2 Different forms of microplastics (Yuan et al., 2022).

terephthalate, and polyvinyl chloride (PVC), along with other categories. PE has been widely used due to its low cost, ease of processing, and recyclability; it is found commonly in shopping bags, microbeads in cosmetics, food packaging films, toys, containers and bottles, and pipes. PS is a rigid thermoplastic produced by the free-radical vinyl polymerization of styrene. Due to its outstanding physical properties and low cost, PS has become one of the most common and abundant plastics used in applications such as food packaging, building insulation, refrigerator liners, and lampshades. PVC is commonly used in the construction and building sectors, including in pipes, because of its tolerance to various chemicals. Although PVC can be recycled multiple times, most countries directly dispose of PVC waste in landfills, which increases the potential risk of PVC plastic weathering into microplastics and accumulating in the environment. PP has strong chemical, fatigue, and heat resistance. It has been widely used in food packaging, pipes, medicine bottles, banknotes, hinged lids, and car parts. PP is resistant to biodegradation but is susceptible to non-biological attack due to its long olefin chains; therefore, PP plastic in the environment may become microplastic contaminants under the influence of environmental factors (Figs. 5.3 and 5.4).

5. Outlook

The accumulation of microplastics in the environment has attracted the attention of researchers to their wide distribution, migration, and removal. As a developing country with a large population base and land area, microplastic pollution in China cannot be underestimated. Research on microplastics in excess sludge is still in its infancy, and there is an urgent need for pollution data collection and analysis. In the future, more attention should be paid to pollution control and resource recycling. Existing wastewater treatment processes should be updated to further reduce the total amount of microplastics discharged into the environment. At the same time, sludge control processes

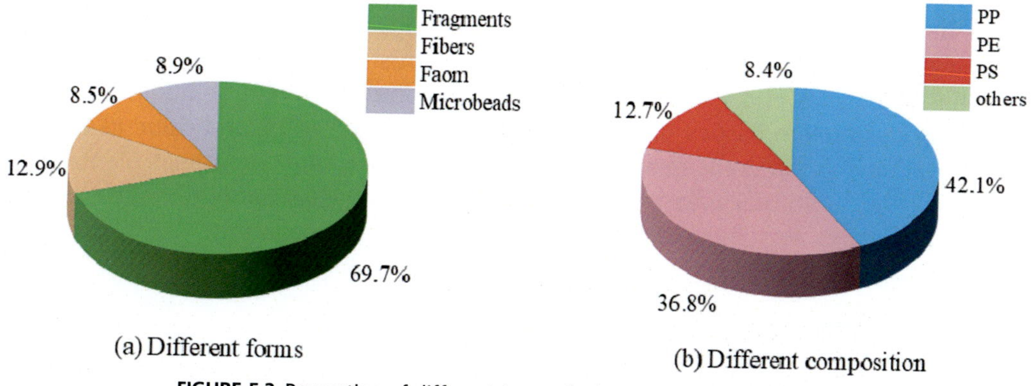

FIGURE 5.3 Proportion of different types of microplastics in excess sludge.

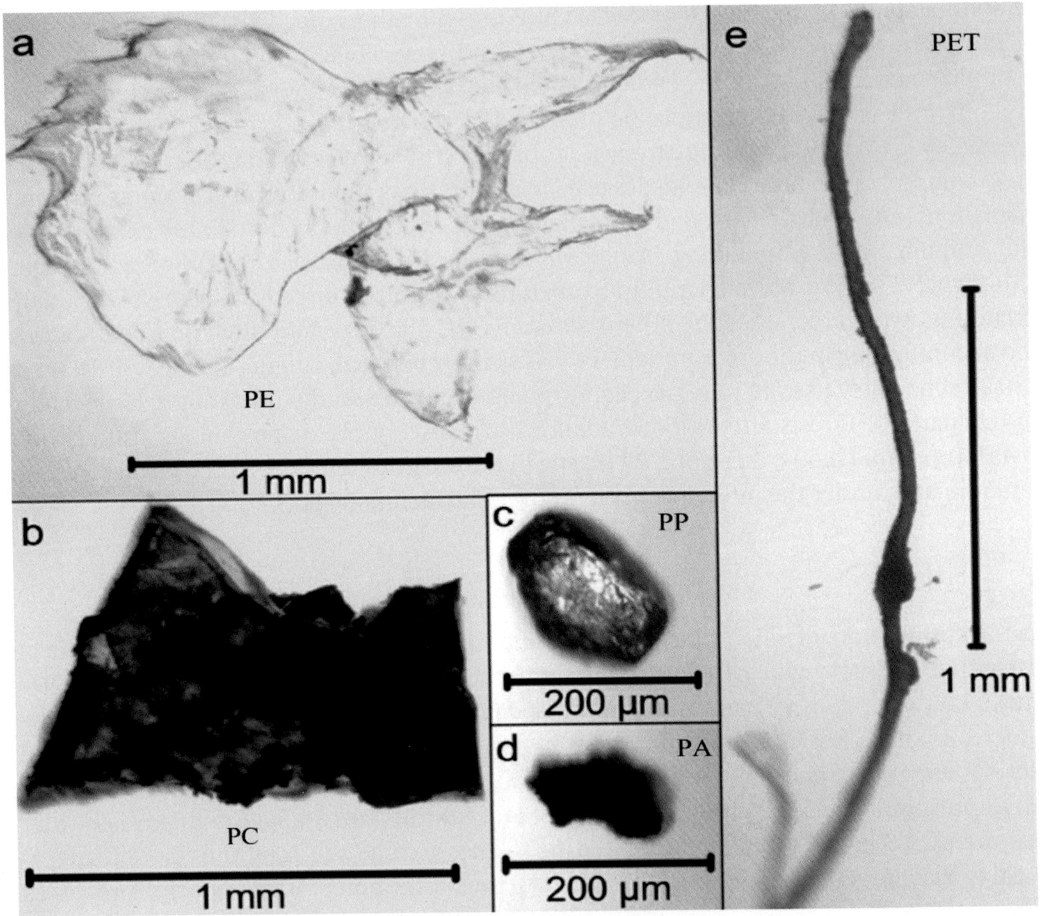

FIGURE 5.4 Microplastics of different form (Yuan et al., 2022).

should be strengthened to prevent secondary microplastic pollution. In the face of increasingly prevalent microplastic pollution, scientists believe that changing the properties of plastics and their recycling methods could provide a solution. Studying characteristics of microplastics such as size, chemical composition, and color, as well as the removal rates of different treatment processes in wastewater treatment plants, will undoubtedly help optimize microplastic design and treatment for better recycling as well as develop effective recycling methods to promote a circular economy.

Acknowledgments

This work was supported by the Science and Technology Program Foundation of Gansu Province (22JR5RA335, 22JR9KA034), Natural Science Foundation of China (52000095, 51868036).

References

Aliabad, M., Nassiri, M., Kor, K., 2019. Microplastics in the surface seawaters of Chabahar Bay, Gulf of Oman (Makran coasts)[J]. Marine Pollution Bulletin 143, 125–133.

Alomar, C., Sureda, A., Capó, X., et al., 2017. Microplastic ingestion by Mullus surmuletus Linnaeus, 1758 fish and its potential for causing oxidative stress. Environmental Research 159, 135–142.

Alvim, C.B., Mendoza-Roca, J.A., Bes-Piá, A., 2020. Wastewater treatment plant as microplastics release source–quantification and identification techniques. Journal of Environmental Management 255, 109739.

Andrady, A., 2017. The plastic in microplastics: a review. Marine Pollution Bulletin 119 (1), 12–22.

Antunes, J., Frias, J., Sobral, P., 2018. Microplastics on the Portuguese coast. Marine Pollution Bulletin 131 (Part A), 294–302.

Browne, A., Crump, P., Niven, S., et al., 2011. Accumulation of microplastic on shorelines woldwide: sources and sinks. Environmental Science & Technology 45 (21), 9175–9179.

Carr, S., Liu, J., Tesoro, A., 2016. Transport and fate of microplastic particles in wastewater treatment plants. Water Research 91, 174–182.

Ding, L., Huang, D., Ouyang, Z., et al., 2022. The effects of microplastics on soil ecosystem: a review. Current Opinion in Environmental Science & Health, 100344.

Fu, Z., Chen, G., Wang, W., et al., 2020. Microplastic pollution research methodologies, abundance, characteristics and risk assessments for aquatic biota in China. Environmental Pollution 266, 115098.

Gao, H., Yan, C., Liu, Q., et al., 2019. Effects of plastic mulching and plastic residue on agricultural production: a meta-analysis. Science of the Total Environment 651, 484–492.

Gouin, T., Avalos, J., Brunning, I., et al., 2015. Use of micro-plastic beads in cosmetic products in Europe and their estimated emissions to the North Sea environment. Seifen Ole Fette Wachse Journal 141 (4), 40–46.

Harley-Nyang, D., Memon, F., Jones, N., et al., 2022. Investigation and analysis of microplastics in sewage sludge and biosolids: a case study from one wastewater treatment works in the UK. Science of the Total Environment 823, 153735.

Kay, P., Hiscoe, R., Moberley, I., et al., 2018. Wastewater treatment plants as a source of microplastics in river catchments. Environmental Science and Pollution Research 25 (20), 20264–20267.

Kedzierski, M., d'Almeida, M., Magueresse, A., et al., 2018. Threat of plastic ageing in marine environment. Adsorption/desorption of micropollutants. Marine Pollution Bulletin 127, 684–694.

Koyuncuoğlu, P., Erden, G., 2021. Sampling, pre-treatment, and identification methods of microplastics in sewage sludge and their effects in agricultural soils: a review. Environmental Monitoring and Assessment 193 (4), 1–28.

Li, J., Zhang, K., Zhang, H., 2018. Adsorption of antibiotics on microplastics. Environmental Pollution 237, 460–467.

Li, Q., Wu, J., Zhao, X., et al., 2019. Separation and identification of microplastics from soil and sewage sludge. Environmental Pollution 254, 113076.

Li, Z., Yang, Y., Chen, X., et al., 2022. A discussion of microplastics in soil and risks for ecosystems and food chains. Chemosphere, 137637.

Liu, X., Yuan, W., Di, M., et al., 2019. Transfer and fate of microplastics during the conventional activated sludge process in one wastewater treatment plant of China. Chemical Engineering Journal 362, 176–182.

Liu, W., Zhang, J., Liu, H., et al., 2021. A review of the removal of microplastics in global wastewater treatment plants: characteristics and mechanisms. Environment International 146, 106277.

Nguyen, M., Hadi, M., Lin, C., et al., 2022. Microplastics in sewage sludge: distribution, toxicity, identification methods, and engineered technologies. Chemosphere 308, 136455.

Nizzetto, L., Futter, M., Langaas, S., 2016. Are agricultural soils dumps for microplastics of urban origin? 50 (20), 10777–10779.

Rolsky, C., Kelkar, V., Driver, E., et al., 2020. Municipal sewage sludge as a source of microplastics in the environment. Current Opinion in Environmental Science & Health 14, 16–22.

Sun, J., Dai, X., Wang, Q., et al., 2019. Microplastics in wastewater treatment plants: detection, occurrence and removal. Water Research 152, 21–37.

Tirkey, A., Upadhyay, L.S.B., 2021. Microplastics: an overview on separation, identification and characterization of microplastics. Marine Pollution Bulletin 170, 112604.

Van-den-Berg, P., Huerta-Lwanga, E., Corradini, F., et al., 2020. Sewage sludge application as a vehicle for microplastics in eastern Spanish agricultural soils. Environmental Pollution 261, 114198.

Waller, C., Griffiths, H., Waluda, C., et al., 2017. Microplastics in the Antarctic marine system: an emerging area of research. Science of the Total Environment 598, 220–227.

Wei, X., Bohlén, M., Lindblad, C., et al., 2021. Microplastics generated from a biodegradable plastic in freshwater and seawater. Water Research 198, 117123.

Yuan, F., Zhao, H., Sun, H., et al., 2022. Investigation of microplastics in sludge from five wastewater treatment plants in Nanjing, China. Journal of Environmental Management 301, 113793.

Zarfl, C., 2019. Promising techniques and open challenges for microplastic identification and quantification in environmental matrices. Analytical and Bioanalytical Chemistry 411 (17), 3743–3756.

Zhang, Z., Chen, Y., 2020. Effects of microplastics on wastewater and sewage sludge treatment and their removal: a review. Chemical Engineering Journal 382, 122955.

Zhou, Q., Zhang, H., Fu, C., et al., 2018. The distribution and morphology of microplastics in coastal soils adjacent to the Bohai Sea and the Yellow Sea. Geoderma 322, 201–208.

Ziccardi, L., Edgington, A., Hentz, K., et al., 2016. Microplastics as vectors for bioaccumulation of hydrophobic organic chemicals in the marine environment: a state-of-the-science review. Environmental Toxicology and Chemistry 35 (7), 1667–1676.

PART II

Behavior of emerging contaminants in organic wastes

6

Occurrence and fate of personal care products and pharmaceuticals in sewage sludge

Muhammad Adil[1] and Pragya Tiwari[2]

[1]PHARMACOLOGY & TOXICOLOGY SECTION, UNIVERSITY OF VETERINARY & ANIMAL SCIENCES, LAHORE, JHANG CAMPUS, JHANG, PAKISTAN; [2]DEPARTMENT OF BIOTECHNOLOGY, YEUNGNAM UNIVERSITY, GYEONGSAN, GYEONGBUK, REPUBLIC OF KOREA

1. Introduction

Personal care products and pharmaceuticals include a variety of substances that are intended for the healthcare of humans or animals. Pharmaceuticals generally encompass cytostatic agents, antihyperlipidemic drugs, antimicrobials, hormones, anti-inflammatory drugs, antiepileptics, antidepressants, anthelmintics, analgesics and antihypertensive drugs (Jelić et al., 2012). Sunscreen, ultraviolet filters, fragrances, insect repellents, antiseptics, disinfectants and preservatives are the most frequently used personal care products (Kosma et al., 2010). Currently, more than 3000 products are being used for the healthcare and well-being of humans and animals (Muthanna and Plósz, 2008). Fig. 6.1 enlists the frequently used classes of personal care products and pharmaceuticals. Personal care products and pharmaceuticals are recognized as emergent contaminants on account of their long-term retention in sewage sludge, water and soil. Despite being primarily meant for beneficial purposes, the indiscriminate use and improper disposal of these products may lead to environmental contamination and consequent ecotoxic effects.

Sewage sludge represents a semi-liquid or liquid mixture of residual substances with 0.25%−12% of solid matter. It is synthesized in wastewater treatment plants during the biochemical process and comprises of a wide range of inorganic and organic elements, along with microorganisms (Verlicchi and Zambello, 2015). Composting, land application, incineration and landfilling are the major disposal routes of sludge. Primary sludge, raw sludge, secondary sludge, mixed sludge and lagoon sludge are the subtypes of untreated sludge (Verlicchi and Zambello, 2015). Raw sludge is obtained through the filtration of raw sewage and consists of suspended solids (Lindberg et al., 2010).

Occurrence and Behavior of Emerging Contaminants in Organic Wastes and Their Control Strategies
https://doi.org/10.1016/B978-0-443-13585-9.00005-7

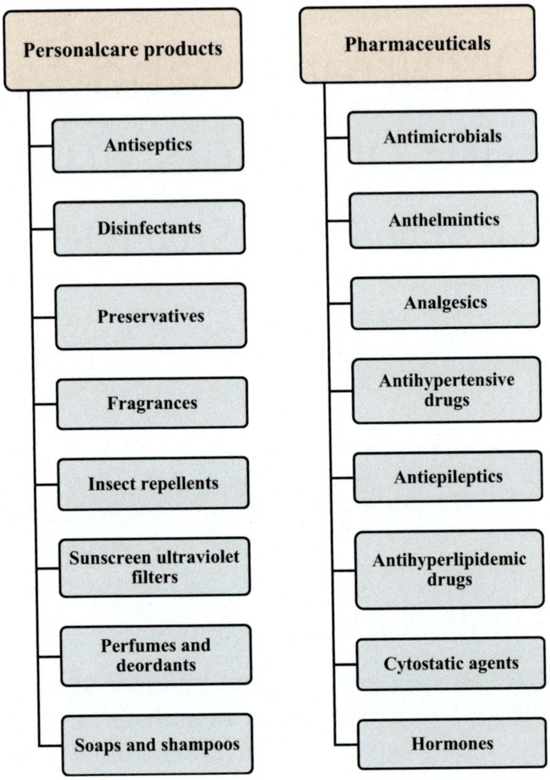

FIGURE 6.1 Typically used classes of personal care products and pharmaceuticals.

Containing around 2%–8% of total dry matter, the primary sludge is collected from primary clarifiers and can be easily subjected to dewatering or thickening. Secondary sludge is synthesized in biological systems such as biological aeriated filters, trickling filters, biological nutrient reactors, membrane biological reactors and conventional activated sludge systems (Verlicchi and Zambello, 2015). Mixed sludge is formed through the mixing of secondary and primary sludges. Lagoon sludge is generated and kept in the aerobic flow basin or anaerobic ponds for stabilization purposes. Treated sludge is subdivided into dried sludge, dewatered sludge, digested sludge, conditioned sludge, biosoilds and composted sludge. Dried sludge is obtained through the reduction of water content via thermal processes. Whereas, reducing the water content through physical/mechanical and chemical ways leads to the production of dewatered/thickened sludge and conditioned sludge, respectively. The anaerobic or aerobic digestion process is carried out to stabilize the sludge and yield the digested sludge. Organic products resulting from the biological stabilization of primary and secondary sludges are referred to as biosoilds. Composted sludge is formed through the aerobic microbial decomposition of organic matter. Ammonia treatment, Fenton's reaction, advanced oxidation, thermal hydrolysis, pasteurization and disinfection are other methods for the production of different types of treated sludge (Verlicchi and Zambello, 2015).

Being rich in diverse nutrients, the sewage sludge is quite useful in restoring the fertility of overexploited land through the enhancement of water-holding capacity and the humus content of the soil (Inglezakis et al., 2014). However, the application of sludge as fertilizer on croplands may lead to contamination of the soil with personal care products and pharmaceuticals or their metabolic derivatives, thereby offering a risk to the entire ecosystem and particularly to the soil-dwelling organisms. Sewage treatment plants represent the chief source of these contaminants in water-containing matrices (Padhye et al., 2014). Despite their relatively low levels in sewage influent, personal care products and pharmaceuticals may deteriorate the elimination efficiency of sewage treatment plants by inhibiting the action of activated sludge bacteria (Thomaidi et al., 2016). So far, the threshold limits of these products in sewage sludge have not been determined. Nevertheless, safety concerns associated with these products have been progressively enhanced during the last 30 years (Schumock et al., 2014). Considering the significant role of sludge-derived personal care products and pharmaceuticals in the context of ecological well-being and public health hazards, this chapter highlights the ubiquity and fate of these pollutants in sewage sludge.

2. Deposition routes of personal care products and pharmaceuticals in sewage sludge

Several pathways, including water treatment plants, inappropriate manufacturer disposal, hospital discharges, sewage treatment plants and domestic wastewater have been identified for the excretion of personal care products and pharmaceuticals into aquatic environments (Liu and Wong, 2013). Manure-based agricultural practices, animal farming, aquaculture facilities and industries also constitute the various pathways for the passage of these pollutants into the environment. Fig. 6.2 illustrates the various deposition routes of personal care products and pharmaceuticals in sewage sludge. Treated hospital-derived and industrial effluents, as well as untreated domestic effluents, may frequently liberate into the water bodies. Some dissolved or unmetabolized pharmaceutical substances are released through the urine and fecal material of animals and/or humans and eventually released into the sewage sludge. Daily use of personal care products for instance creams, skin lotions, sunscreens, toothpastes, shampoos and other types of cosmetics can cause contamination of sewage systems and surface water. Besides, the release of personal care products also occurs during their utilization for recreational purposes such as swimming and sunbathing (Brausch and Rand, 2011). The utilization of reclaimed water for irrigation, improper disposal of previously-treated animal carcasses and surplus drugs to landfills, and runoff of medical products in farmyards are other pathways of exposure to personal care products and pharmaceuticals (Awad et al., 2014).

Sewage sludge represents a significant source of personal care products and pharmaceuticals for prospective ecological contamination. Human and animal drugs can

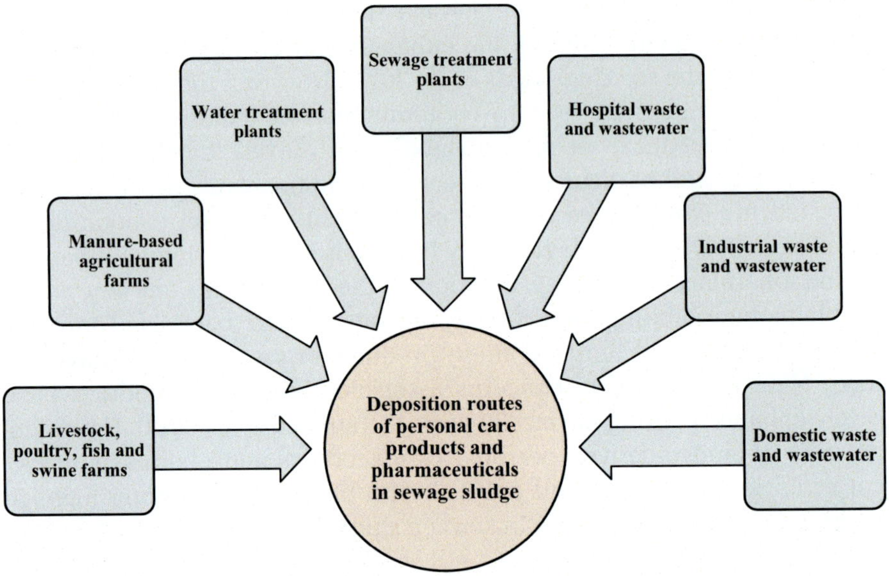

FIGURE 6.2 Deposition routes of personal care products and pharmaceuticals in sewage sludge.

give rise to environmental contamination via direct or indirect ways. The hydrologic cycle is the primary route for environmental dissemination of residual personal care products and pharmaceuticals (Mompelat et al., 2009). The quantities of these products in sewage, drinking water, groundwater and surface water usually vary from ng/L to μg/L (Dai et al., 2015). The liberated substances may either maintain their typical physico-chemical forms and concentrations or, alternatively, be converted into metabolic products inside the aquatic matrices. Many compounds are efficiently converted into their hydrophilic metabolites (Reddersen et al., 2002), whereas, others are primarily excreted and unmetabolized by animals and humans. Generally, the concentrations of metabolic products exceed those of the original drugs due to their effective consumption. Conversely, the concentrations of certain pharmaceutical excipients may virtually remain unaltered.

Biosoilds are frequently utilized as fertilizer on account of their considerable nutrient contents. Thirty-nine percent of the generated sewage sludge was consumed for agricultural applications in European countries during 2010 (Lamastra et al., 2018). The recycling of wastewater treatment by-products and biosolids as organic fertilizer was started as a viable strategy for optimizing waste disposal and reducing the consumption of chemical fertilizers. Nevertheless, the detection of personal care products and pharmaceuticals in sewage sludge immediately raised serious concerns about the potential ecotoxic effects. The ubiquity of personal care products and pharmaceuticals in sewage sludge depends upon their physico-chemical attributes as well as the effectiveness of sewage and wastewater treatment plants (Dong et al., 2016). Carbamazepine, ibuprofen and diclofenac are predominantly found in biosolids and sewage sludge. Hydrophobic compounds exhibit more affinity for organic matter due to their high octanol/water partition co-efficients. Conversely, lipophobic substances have a poor capacity of binding with suspended solids and sludge particles (Morissette et al., 2015).

3. Occurrence of personal care products and pharmaceuticals in sewage sludge

Several antibiotics are usually excreted into the environment as parent drugs without undergoing any biotransformation. Sewage sludge primarily contains macrolides, fluoroquinolones and sulphonamides (Jelić et al., 2012). Tylosin, azithromycin, ciprofloxacin and sulphamethoxazole are commonly found in sewage treatment plant effluent (Huang et al., 2011). Whereas, high amounts of tetracycline and trimethoprim were documented in the effluent and influent of sewage treatment plants (Brown et al., 2006; Leung et al., 2012). β-blockers, including metoprolol, propranolol and atenolol (used to treat hypertension, migraine headache and angina pectoris) have also been recorded in the effluents and influents of sewage treatment plants (Vieno et al., 2007). Natural estrogens, including 17α-ethinylestradiol, estriol, estrone and 17β-estradiol, represent another class of sewage sludge contaminants (Citulski and Farahbakhsh, 2010). Hormones contained in the sewage treatment plants may be biotransformed, incorporated into secondary effluents, and released with the sludge material (Belhaj et al., 2015). Thus, bio-deconjugation is an efficient way to eliminate natural hormones from sewage treatment plants (Liu et al., 2015). Besides, sewage sludge also contained small quantities of anticonvulsants and antihyperlipidemic drugs (Kostich et al., 2014). Other pharmaceuticals detected in sewage above the minimum regulatory levels in sludge include gemfibrozil, flufenamic acid and tetrahydrocannabinol (Carmona et al., 2014).

Parabens are commonly used for the preservation of pharmaceuticals, foodstuffs and cosmetics (Li et al., 2015). Propyl paraben and methyl paraben have been widely recorded in the influents of sewage treatment plants (Carmona et al., 2014; Kasprzyk-Hordern et al., 2008). Nevertheless, parabens are biodegradable and can be adequately eliminated through the activated sludge process (Hernández Leal et al., 2010). Seasonal as well as diurnal variations have been documented in the detected level of parabens in raw sewage samples (Pedrouzo et al., 2009). The bactericidal agent triclosan, which is often added to cosmetic products such as toothpastes, soaps, shampoos, mouth rinses, skin-care lotions and deodorants, has been recorded in the effluents and influents of sewage treatment plants. Nearly 50% of the triclosan is transformed into metabolic products or undergoes the methylation reaction and is ultimately liberated into aquatic environments via effluent discharge (Bester, 2007). Table 6.1 enlists the recorded levels of different personal care products and pharmaceuticals in sewage sludge.

Table 6.1 Recorded levels of different personal care products and pharmaceuticals in sewage sludge.

Product	Detected concentration (ng/g)	References
Azithromycin	830	Clarke and Smith (2011)
	749	Ben et al. (2018)
Carbamazepine	44–11,060	Dong et al. (2016)
	22	Subedi et al. (2017)
Ciprofloxacin	1400–2030	Golet et al. (2002)
	5700–7700	Lindberg et al. (2006)
Diclofenac	30	Samaras et al. (2013)
	1720–11,060	Dong et al. (2016)
Estrone	27.7	Gabet-Giraud et al. (2010)
	39.2	Ben et al. (2018)
Galaxolide	26	Stevens et al. (2003)
	5.4–8.5	Ternes et al. (2005)
Ibuprofen	180	Samaras et al. (2013)
	18.3–145	Subedi et al. (2017)
Naproxen	11.1–35.1	Yu and Wu (2012)
	23.8–72.2	Martín et al. (2012)
Norfloxacin	1540–1960	Golet et al. (2002)
	4700–5800	Lindberg et al. (2006)
Sulfamethoxazole	4.2	Ben et al. (2018)
	31	Subedi et al. (2017)
Tonalide	4	Stevens et al. (2003)
	2.3–4.3	Ternes et al. (2005)
Triclosan	4400	Chen and Bester (2009)
	944	Bourdat-Deschamps et al. (2017)
Trimethoprim	26	McClellan and Halden (2010)
	7.6–13	Subedi et al. (2017)
17β-estradiol	9600	Gabet-Giraud et al. (2010)
	2500	Ben et al. (2018)

N, N-diethyl-m-toluamide, the highly frequent bioactive component of insect repellents, globally persists in relatively low quantities in both the effluents and influents of sewage treatment plants (Yang et al., 2017). However, its concentration is substantially reduced in the winter season on account of diminished consumption (Costanzo et al., 2007). Synthetic musks, for example, polycyclic musks (e.g., toxalide and galaxolide fragrances) and nitro musks (for instance, musk ketone and musk xylene), are generally used in various products, including detergents, deodorants and soaps (Daughton and Ternes, 1999). Musk xylene and musk ketone were detected in lesser concentrations in 83%–90% of the tested sewage treatment plant effluents (Brausch and Rand, 2011). Likewise, high levels of toxalide and galaxolide fragrances are usually found in the influents of sewage treatment plants. In contrast to synthetic musks, the concentrations of hydrophobic personal care products, including galaxolide, triclocarbon and triclosan, were comparatively higher in sewage sludge (Stasinakis, 2012). Ultraviolet filters are used in lotions, sunscreens and other types of cosmetics for protecting the skin from the harmful effects of ultraviolet radiation. Low quantities of ultraviolet filters were found in all the examined sewage treatment plant effluents (Balmer et al., 2005). Moreover, relatively high levels of ultraviolet filters were noticed during hot and humid climatic conditions (Bester, 2007; Tsui et al., 2014). More than 80% of the effluents and influents of sewage treatment plants also contained other products such as 2-ethyl-hexyl-4-trimethoxycinnamate, 2,4-dihydroxybenzophenone, benzophenone-4 and benzophenone-3 (Tsui et al., 2014).

4. Analytical techniques for quantification of personal care products and pharmaceuticals in sewage sludge

Various kinds of sludge samples should be collected to effectively analyze the distribution of personal care products and pharmaceuticals in the bulk of sewage sludge. Inappropriate collection of samples or prolonged storage may lead to false measurements (Kot-Wasik et al., 2007). Sampling locations are based on the nature of the required test samples. Non-transparent vessels are recommended for collecting samples of photosensitive substances such as metoprolol, propranolol and triclosan (Andreozzi et al., 2003). The collected samples may be consequently subjected to freezing, lyophilization, air drying, membrane filtration, homogenization or maceration for a few minutes (Abril et al., 2018; Cerqueira et al., 2018). Eventually, the samples should be kept at −20°C for subsequent analysis (Azzouz and Ballesteros, 2012). Delayed analysis of stored samples may cause an alteration in the composition.

Solid-liquid extraction is primarily used for separating the components of interest from other high-molecular-weight substances of the solid sludge. Soxhlet extraction represents a typical example of this frequently employed technique. Nevertheless, it is time-consuming and labor-intensive, and necessitates larger volumes of samples and organic solvents. Alternatively, ultrasound-assisted extraction can be applied as a time-saving and eco-friendly procedure. Besides, microwave-assisted extraction can be utilized for isolating the constituents of interest from the complex sludge samples. In addition to its need for relatively smaller quantities of solvents, this method is also characterized by a rapid and controlled extraction process (Eskilsson and Björklund, 2000). However, it essentially requires a filtration step (following extraction) and expensive equipment. Accelerated solvent extraction also referred to as pressurized liquid extraction is a fully automated technique that utilizes small amounts of liquid extractants like acetone, ethanol and hexane at high temperature and pressure conditions (Nieto et al., 2010). Despite its efficient extraction capacity, a clean-up process is often essential for the removal of complex matrix from the extracts. Pressurized hot water extraction is another cost-effective and eco-friendly method for separating the target analytes from sludge samples using pressurized water at high temperatures.

Quick, easy, cheap, effective, rugged and safe extraction is a novel extraction method that requires small quantities of organic solvents and provides high selectivity as well as good recovery (Pérez-Lemus et al., 2019). Apart from being inexpensive, it needs less time and amounts of solvent in comparison with accelerated solvent extraction. In matrix solid-phase dispersion, the sample is homogenized on a solid surface using a dispersing agent that helps to disrupt the sample and extract the ingredients of interest through an elution solvent (Capriotti et al., 2010). Various methods for the extraction of personal care products and pharmaceuticals in sewage sludge have been enlisted in Fig. 6.3. The majority of the aforementioned extraction methods are devoid of adequate selectivity and therefore necessitate the subsequent conduction of a clean-up process. Typically used clean-up agents include primary secondary amine and C18. Personal care products and pharmaceuticals extracted from sewage sludge are commonly subjected to a clean-up step using solid-phase extraction (López-Serna et al., 2018). This technique is simple, rapid and can be conveniently linked with liquid chromatography. Substances having different solubility in water and non-polar organic solvents can be easily segregated through the process of liquid-liquid extraction.

Latest assays for the quantification of contaminants in environmental samples are predominantly based upon the application of high-performance liquid chromatography and gas chromatography, which are capable of efficiently segregating the complex

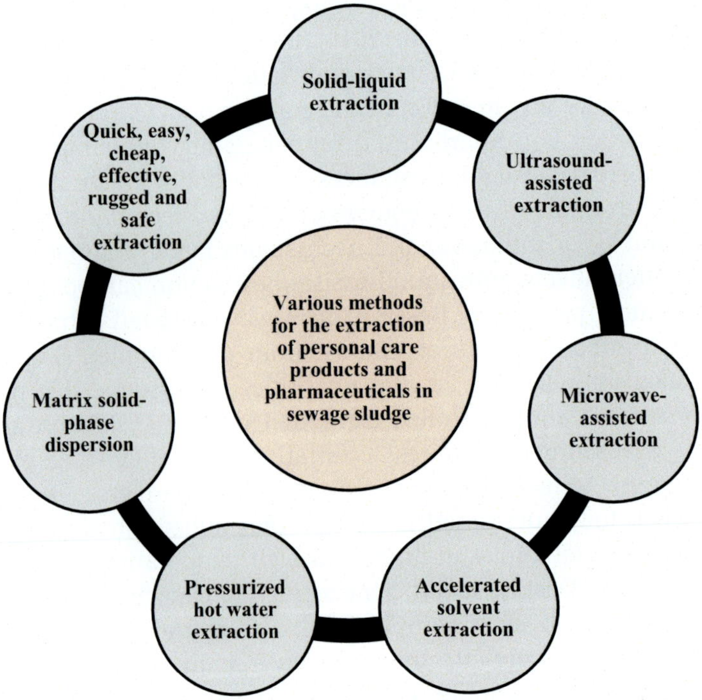

FIGURE 6.3 Various methods for the extraction of personal care products and pharmaceuticals in sewage sludge.

mixtures. Both methanol and acetonitrile are utilized in terms of organic mobile phases in the liquid chromatographic assays. The pH of the mobile phase is usually adjusted to neutral or acidic values for multi-residual analyses. Tri-n-butylamine or other volatile ion-pairing agents can be used for separating acidic drugs by means of ion-pair liquid chromatography. Formic acid, acetic acid, oxalic acid, methyl ammonium acetate or ammonium acetate can be consumed to enhance the sensitivity of mass spectrometry-based detection of antibiotics (Debska et al., 2005; Hilton and Thomas, 2003). Liquid chromatography-based separation methods are frequently carried out for the estimation of personal care products and pharmaceuticals in sludge samples. Biomacromolecules, including proteins, peptides, sugars and rubbers, can be efficiently separated and analyzed by means of size-exclusion chromatography. Gel permeation chromatography is also employed in conjunction with other clean-up procedures. Ultra-high-performance liquid chromatography is characterized by significantly improved sensitivity, speed and resolution. Chromatographic separation is also employed in tandem with mass spectrometry for the assessment of personal care products and pharmaceuticals in sewage sludge. In particular, the relatively less expansive method of gas chromatography is widely performed throughout the world (Schoeman et al., 2017). Gas chromatography can be combined with other modalities such as electron capture detection or mass spectrometry (Daso et al., 2012). Unlike gas chromatography-mass spectrometry, the derivatization step is bypassed in the case of liquid chromatography-mass spectrometry. Versatility and convenience of sample preparation are additional advantages of liquid chromatography-mass spectrometry.

5. Elimination of personal care products and pharmaceuticals from sewage sludge

The excretion of personal care products and pharmaceuticals from sewage treatment plants is a multifaceted process, governed by the adsorption capacity of activated sludge as well as the biodegradability, volatility, solubility and hydrophilicity of pollutants (Evgenidou et al., 2015). More than 90% of some products (such as parabens) can be appropriately removed, whereas, only partial excretion of most substances occurs by means of primary and secondary treatment practices of the conventional sewage plants (González-Mariño et al., 2011). Hydrophilic personal care products and pharmaceuticals are partially removed by means of the primary treatment processes (Luo et al., 2014). Owing to their hydrophilic features, the elimination efficacies of sulpiride, trimethoprim, metoprolol, carbamazepine, N, N-diethylm-toluamide and caffeine were less than 20% in the primary sedimentation tanks (Wang et al., 2014). Conversely, around 40% of calaxilid and toxalide fragrances can be adequately eliminated through primary treatment on account of greater partition coefficients between the solid and liquid media (Stamatis and Konstantinou, 2013). Sewage treatment methods are categorized into various types (Poerio et al., 2019), as shown in Fig. 6.4.

Preliminary treatment	Primary treatment	Secondary treatment	Tertiary treatment
• Aeration	• Adsorption	• Activated sludge process	• Activated carbon treatment
• Degasification	• Chorination	• Aerobic digestion	• Disinfection
• Filtration	• Coagulation	• Anaerobic digestion	• Filtration
• Floatation	• Neutralization	• Lagooning	• Ion exchange
• Sedimentation	• Ozonation	• Oxidation ponds	• Membrane techniques
• Screening		• Trickling filtration	• Oxidation

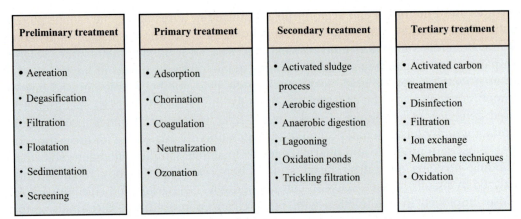

FIGURE 6.4 Various types of sewage sludge treatment methods.

Secondary treatment predominantly includes biological processes for the elimination of personal care products and pharmaceuticals via biodegradation, biotransformation, adsorption and partition (Miao et al., 2005). More than 80% of triclosan was eliminated in the activated sludge process via biotransformation (Federle et al., 2002). Only 25% of diclofenac, whereas, 75%–87% of ketoprofen, ibuprofen and caffeine were biodegraded through the secondary treatment process (Salgado et al., 2012). Conversion to metabolites or by-products during secondary treatment can impair the elimination of personal care products and pharmaceuticals (Miao et al., 2005). Exposure to β-blockers and antimicrobial agents can inhibit the action of activated sludge bacteria and thereby reduce the elimination capacity of sewage treatment plants (Dann and Hontela, 2011; Drury et al., 2013).

Triclosan was effectively broken down by different bacteria to yield 2,4-dichlorophenol, catechol and phenol, following the co-utilization of acetate under methanogenic conditions. Thermophilic and mesophilic digestion processes resulted in over 60% removal of tonalide and galaxolide (Carballa et al., 2006). Similarly, 50% removal of cashmeran, phantolide, celestolide, tonalide and galaxolide from the sewage treatment plant was achieved by means of anaerobic digestion (Kupper et al., 2006). Anaerobic digestion of sludge led to 100% and 14% removal of naproxen and bisoprol during 14 and 161 days, respectively (Lahti and Oikari, 2011). Conversely, slight or no removal of traesolide, phantolide, celestolide, cashmerane, tonalide and galaxolide has also been linked with the anaerobic digestion of sludge (Clara et al., 2011). Fluoroquinolones were not efficiently removed through the digestion of sludge under methanogenic conditions (Golet et al., 2003). Likewise, several investigations have described the recalcitrance of carbamazepine and diclofenac during anaerobic sludge digestion (Carballa et al., 2006; Stamatelatou et al., 2003). Mass removal of progesterone, androstenedione, testosterone, 17α-ethinyl estradiol, estriol, 17β-estradiol and estrone was carried out through anaerobic sludge digestion (Esperanza et al., 2007).

Nevertheless, less than 40% removal of estriol, estrone and 17β-estradiol was achieved following the anaerobic sludge digestion (Muller et al., 2010). Despite causing more than 50% biodegradation of the total steroid estrogens (comprising of 17α-ethinyl estradiol, estrone sulfate conjugate, 17β-estradiol, estriol and estrone) in primary sludge, thermophilic and mesophilic digestion techniques failed to significantly remove the same substances from mixed sludge samples (Paterakis et al., 2012).

The inefficient elimination of various personal care products and pharmaceuticals via secondary treatment processes necessitates the application of tertiary treatment in terms of membrane separation, sand filtration and advanced oxidation processes before the conduction of ultraviolet or chemical disinfection (Yang et al., 2017). The sand filtration system failed to eliminate triclosan in the sewage treatment plant (McAvoy et al., 2002). Moreover, chemical treatment failed to properly remove the light-resistant ultraviolet filters (4-methyl-benzylidene-camphor) and clarithromycin (Brausch and Rand, 2011; Zhang et al., 2013). The majority of parabens, excluding the di-chlorinated substances, were properly eliminated by means of ozonation (Tay et al., 2010). The concurrent use of ultraviolet systems and biological processes can substantially augment the excretion of personal care products and pharmaceuticals in sewage treatment plants (Wang et al., 2014). Over 90% of the tested personal care products and pharmaceuticals were efficiently removed through the integration of sand filtration and ozonation with the activated sludge technique (Nakada et al., 2007). Contrary to the application of a single biological process, the combination of an advanced oxidation process and a bioreactor significantly improved the elimination of ifosfamide and cyclophosphamide.

6. Conclusions and future perspectives

Despite the strict regulatory measures, the consumption of personal care products and pharmaceuticals cannot be restricted due to their widespread applications in animals and humans. Different substances belonging to the same chemical group may considerably differ in terms of their biodegradation capacities. The elimination capacity of personal care products is usually greater than that of pharmaceuticals. Bioavailability, temperature and characteristics of sludge influence the fate of personal care products and pharmaceuticals. Although these products are extensively excreted into aquatic environments, their removal mechanisms and degradation products have not been precisely elucidated. The ecotoxic impacts of various personal care products and pharmaceuticals on different organisms significantly vary. There is a paucity of data about the bioaccumulation capacity of these contaminants in animal tissues and plants. Advancements in environmental risk assessment, especially the estimation of predicted no-effect concentrations, are critically required.

Advanced oxidation processes, membrane bioreactors, tertiary treatment along with nutrient retrieval and activated sludge processes are typically used techniques for the exclusion of personal care products and pharmaceuticals. Nevertheless, some products, such as antimicrobial drugs, are not efficiently removed using conventional sewage treatment plants. Despite the better removal efficiency of chemical methods such as ozonation and chlorination, chemical-induced toxicity can be a possible drawback. The concurrent use of other methods with anaerobic digestion has shown promising results. Although the ubiquity and fate of personal care products and pharmaceuticals in sewage sludge and the removal efficacy of various elimination techniques have been extensively investigated, comprehensive studies on their removal mechanisms and antagonistic effects on biological processes are still lacking. Hydrothermal liquefaction or biodrying and hydro-thermal carbonization have been examined as prospective technologies for diminishing the concentrations of contaminants before the administration of sewage sludge on agricultural land. The consumption of treated sludge as fertilizer necessitates thorough caution on account of associated likely ecotoxicological concerns.

Traditional sewage treatment plants are modeled for the excretion of suspended solids and organic matter. Accordingly, high levels of personal care products and pharmaceuticals contained in reclaimed water and sewage sludge may possibly enter the aquatic system and food chain. Presently, insufficient information is available regarding the elimination mechanisms of personal care products and pharmaceuticals, and their antagonistic effects on biological processes in the treatment plants for sewage and water. Latest techniques for the elimination of personal care products and pharmaceuticals, including advanced oxidation processes, carbon absorption and membrane filtration, have been introduced. Nevertheless, various elimination technologies exhibit considerable variations in terms of cost and performance. Consequently, it is essential to ascertain the influence of personal care products and pharmaceuticals on the microbial community, process stability and treatment performance of biological phenomena in water and sewage treatment plants. The exclusion of many products and/or their metabolites is difficult to achieve by means of sludge treatment processes. Specific conditions that are necessary for reducing the accumulation of personal care products and pharmaceuticals in the sewage sludge should be properly determined.

References

Abril, C., Santos, J.L., Malvar, J.L., Martín, J., Aparicio, I., Alonso, E., 2018. Determination of perfluorinated compounds, bisphenol A, anionic surfactants and personal care products in digested sludge, compost and soil by liquid-chromatography-tandem mass spectrometry. Journal of Chromatography A 1576, 34–41.

Andreozzi, R., Raffaele, M., Nicklas, P., 2003. Pharmaceuticals in STP effluents and their solar photodegradation in aquatic environment. Chemosphere 50 (10), 1319–1330.

Awad, Y.M., Kim, S.-C., Abd El-Azeem, S.A., Kim, K.-H., Kim, K.-R., Kim, K., Ok, Y.S., 2014. Veterinary antibiotics contamination in water, sediment, and soil near a swine manure composting facility. Environmental Earth Sciences 71, 1433–1440.

Azzouz, A., Ballesteros, E., 2012. Combined microwave-assisted extraction and continuous solid-phase extraction prior to gas chromatography–mass spectrometry determination of pharmaceuticals, personal care products and hormones in soils, sediments and sludge. Science of the Total Environment 419, 208–215.

Balmer, M.E., Buser, H.-R., Müller, M.D., Poiger, T., 2005. Occurrence of some organic UV filters in wastewater, in surface waters, and in fish from Swiss lakes. Environmental Science and Technology 39 (4), 953–962.

Belhaj, D., Baccar, R., Jaabiri, I., Bouzid, J., Kallel, M., Ayadi, H., Zhou, J.L., 2015. Fate of selected estrogenic hormones in an urban sewage treatment plant in Tunisia (North Africa). Science of the Total Environment 505, 154–160.

Ben, W., Zhu, B., Yuan, X., Zhang, Y., Yang, M., Qiang, Z., 2018. Occurrence, removal and risk of organic micropollutants in wastewater treatment plants across China: comparison of wastewater treatment processes. Water Research 130, 38–46.

Bester, K., 2007. Personal care compounds in the environment pathways, fate, and methods for determination. Environmental Engineering and Management Journal 6 (3), 259–260.

Bourdat-Deschamps, M., Ferhi, S., Bernet, N., Feder, F., Crouzet, O., Patureau, D., Benoit, P., 2017. Fate and impacts of pharmaceuticals and personal care products after repeated applications of organic waste products in long-term field experiments. Science of the Total Environment 607, 271–280.

Brausch, J.M., Rand, G.M., 2011. A review of personal care products in the aquatic environment: environmental concentrations and toxicity. Chemosphere 82 (11), 1518–1532.

Brown, K.D., Kulis, J., Thomson, B., Chapman, T.H., Mawhinney, D.B., 2006. Occurrence of antibiotics in hospital, residential, and dairy effluent, municipal wastewater, and the Rio Grande in New Mexico. Science of the Total Environment 366 (2–3), 772–783.

Capriotti, A.L., Cavaliere, C., Giansanti, P., Gubbiotti, R., Samperi, R., Laganà, A., 2010. Recent developments in matrix solid-phase dispersion extraction. Journal of Chromatography A 1217 (16), 2521–2532.

Carballa, M., Omil, F., Alder, A., Lema, J., 2006. Comparison between the conventional anaerobic digestion of sewage sludge and its combination with a chemical or thermal pre-treatment concerning the removal of pharmaceuticals and personal care products. Water Science and Technology 53 (8), 109–117.

Carmona, E., Andreu, V., Picó, Y., 2014. Occurrence of acidic pharmaceuticals and personal care products in Turia River Basin: from waste to drinking water. Science of the Total Environment 484, 53–63.

Cerqueira, M.B., Soares, K.L., Caldas, S.S., Primel, E.G., 2018. Sample as solid support in MSPD: a new possibility for determination of pharmaceuticals, personal care and degradation products in sewage sludge. Chemosphere 211, 875–883.

Chen, X., Bester, K., 2009. Determination of organic micro-pollutants such as personal care products, plasticizers and flame retardants in sludge. Analytical and Bioanalytical Chemistry 395, 1877–1884.

Citulski, J.A., Farahbakhsh, K., 2010. Fate of endocrine-active compounds during municipal biosolids treatment: a review. Environmental Science and Technology 44 (22), 8367–8376.

Clara, M., Gans, O., Windhofer, G., Krenn, U., Hartl, W., Braun, K., Scheffknecht, C., 2011. Occurrence of polycyclic musks in wastewater and receiving water bodies and fate during wastewater treatment. Chemosphere 82 (8), 1116–1123.

Clarke, B.O., Smith, S.R., 2011. Review of 'emerging' organic contaminants in biosolids and assessment of international research priorities for the agricultural use of biosolids. Environment International 37 (1), 226–247.

Costanzo, S., Watkinson, A., Murby, E., Kolpin, D.W., Sandstrom, M.W., 2007. Is there a risk associated with the insect repellent DEET (N, N-diethyl-m-toluamide) commonly found in aquatic environments? Science of the Total Environment 384 (1–3), 214–220.

Dai, G., Wang, B., Huang, J., Dong, R., Deng, S., Yu, G., 2015. Occurrence and source apportionment of pharmaceuticals and personal care products in the Beiyun River of Beijing, China. Chemosphere 119, 1033–1039.

Dann, A.B., Hontela, A., 2011. Triclosan: environmental exposure, toxicity and mechanisms of action. Journal of Applied Toxicology 31 (4), 285–311.

Daso, A.P., Fatoki, O.S., Odendaal, J.P., Olujimi, O.O., 2012. Occurrence of selected polybrominated diphenyl ethers and 2, 2′, 4, 4′, 5, 5′-hexabromobiphenyl (BB-153) in sewage sludge and effluent samples of a wastewater-treatment plant in Cape Town, South Africa. Archives of Environmental Contamination and Toxicology 62, 391–402.

Daughton, C.G., Ternes, T.A., 1999. Pharmaceuticals and personal care products in the environment: agents of subtle change? Environmental Health Perspectives 107 (Suppl. 6), 907–938.

Debska, J., Kot-Wasik, A., Namiesnik, J., 2005. Determination of nonsteroidal antiinflammatory drugs in water samples using liquid chromatography coupled with diode-array detector and mass spectrometry. Journal of Separation Science 28 (17), 2419–2426.

Dong, R., Yu, G., Guan, Y., Wang, B., Huang, J., Deng, S., Wang, Y., 2016. Occurrence and discharge of pharmaceuticals and personal care products in dewatered sludge from WWTPs in Beijing and Shenzhen. Emerging Contaminants 2 (1), 1–6.

Drury, B., Scott, J., Rosi-Marshall, E.J., Kelly, J.J., 2013. Triclosan exposure increases triclosan resistance and influences taxonomic composition of benthic bacterial communities. Environmental Science and Technology 47 (15), 8923–8930.

Eskilsson, C.S., Björklund, E., 2000. Analytical-scale microwave-assisted extraction. Journal of Chromatography A 902 (1), 227–250.

Esperanza, M., Suidan, M.T., Marfil-Vega, R., Gonzalez, C., Sorial, G.A., McCauley, P., Brenner, R., 2007. Fate of sex hormones in two pilot-scale municipal wastewater treatment plants: conventional treatment. Chemosphere 66 (8), 1535–1544.

Evgenidou, E.N., Konstantinou, I.K., Lambropoulou, D.A., 2015. Occurrence and removal of transformation products of PPCPs and illicit drugs in wastewaters: a review. Science of the Total Environment 505, 905–926.

Federle, T.W., Kaiser, S.K., Nuck, B.A., 2002. Fate and effects of triclosan in activated sludge. Environmental Toxicology and Chemistry: International Journal 21 (7), 1330–1337.

Gabet-Giraud, V., Miege, C., Herbreteau, B., Hernandez-Raquet, G., Coquery, M., 2010. Development and validation of an analytical method by LC-MS/MS for the quantification of estrogens in sewage sludge. Analytical and Bioanalytical Chemistry 396, 1841–1851.

Golet, E.M., Strehler, A., Alder, A.C., Giger, W., 2002. Determination of fluoroquinolone antibacterial agents in sewage sludge and sludge-treated soil using accelerated solvent extraction followed by solid-phase extraction. Analytical Chemistry 74 (21), 5455–5462.

Golet, E.M., Xifra, I., Siegrist, H., Alder, A.C., Giger, W., 2003. Environmental exposure assessment of fluoroquinolone antibacterial agents from sewage to soil. Environmental Science and Technology 37 (15), 3243–3249.

González-Mariño, I., Quintana, J.B., Rodríguez, I., Cela, R., 2011. Evaluation of the occurrence and biodegradation of parabens and halogenated by-products in wastewater by accurate-mass liquid chromatography-quadrupole-time-of-flight-mass spectrometry (LC-QTOF-MS). Water Research 45 (20), 6770–6780.

Hernández Leal, L.a., Vieno, N., Temmink, H., Zeeman, G., Buisman, C.J., 2010. Occurrence of xenobiotics in gray water and removal in three biological treatment systems. Environmental Science and Technology 44 (17), 6835–6842.

Hilton, M.J., Thomas, K.V., 2003. Determination of selected human pharmaceutical compounds in effluent and surface water samples by high-performance liquid chromatography–electrospray tandem mass spectrometry. Journal of Chromatography A 1015 (1–2), 129–141.

Huang, C.-H., Renew, J.E., Smeby, K.L., Pinkston, K., Sedlak, D.L., 2011. Assessment of potential antibiotic contaminants in water and preliminary occurrence analysis. Journal of contemporary water research and education 120 (1), 4.

Inglezakis, V.J., Zorpas, A.A., Karagiannidis, A., Samaras, P., Voukkali, I., Sklari, S., 2014. European Union legislation on sewage sludge management. Fresenius Environmental Bulletin 23 (2), 635–639.

Jelić, A., Gros, M., Petrović, M., Ginebreda, A., Barceló, D., 2012. Occurrence and elimination of pharmaceuticals during conventional wastewater treatment. In: Guasch, H., Ginebreda, A., Geiszinger, A. (Eds.), Emerging and Priority Pollutants in Rivers: Bringing Science into River Management Plans. Springer, Berlin, Heidelberg, pp. 1–23. https://doi.org/10.1007/978-3-642-25722-3_1.

Kasprzyk-Hordern, B., Dinsdale, R.M., Guwy, A.J., 2008. Multiresidue methods for the analysis of pharmaceuticals, personal care products and illicit drugs in surface water and wastewater by solid-phase extraction and ultra performance liquid chromatography–electrospray tandem mass spectrometry. Analytical and Bioanalytical Chemistry 391, 1293–1308.

Kosma, C.I., Lambropoulou, D.A., Albanis, T.A., 2010. Occurrence and removal of PPCPs in municipal and hospital wastewaters in Greece. Journal of Hazardous Materials 179 (1–3), 804–817.

Kostich, M.S., Batt, A.L., Lazorchak, J.M., 2014. Concentrations of prioritized pharmaceuticals in effluents from 50 large wastewater treatment plants in the US and implications for risk estimation. Environmental Pollution 184, 354–359.

Kot-Wasik, A., Dębska, J., Namieśnik, J., 2007. Analytical techniques in studies of the environmental fate of pharmaceuticals and personal-care products. TrAC, Trends in Analytical Chemistry 26 (6), 557–568.

Kupper, T., Plagellat, C., Brändli, R., De Alencastro, L., Grandjean, D., Tarradellas, J., 2006. Fate and removal of polycyclic musks, UV filters and biocides during wastewater treatment. Water Research 40 (14), 2603–2612.

Lahti, M., Oikari, A., 2011. Microbial transformation of pharmaceuticals naproxen, bisoprolol, and diclofenac in aerobic and anaerobic environments. Archives of Environmental Contamination and Toxicology 61, 202–210.

Lamastra, L., Suciu, N.A., Trevisan, M., 2018. Sewage sludge for sustainable agriculture: contaminants' contents and potential use as fertilizer. Chemical and Biological Technologies in Agriculture 5 (1), 1–6.

Leung, H.W., Minh, T., Murphy, M.B., Lam, J.C., So, M.K., Martin, M., Richardson, B.J., 2012. Distribution, fate and risk assessment of antibiotics in sewage treatment plants in Hong Kong, South China. Environment International 42, 1–9.

Li, W., Shi, Y., Gao, L., Liu, J., Cai, Y., 2015. Occurrence, fate and risk assessment of parabens and their chlorinated derivatives in an advanced wastewater treatment plant. Journal of Hazardous Materials 300, 29–38.

Lindberg, R.H., Fick, J., Tysklind, M., 2010. Screening of antimycotics in Swedish sewage treatment plants–Waters and sludge. Water Research 44 (2), 649–657.

Lindberg, R.H., Olofsson, U., Rendahl, P., Johansson, M.I., Tysklind, M., Andersson, B.A., 2006. Behavior of fluoroquinolones and trimethoprim during mechanical, chemical, and active sludge treatment of sewage water and digestion of sludge. Environmental Science and Technology 40 (3), 1042–1048.

Liu, J.-L., Wong, M.-H., 2013. Pharmaceuticals and personal care products (PPCPs): a review on environmental contamination in China. Environment International 59, 208–224.

Liu, Z.-h., Lu, G.-n., Yin, H., Dang, Z., Rittmann, B., 2015. Removal of natural estrogens and their conjugates in municipal wastewater treatment plants: a critical review. Environmental Science and Technology 49 (9), 5288–5300.

López-Serna, R., Marín-de-Jesús, D., Irusta-Mata, R., García-Encina, P.A., Lebrero, R., Fdez-Polanco, M., Muñoz, R., 2018. Multiresidue analytical method for pharmaceuticals and personal care products in sewage and sewage sludge by online direct immersion SPME on-fiber derivatization–GCMS. Talanta 186, 506–512.

Luo, Y., Guo, W., Ngo, H.H., Nghiem, L.D., Hai, F.I., Zhang, J., Wang, X.C., 2014. A review on the occurrence of micropollutants in the aquatic environment and their fate and removal during wastewater treatment. Science of the Total Environment 473, 619–641.

Martín, J., Camacho-Muñoz, D., Santos, J., Aparicio, I., Alonso, E., 2012. Occurrence of pharmaceutical compounds in wastewater and sludge from wastewater treatment plants: removal and ecotoxicological impact of wastewater discharges and sludge disposal. Journal of Hazardous Materials 239, 40–47.

McAvoy, D.C., Schatowitz, B., Jacob, M., Hauk, A., Eckhoff, W.S., 2002. Measurement of triclosan in wastewater treatment systems. Environmental Toxicology and Chemistry: International Journal 21 (7), 1323–1329.

McClellan, K., Halden, R.U., 2010. Pharmaceuticals and personal care products in archived US biosolids from the 2001 EPA national sewage sludge survey. Water Research 44 (2), 658–668.

Miao, X.-S., Yang, J.-J., Metcalfe, C.D., 2005. Carbamazepine and its metabolites in wastewater and in biosolids in a municipal wastewater treatment plant. Environmental Science and Technology 39 (19), 7469–7475.

Mompelat, S., Le Bot, B., Thomas, O., 2009. Occurrence and fate of pharmaceutical products and by-products, from resource to drinking water. Environment International 35 (5), 803–814.

Morissette, M.-F., Duy, S.V., Arp, H., Sauvé, S., 2015. Sorption and desorption of diverse contaminants of varying polarity in wastewater sludge with and without alum. Environmental Science: Processes & Impacts 17 (3), 674–682.

Muller, M., Combalbert, S., Delgenès, N., Bergheaud, V., Rocher, V., Benoît, P., Hernandez-Raquet, G., 2010. Occurrence of estrogens in sewage sludge and their fate during plant-scale anaerobic digestion. Chemosphere 81 (1), 65–71.

Muthanna, T., Plósz, B., 2008. The impact of hospital sewage discharge on the assessment of environmental risk posed by priority pharmaceuticals: hydrodynamic modelling and measurements. International Conference on Urban Drainage, Edinburgh, Scotland, UK.-2008.-C.

Nakada, N., Shinohara, H., Murata, A., Kiri, K., Managaki, S., Sato, N., Takada, H., 2007. Removal of selected pharmaceuticals and personal care products (PPCPs) and endocrine-disrupting chemicals (EDCs) during sand filtration and ozonation at a municipal sewage treatment plant. Water Research 41 (19), 4373–4382.

Nieto, A., Borrull, F., Pocurull, E., Marcé, R.M., 2010. Pressurized liquid extraction: a useful technique to extract pharmaceuticals and personal-care products from sewage sludge. TrAC, Trends in Analytical Chemistry 29 (7), 752–764.

Padhye, L.P., Yao, H., Kung'u, F.T., Huang, C.-H., 2014. Year-long evaluation on the occurrence and fate of pharmaceuticals, personal care products, and endocrine disrupting chemicals in an urban drinking water treatment plant. Water Research 51, 266–276.

Paterakis, N., Chiu, T., Koh, Y., Lester, J., McAdam, E., Scrimshaw, M., Cartmell, E., 2012. The effectiveness of anaerobic digestion in removing estrogens and nonylphenol ethoxylates. Journal of Hazardous Materials 199, 88–95.

Pedrouzo, M., Borrull, F., Marcé, R.M., Pocurull, E., 2009. Ultra-high-performance liquid chromatography–tandem mass spectrometry for determining the presence of eleven personal care products in surface and wastewaters. Journal of Chromatography A 1216 (42), 6994–7000.

Pérez-Lemus, N., López-Serna, R., Pérez-Elvira, S.I., Barrado, E., 2019. Analytical methodologies for the determination of pharmaceuticals and personal care products (PPCPs) in sewage sludge: a critical review. Analytica Chimica Acta 1083, 19–40.

Poerio, T., Piacentini, E., Mazzei, R., 2019. Membrane processes for microplastic removal. Molecules 24 (22), 4148.

Reddersen, K., Heberer, T., Dünnbier, U., 2002. Identification and significance of phenazone drugs and their metabolites in ground-and drinking water. Chemosphere 49 (6), 539–544.

Salgado, R., Marques, R., Noronha, J., Carvalho, G., Oehmen, A., Reis, M., 2012. Assessing the removal of pharmaceuticals and personal care products in a full-scale activated sludge plant. Environmental Science and Pollution Research 19, 1818–1827.

Samaras, V.G., Stasinakis, A.S., Mamais, D., Thomaidis, N.S., Lekkas, T.D., 2013. Fate of selected pharmaceuticals and synthetic endocrine disrupting compounds during wastewater treatment and sludge anaerobic digestion. Journal of Hazardous Materials 244, 259–267.

Schoeman, C., Dlamini, M., Okonkwo, O., 2017. The impact of a wastewater treatment works in Southern Gauteng, South Africa on efavirenz and nevirapine discharges into the aquatic environment. Emerging Contaminants 3 (2), 95–106.

Schumock, G.T., Li, E.C., Suda, K.J., Matusiak, L.M., Hunkler, R.J., Vermeulen, L.C., Hoffman, J.M., 2014. National trends in prescription drug expenditures and projections for 2014. American Journal of Health-System Pharmacy 71 (6), 482–499.

Stamatelatou, K., Frouda, C., Fountoulakis, M., Drillia, P., Kornaros, M., Lyberatos, G., 2003. Pharmaceuticals and health care products in wastewater effluents: the example of carbamazepine. Water Science and Technology: Water Supply 3 (4), 131–137.

Stamatis, N.K., Konstantinou, I.K., 2013. Occurrence and removal of emerging pharmaceutical, personal care compounds and caffeine tracer in municipal sewage treatment plant in Western Greece. Journal of Environmental Science and Health, Part B 48 (9), 800–813.

Stasinakis, A.S., 2012. Review on the fate of emerging contaminants during sludge anaerobic digestion. Bioresource Technology 121, 432–440.

Stevens, J.L., Northcott, G.L., Stern, G.A., Tomy, G.T., Jones, K.C., 2003. PAHs, PCBs, PCNs, organochlorine pesticides, synthetic musks, and polychlorinated n-alkanes in UK sewage sludge: survey results and implications. Environmental Science and Technology 37 (3), 462–467.

Subedi, B., Balakrishna, K., Joshua, D.I., Kannan, K., 2017. Mass loading and removal of pharmaceuticals and personal care products including psychoactives, antihypertensives, and antibiotics in two sewage treatment plants in southern India. Chemosphere 167, 429–437.

Tay, K.S., Rahman, N.A., Radzi Bin Abas, M., 2010. Kinetic studies of the degradation of parabens in aqueous solution by ozone oxidation. Environmental Chemistry Letters 8, 331–337.

Ternes, T.A., Bonerz, M., Herrmann, N., Löffler, D., Keller, E., Lacida, B.B., Alder, A.C., 2005. Determination of pharmaceuticals, iodinated contrast media and musk fragrances in sludge by LC tandem MS and GC/MS. Journal of Chromatography A 1067 (1–2), 213–223.

Thomaidi, V.S., Stasinakis, A.S., Borova, V.L., Thomaidis, N.S., 2016. Assessing the risk associated with the presence of emerging organic contaminants in sludge-amended soil: a country-level analysis. Science of the Total Environment 548, 280–288.

Tsui, M.M., Leung, H., Lam, P.K., Murphy, M.B., 2014. Seasonal occurrence, removal efficiencies and preliminary risk assessment of multiple classes of organic UV filters in wastewater treatment plants. Water Research 53, 58–67.

Verlicchi, P., Zambello, E., 2015. Pharmaceuticals and personal care products in untreated and treated sewage sludge: occurrence and environmental risk in the case of application on soil-a critical review. Science of the Total Environment 538, 750–767.

Vieno, N., Tuhkanen, T., Kronberg, L., 2007. Elimination of pharmaceuticals in sewage treatment plants in Finland. Water Research 41 (5), 1001–1012.

Wang, D., Sui, Q., Lu, S.-G., Zhao, W.-T., Qiu, Z.-F., Miao, Z.-W., Yu, G., 2014. Occurrence and removal of six pharmaceuticals and personal care products in a wastewater treatment plant employing anaerobic/anoxic/aerobic and UV processes in Shanghai, China. Environmental Science and Pollution Research 21, 4276–4285.

Yang, Y., Ok, Y.S., Kim, K.-H., Kwon, E.E., Tsang, Y.F., 2017. Occurrences and removal of pharmaceuticals and personal care products (PPCPs) in drinking water and water/sewage treatment plants: a review. Science of the Total Environment 596, 303–320.

Yu, Y., Wu, L., 2012. Analysis of endocrine disrupting compounds, pharmaceuticals and personal care products in sewage sludge by gas chromatography–mass spectrometry. Talanta 89, 258–263.

Zhang, R., Tang, J., Li, J., Zheng, Q., Liu, D., Chen, Y., Zhang, G., 2013. Antibiotics in the offshore waters of the Bohai Sea and the Yellow Sea in China: occurrence, distribution and ecological risks. Environmental Pollution 174, 71–77.

7

Environmental behaviors of exogenous emerging contaminants on the anaerobic digestion of waste activated sludge

Jingyang Luo[1,2] and Yang Wu[3]

[1]KEY LABORATORY OF INTEGRATED REGULATION AND RESOURCE DEVELOPMENT ON SHALLOW LAKES, MINISTRY OF EDUCATION, HOHAI UNIVERSITY, NANJING, CHINA; [2]COLLEGE OF ENVIRONMENT, HOHAI UNIVERSITY, NANJING, CHINA; [3]STATE KEY LABORATORY OF POLLUTION CONTROL AND RESOURCE REUSE, SCHOOL OF ENVIRONMENTAL SCIENCE AND ENGINEERING, TONGJI UNIVERSITY, SHANGHAI, CHINA

1. Introduction

Over the past decades, the exponential increase of diverse synthetic chemicals and associated consumption in domestic and industrial sectors has resulted in the massive release of emerging contaminants (ECs) into wastewater (Richmond et al., 2018), such as nanoparticles (NPs), pharmaceuticals and personal care products (PPCPs), persistent organic pollutants (POPs), microplastics (MPs), etc., which has become a global issue with increasing environmental concern. However, the limited removal efficacy of conventional wastewater treatment processes in wastewater treatment plants results in the large accumulation of ECs in waste activated sludge (WAS), with concentrations reaching up to tens of mg/g dry matter (Rathi et al., 2021). In addition to its pollution characteristics, WAS also exhibited resource features due to the high organic proportion, which could be a promising feedstock for resource recovery (Wu et al., 2023). Presently, anaerobic digestion is widely regarded as a sustainable method for WAS treatment with the production of renewable fuels and chemicals (e.g., volatile fatty acids (VFAs), hydrogen (H_2), and methane (CH_4)), which serves to mitigate greenhouse gas emissions and reduce dependence on fossil fuels (Shi et al., 2022).

WAS digestion typically starts from solubilization and hydrolysis, which involve the conversion of complex fermentation substrates into soluble organic compounds and eventually into soluble monomers. Then it is followed by acidogenesis and methanogenesis, in which various intermediate products are fermented to methane with the

Occurrence and Behavior of Emerging Contaminants in Organic Wastes and Their Control Strategies
https://doi.org/10.1016/B978-0-443-13585-9.00003-3

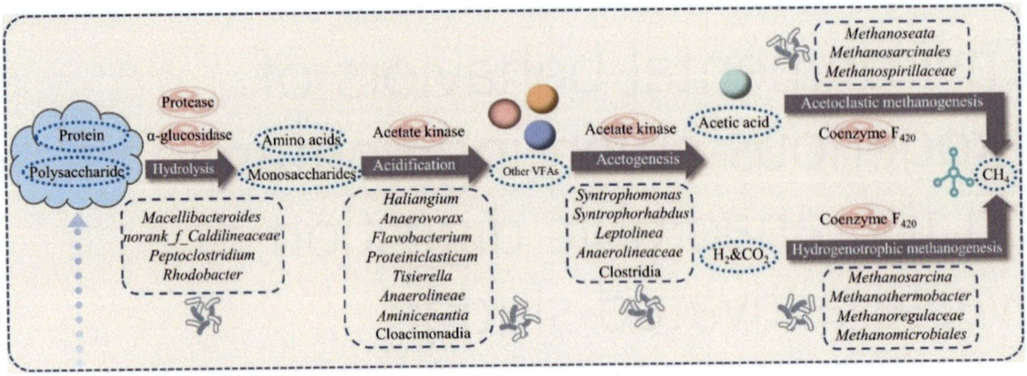

FIGURE 7.1 Proposed metabolic pathway of WAS digestion process (Luo et al., 2020).

participation of functional microbial consortia (Zhang et al., 2021) (Fig. 7.1). The active involvement of functional microorganisms plays a key role in the process of anaerobic digestion and is directly correlated with digestion efficiency (Wang et al., 2021). However, microbial activity and metabolic functions are associated with various factors, including the operational conditions and the potential toxic compounds in digesters. Consequently, the presence of undesirable ECs has the potential to negatively impact the performance of WAS anaerobic digestion. Previous studies have demonstrated the detrimental effects of exogenous ECs on WAS digestion, which vary greatly based on the species and concentrations involved (Luo et al., 2016; Wang et al., 2017; Xie et al., 2019; Xu et al., 2017). Despite this, the impact of pollutants on WAS digestion is often overlooked in research aimed at manipulating and optimizing the process.

The chapter aims to assess the impacts of frequently detected ECs with high concentrations, including inorganic NPs, PPCPs, MPs, etc., on the performance of WAS digestion. Also, a comprehensive understanding of the underlying mechanisms and potential mitigating strategies is proposed. It would help to gain deeper insights into the influence of contaminants on WAS digestion and raise awareness about the potential risks posed by ECs in WAS.

2. Potential impacts of exogenous ECs on WAS digestion and the underlying mechanisms

2.1 Inorganic nanoparticles pollutants

The augmented use of NPs-enriched products has led to a substantial discharge of various NPs into WWTPs, where they are subjected to adsorption and subsequently concentration in WAS (Walser et al., 2012). Predicted environmental concentrations of NPs range from μg/L to mg/L (Wang and Chen, 2016). Various researches have investigated the environmental behaviors of NPs in WAS digestion processes, with controversial outcomes contingent upon the NP types (Chen et al., 2014; Mu and Chen, 2011; Wang et al., 2016).

In terms of Fe NPs, they are hypothesized to enhance the performance of WAS digestion by serving as electron carriers between electron donors and acceptors. By improving direct interspecies electron transfer (DIET) during WAS anaerobic digestion (Kato et al., 2012), Fe NPs could accelerate the methanogenesis process. Meanwhile, Fe NPs positively impact the WAS solubilization and hydrolysis processes by fostering micro-electrolysis between the sludge and the NPs. This micro-electrolysis disrupts the cemented structure of the sludge matrix, thereby releasing more soluble substrates that are amenable to further biodegradation (Luo et al., 2014). In addition, the Fe^{2+}, which is released from Fe NPs, serves as a crucial component in the assembly of Fe—S clusters in enzymes and is crucial in the electron transport process of cellular redox activity (Luo et al., 2014; Zandvoort et al., 2006). Also, the precipitation of Fe^{2+} with S^{2}- has been shown to mitigate the detrimental impact of S^{2-} on microbial populations (Li et al., 2007). However, Yang et al. demonstrated that the utilization of Fe NPs in anaerobic digestion processes resulted in beneficial impacts on solubilization and acidogenesis but had an inhibitory effect on methanogenesis. A 20% decrease of methane production was observed with 1 mM Fe NPs. The authors proposed that the strong reducing conditions at the surface of Fe NPs resulted in the rapid inactivation and damage of cell integrity and activity. The dissolution of Fe NPs also led to the promotion of rapid hydrogen production and accumulation, resulting in an unfavorable high H_2 partial pressure for acidogenesis and the destruction of the syntrophic relationship between bacteria and methanogens in anaerobic digestion. Additionally, the high concentration of soluble Fe^{2+} has the potential to complex with PO_4^{3-}, resulting in a deficiency of this essential nutrient for methanogens and further inhibiting methanogenesis (Yang et al., 2013b).

The impacts of TiO_2 NPs, Al_2O_3 NPs, SiO_2 NPs, and Au NPs on WAS digestion were commonly considered to be negligible (García et al., 2012; Mu et al., 2011). However, it should be pointed out that the effects of these NPs might be contingent upon environmental conditions. García et al. observed that under mesophilic conditions, TiO_2 NPs had no effect on anaerobic digestion but led to a 10% increase in biogas production under thermophilic conditions. Furthermore, TiO_2 NPs have been shown to enhance H_2 production through photofermentation after anaerobic dark fermentation, with a 46.1% improvement in H_2 production at a concentration of 100 mg/L. TiO_2 NPs play a crucial role in facilitating the hydrolysis of macro-molecular substrates (such as proteins and polysaccharides) in WAS, promoting the growth rate of photosynthetic bacteria, but inhibiting hydrogen-uptake enzymes (Zhao and Chen, 2011).

Moreover, some NPs, such as ZnO NPs, Ag NPs, MgO NPs, graphene oxide (GO) NPs, and CeO_2 NPs, had adverse effects on WAS anaerobic digestion, particularly at high concentrations (Table 7.1). For example, low levels of ZnO NPs (1 or 6 mg/g TSS) had no noticeable impact on WAS anaerobic digestion but resulted in 22.8% and 81.1% reductions in methane production after short-term exposure and 18.3% and 75.1% reductions after long-term exposure at concentrations of 30 and 150 mg/g TSS,

Table 7.1 Influences of nanoparticles on the WAS anaerobic digestion.

NPs category	Concentration	Influences on WAS digestion	References
nZVI	10 mg/kg TSS	Increase CH_4 production by 20%	Wang et al. (2016)
nZVI	1 g/L	Increase CH_4 production by 374%	Xiu et al. (2010)
nZVI	5 g/L	Increase VFAs production by 512%	Luo et al. (2014)
nZVI	1–30 mM	Decrease CH_4 production by 20%–69%	Yang et al. (2013)
Fe_2O_3 NPs	5 mg/L	Reduce CH_4 production by 4%	Zhang et al. (2020)
(30–60 nm)	50 mg/L	Reduce CH_4 production by 4%	
(48 days)	150 mg/L	Reduce CH_4 production by 10.9%	
	250 mg/L	Reduce CH_4 production by 18.2%	
	500 mg/L	Reduce CH_4 production by 28.9%	
Fe_2O_3 NPs	5 mg/L	Increase CH_4 production by 8.2%	
(30–60 nm)	50 mg/L	Increase CH_4 production by 1.4%	
(72 h)	150 mg/L	Increase CH_4 production by 5.6%	
	250 mg/L	Increase CH_4 production by 4%	
	500 mg/L	Increase CH_4 production by 6.4%	
	750 mg/L	Increase CH_4 production by 1.4%	
	1000 mg/L	Increase CH_4 production by 8.7%	
Fe_2O_3 NPs	100 mg/kg TSS	Increase CH_4 production by 17%	Wang et al. (2016)
FeT	200 mg/L	Increase CH_4 production by 22%	Baniamerian et al. (2021)
Fe_3O_4 NPs	231 mg/g VSS	Increase CH_4 production by 56.9%	Wang and Jeong (2018)
Fe_3O_4 NPs	100 mg/L	Increase CH_4 production by 234%	Huang et al. (2019)
MgO NPs	500 mg/kg TSS	Reduce CH_4 production by 99%	Wang et al. (2016)
Ag NPs	500 mg/kg TSS	Reduce CH_4 production by 27%	Wang et al. (2016)
Ag NPs	40 mg/kg TSS	Non-effect on CH_4 production	Yang et al. (2012)
Ag NPs	0.85 g/L	Intermediate inhibition within 33%–50% CH_4 reduction	García et al. (2012)
TiO_2 NPs	150 mg/kg TSS	Non-effect on CH_4 production	Mu et al. (2011)
Al_2O_3 NPs	150 mg/kg TSS	Non-effect on CH_4 production	Mu et al. (2011)
Al_2O_3 NPs	50 mg/L	Increase CH_4 production by 13.5%	Kökdemir Ünşar and Perendeci (2018)
(40-50 nm)	250 mg/L	Increase CH_4 production by 14.8%	
48 days	500 mg/L	Increase CH_4 production by 8.4%	
Al_2O_3 NPs	250 mg/g TSS	Increase CH_4 production by 14.8%	Kökdemir Ünşar and Perendeci (2018)
Co NPs	0.16 mg/g TSS	Increase CH_4 production by 42%	Zaidi et al. (2019)
Ni NPS	0.16 mg/g TSS	Increase CH_4 production by 31.7%	Zaidi et al. (2019)
NiT	23.5 mg/L	Increase CH_4 production by 24%	Baniamerian et al. (2021)
SiO_2 NPs	150 mg/kg TSS	Non-effect on CH_4 production	Mu et al. (2011)
ZnO NPs	1–150 mg/kg TSS	Dosage dependent: non-effect at low dose (1 or 6 mg/g TSS); but inhibit at high level (induced 18.3%–22.8% and 75.1%–81.1% of CH_4 inhibition at 30 and 150 mg/g TSS, respectively.	Mu et al. (2011)
ZnO NPs	–	EC_{50} for CH_4 production: 57.4 mg Zn/L for ZnO NPs	Luna-delRisco et al. (2011)

Table 7.1 Influences of nanoparticles on the WAS anaerobic digestion.—cont'd

NPs category	Concentration	Influences on WAS digestion	References
ZnO NPs	15 mg/kg TSS	Reduce CH_4 production (76.8% of the control) and delay peak production rate	Zhao et al. (2018)
ZnO NPs	30 mg/kg	Reduce CH_4 production by 45.9% at the initial stage; but showed partial recovery after adaption	Zhao et al. (2019a)
ZnO NPs	30 mg/g dry sludge	Reduce CH_4 production by 74.9%	Zhao et al. (2019b)
Cu NPs	14.53 mg/g TSS	Decrease the VFAs production from 223.5 to 120.5 mg COD/g VSS.	Chen et al. (2014)
CuO NPs	10.7 mg Cu/L	EC_{50} concentrations for methane inhibition	Luna-delRisco et al. (2011)
TiO_2 NPs	100 mg/L	46.1% H_2 promotion during anaerobic dark fermentation	Zhao and Chen (2011)
TiO_2 NPs (4-8 nm)	500, 1000, 1500, 2000 mg/L	14.9% (average) increase in methane production	Cervantes-Avilés et al. (2018)
GO NPs	300 mg/L	Increase CH_4 production by up to 23-fold	Bueno-Lopez et al. (2018)
GO NPs	0.054 and 0.108 mg/mg VSS	Reduce CH_4 production by 7.6% and 12.6%, respectively	Dong et al. (2019)
Graphene	30 and 120 mg/L	Increase CH_4 production by 17.0% and 51.4%, respectively.	Tian et al. (2017)
CeO_2	640 mg/L	Reduce CH_4 production by 90%.IC_{50}: 60 mg/L (mesophilic), IC_{50}: <320 mg/L (thermophilic)	García et al. (2012)
CeO_2	5–150 mg/g VSS	Reduce 15%–19% VFAs production for flocculent sludge and 35% for the granular sludge; But no inhibition on methanogenesis	Ma et al. (2013)

respectively. ZnO NPs inhibited the hydrolysis, acidification, and methanogenesis steps during short-term exposure while only impacting the hydrolysis and methanogenesis processes during long-term exposure (Mu et al., 2011). The varying responses of methanogenic archaea and acidogenic bacteria to ZnO NPs could be due to differences in their cell structures and metabolic pathways (Nanninga, 2009). Nevertheless, it is worth noting that anaerobic microorganisms have been observed to possess self-recovering abilities in response to toxic NP exposure through adaptation. For instance, Zhao et al. observed that ZnO NPs reduced peak production and inhibited methane production by 45.9% initially (14 h). However, the methanogenic capacity of microbes in WAS recovered, and the inhibition effects decreased to 31.4% as they adapted to the toxic environment (at 36 h) and continued to recover with a final inhibition of 26.7% after 35 days (Zhao et al., 2018). Mu et al. also reported the alleviation of ZnO NPs' toxic effects on methane production after prolonged exposure (Mu et al., 2011).

The mechanisms behind the negative effects of NPs on anaerobic digestion are yet to be fully understood (Fig. 7.2). Generally, the generation of reactive oxygen species (ROS) is widely accepted as the most probable mechanism for NP-induced toxicity. The small size of NPs allows for their penetration of cell membranes and subsequent entry into cells, leading to the production of ROS. This excess of ROS results in oxidative stress on the membrane, peroxidation of various organelle components, and disruption of homeostasis. Elevated levels of lactate dehydrogenase in NP reactors provide clear evidence of cell membrane damage caused by NPs, which leads to the dysfunction of cellular metabolism crucial to anaerobic digestion (Wang and Chen, 2016; Wang et al., 2016). For instance, the activities of protease, acetate kinase (AK), and coenzyme F420 were observed to decline by 25.3%, 22.9%, and 40.9%, respectively, at a ZnO NPs concentration of 150 mg/g TSS (Mu et al., 2011). Furthermore, NPs can alter the microbial community. Wang et al. reported that the gene copies of bacteria and archaea decreased to 84% and 32%, 79% and 31% of control with the exposure of 500 mg/g TSS Ag NPs and MgO NPs, respectively. In particular, a decreased abundance of active *α-Proteobacteria*, *β-Proteobacteria*, *Bacteroidetes*, and *Methanosaeta*, which are typically abundant in conventional digesters, was observed, especially the *Methanosaeta* (Wang et al., 2016). Commonly, the methanogenic archaea are less tolerant to the toxic effects of NPs.

Additionally, the release of toxic substances from NPs may contribute to their adverse impact on anaerobic digestion (Mu et al., 2011). As posited by Wang et al. the liberation of Fe^{2+}, Ag^{+}, and Mg^{2+} appears to play a primary role in the facilitating and/or inhibitory effects of Fe NPs, Ag NPs, and MgO NPs (Wang et al., 2016). Yang et al. observed that, under anaerobic conditions, exposure to 40 mg/L of Ag NPs had no significant impact on

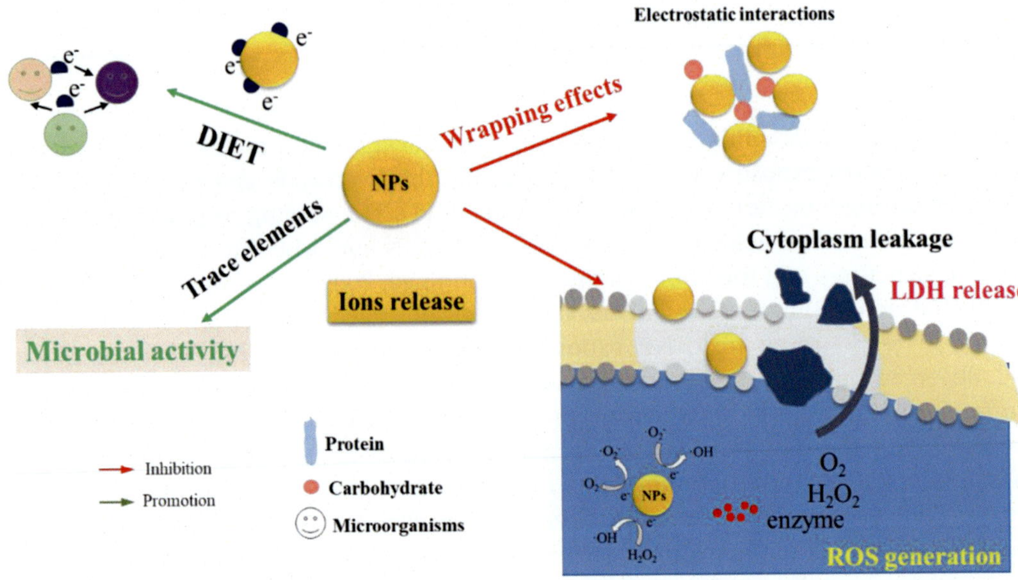

FIGURE 7.2 Main mechanisms of NPs influences on anaerobic digestion process (Luo et al., 2020).

methane production (Yang et al., 2012). However, when Ag NPs were initially subjected to a microaerobic environment, leading to the release of Ag^{+}, a remarkable reduction in total methanogenic gene copies was observed, with an alteration in the competition between hydrogenotrophic methanogens and acetoclastic methanogens (Yang et al., 2013a). The release of toxic substances from NPs is dependent upon both their intrinsic physical and chemical properties (such as chemical composition, particle size, etc.) and the environmental conditions they encounter (such as pH and temperature).

In addition, the toxicity of NPs in anaerobic digestion systems can also be influenced by various physicochemical parameters. The effects of electrostatic interactions between NPs and fermentation substrates also contribute to the NPs' toxicity. For instance, GO has been shown to negatively impact methane production due to its strong adsorption of organic substrates (Bueno-Lopez et al., 2018; Dong et al., 2019). This effect can be mitigated through thermal treatment of the GO-WAS mixture or reduction of the GO to graphene (Dong et al., 2019). CeO_2 NPs caused a reduction in VFA production in anaerobic sludge, which was attributed to physical penetration and membrane reduction rather than the production of ROS (Ma et al., 2013). It is also impacted by the rupture of outer membrane proteins and phospholipids (Wu and Narsimhan, 2008). Furthermore, CeO_2 NPs possess catalytic properties that may perturb the respiration mechanisms of microorganisms (García et al., 2012).

Adsorption processes are a promising technology to mitigate the toxicity of NPs toward WAS digestion. In WWTPs, activated carbons, including powdered and granulated activated carbons, are frequently utilized as the sorbent material for NPs due to their high specific surface area and low preparation cost (Park et al., 2017). Previous work exhibited that granulated activated carbons could be an efficient adsorbent to adsorb Ag NPs, and the efficiency was related to the carbon type and solution chemistry (Gicheva & Yordanov, 2013). Meanwhile, other studies also demonstrated that, due to the positively charged surface, granulated activated carbon was preferred to adsorb negatively charged TiO_2 NPs via electrostatic interactions (Salih et al., 2012). Positively charged activated alumina also exhibited high adsorption capacity to negatively charged NPs at near-neutral condition (Park et al., 2017). Notably, activated carbons could also accelerate microbial electron transfer (mainly DIET) due to their abundant functional groups, which was beneficial to improve the ultimate methane production (Liu et al., 2023a). Nevertheless, in a series of treatment steps, different adsorption mechanisms affecting the adsorption capacity of NPs should be studied to further characterize and avoid the factors leading to the breakthrough of NPs in WWTPs. It is also necessary to investigate the impacts of existing widely utilized combination technologies such as coagulation and flocculation plus sand filtration processes for NP removal (Troester et al., 2016).

2.2 Xenobiotic organic pollutants

Xenobiotics mainly refer to man-made organic compounds that are not naturally occurring in the environment, such as PPCPs and POPs (Östman et al., 2017; Symsaris et al., 2015; Wieck et al., 2018). These pollutants, especially hydrophobic compounds, tend to be removed through sorption to WAS during biological treatment processes in WWTPs (Mailler et al., 2017; Montes-Grajales et al., 2017). The levels of xenobiotic pollutants in wastewater solids have been reported to vary from μg/kg to mg/kg, and their levels are projected to continue rising in the future, given their widespread use (Östman et al., 2017; Wang et al., 2017; Xie et al., 2019).

2.2.1 Influences of PPCPs on the WAS anaerobic digestion

The frequent detection of PPCPs in WWTPs and their continued presence in WAS have raised concerns about the potential impact of these compounds on subsequent biological processes (Luo et al., 2019).

Among them, antibiotics have attracted significant attention in recent years due to their indiscriminate and excessive use. Li et al. found that low levels of fluoroquinolone antibiotics (FQs: ofloxacin, norfloxacin, ciprofloxacin, and lomefloxacin) added (2 mg/L) slightly increased methane production by 6.3% from the thermally hydrolyzed sludge compared to the control. However, high dosages of FQs (100 mg/L) showed no improvement (20 mg/L) or even led to an approximate 8% reduction in methane production (Li et al., 2017). The sulfonamide antibiotics (sulfadiazine and sulfamethazine) enhanced the production of VFAs, particularly acetic acid, by promoting the solubilization, hydrolysis, and acidification of WAS (Hu et al., 2018; Xie et al., 2019). The activities of protease, α-glucosidase, and AK were improved by sulfadiazine, while the methanogenesis process was suppressed (Xie et al., 2019). Similarly, roxithromycin (ROX) resulted in an increase of VFAs from 295 to 610 mg COD/L when the ROX level rose from 0 to 100 mg/kg TSS. ROX was found to enhance the activity of AK and inhibit the activities of α-glucosidase and coenzyme F420. But a stronger inhibitory effect on methane production than the hydrolysis process was observed, resulting in an increase in VFA accumulation (Chen et al., 2020a,b). These results suggest that antibiotics might mainly impact the activities of methanogens (Hu et al., 2018). Some researchers have posited that the inhibition of pharmaceuticals on methanogens may be correlated with their ability to adsorb onto cellular membranes. For example, Fountoulakis et al. observed that the toxicities of pharmaceuticals on methanogens as well as the methane production follow the order of propranolol hydrochloride > diclofenac sodium > carbamazepine > ofloxacin > clofibric acid > sulfamethoxazole (Fountoulakis et al., 2004). Furthermore, it was noted that hydrogenotrophic methanogens appear to be more resilient to the inhibitory effects of these pharmaceuticals compared to acetoclastic methanogens (Symsaris et al., 2015).

The widespread use of antimicrobial and biocidal agents in PPCPs has resulted in their widespread detection as xenobiotics in wastewater. These agents also have negative impacts on anaerobic digestion and the production of methane (Fig. 7.3). For example, benzalkonium chloride has been found to gradually decrease methane production, which leads to complete inhibition at a concentration of 100 mg/L (Flores et al., 2015). Additionally, the surfactant-like characteristics of these agents can lead to a vulnerability to methanogenic archaea due to their extraordinary hydrophobic cytoplasmic

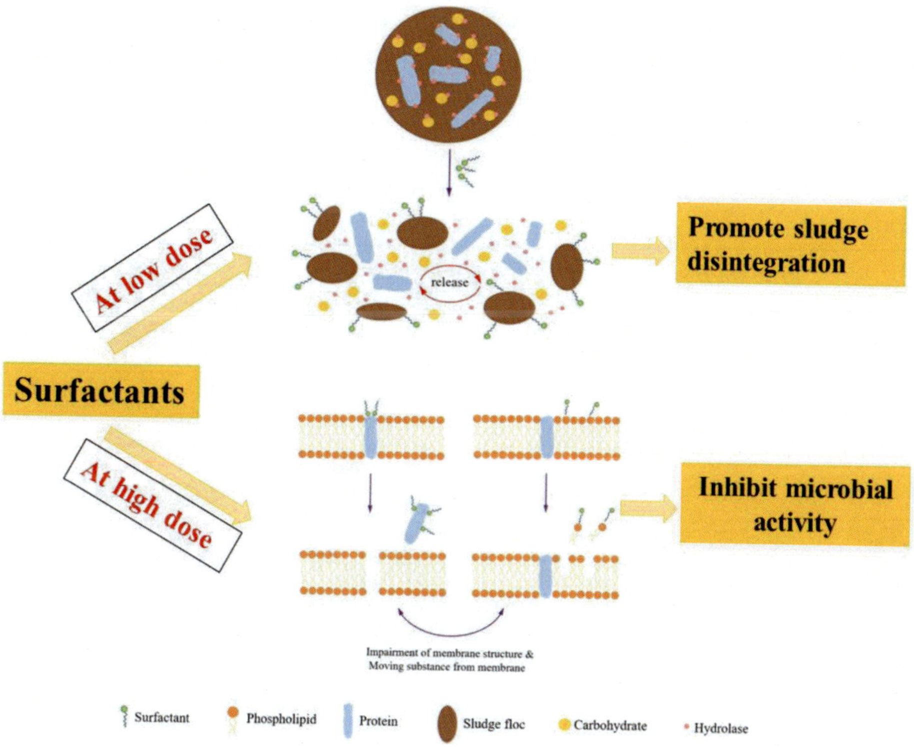

FIGURE 7.3 Proposed mechanisms of surfactants on WAS digestion process at different dose (Luo et al., 2020).

membranes and potential negative impact on the proton motive force for ATP generation (Flores et al., 2015; Tezel, 2009). On the other hand, the surfactant properties can also lead to the disruption of extracellular polymeric substances (EPS) in the anaerobic digestion system, resulting in an increase in bioavailable substrates for acetogenic bacteria metabolism and promoting the production of VFAs (Luo et al., 2019; Wang et al., 2017). Luo et al. found that low-level exposure to biocides such as chlorhexidine and hexadecyltrimethylammonium bromide can lead to a significant increase in VFA accumulation and the enhancement of hydrolases and acidogenic enzymes involved in their production. However, the promoting effects weaken in high loads or the coexistence of chlorhexidine and HTAB (Luo et al., 2019). These findings support the conclusion that the toxicity of biocides is directly proportional to their concentration (Durham and Young, 2009).

2.2.2 Influences of persistent organic pollutants on WAS anaerobic digestion

It has been established that numerous POPs are frequently detected in WAS at elevated levels (Luo et al., 2016; Symsaris et al., 2015). For example, the concentration of polycyclic aromatic hydrocarbons (PAHs) in sludge has been reported to range between 1 and 2000 mg/kg dry weight (Cai et al., 2007; Hua et al., 2008). Despite the presence of POPs in WAS, research into their impact on WAS digestion has been limited, especially their acute toxicities, which might be due to the "persistent" and "chronic" nature of these pollutants (Table 7.2).

Table 7.2 Influences of xenobiotic organic pollutants on the WAS anaerobic digestion.

Names	Classifications	Concentration	Influences on WAS digestion	References
Sulfadiazine	Antibiotics	50 mg/kg TSS	Increase VFAs production by 32%	Xie et al. (2019)
Sulfamethazine	Antibiotics	24 mg/kg VSS	Increase VFAs production by 173%	Hu et al. (2018a)
Ciprofloxacin	Antibiotics	0.05–50 mg/L	CH_4 production decrease from 115.3 mL in control to 66.7 mL at 50 mg/L ciprofloxacin Non-effect at low dose (0.05 mg/L)	Mai et al. (2018)
Ciprofloxacin	Antibiotics	0–500 mg/kg TSS	Delay the peak CH_4 production rate by 1 and 2 h with a 42.9% and 68.7% lower peak value at 100 and 500 mg/kg TSS, respectively	Zhao et al. (2018)
Ciprofloxacin Ofloxacin NOR Lomefloxacin	Antibiotics	2–100 mg/L	Increase CH_4 production by 106.3% at 2 mg/L. Non-effect at 20 mg/L; but reduce around 8% at 100 mg/L	Li et al. (2017)
Cefamandole nafate	Antibiotics	1–5 mg/g TSS	Reduce VFAs production by 8.9%–12.7%	Luo et al. (2023)
Cefpirome sulfate	Antibiotics	1–5 mg/g TSS	Non-effects on VFAs production	Luo et al. (2023)
Sulfamethoxazole	Antibiotics	100 mg/kg TSS	Increase VFAs production by 80%	Wei et al. (2022)
Roxithromycin	Antibiotics	100 mg/kg TSS	Increase VFAs production by 110% The maximum VFA production time decreased from 6 to 3 days	Chen et al. (2020a,b)
Tetracycline	Antibiotics	0–60 mg/kg TSS	VFAs yield decreased from 125.1 in control to 90.8 mg COD/g VSS at 60 mg/kg TSS	He et al. (2021)
Chlortetracycline	Antibiotics	10 mg/kg TSS	Increase VFAs production by 21.1%	Tang et al. (2021)
Sulfaquinoxaline	Antibiotics	100 mg/kg TSS	Increase VFAs production by 70%	Wei et al. (2022)
Rifampicin	Antibiotics	[a] IC_{20}: 100 mg/L; IC_{50}: >250 mg/L		Sanz et al. (1996)
Ampicillin	Antibiotics	IC_{20}: 10 mg/L;		
Novobiocin	Antibiotics	IC_{20}: 10 mg/L;		
Penicillin	Antibiotics	IC_{20}: 10 mg/L;		
Gentamicin	Antibiotics	IC_{20}: 35 mg/L;		
Hygromycin B	Antibiotics	IC_{20}: 64 mg/L; IC_{50}: 210 mg/L; IC_{80}: >300 mg/L		
Kanamycin	Antibiotics	IC_{20}: 100 mg/L;		

Table 7.2 Influences of xenobiotic organic pollutants on the WAS anaerobic digestion.—cont'd

Names	Classifications	Concentration	Influences on WAS digestion	References
Neomycin	Antibiotics	IC_{20}:20 mg/L; IC_{50}: >500 mg/L		
Spectionomycin	Antibiotics	IC_{20}: >20 mg/L		
Streptomycin	Antibiotics	IC_{20}: 18 mg/L		
Chlortetracycline	Antibiotics	IC_{20}: 5 mg/L; IC_{50}: 40 mg/L; IC_{80}: 152 mg/L		
Doxycycline	Antibiotics	IC_{20}: 8 mg/L;		
Tylosin	Antibiotics	IC_{20}: 15 mg/L; IC_{50}: 334 mg/L; IC_{80}: >400 mg/L		
Chloramphenicol	Antibiotics	IC_{20}:5 mg/L; IC_{50}: 40 mg/L; IC_{80}: 152 mg/L		
Propranolol	Antibiotics	IC_{20}: 33 mg/L; IC_{50}: 334 mg/L; IC_{80}: >400 mg/L		
Ofloxacin	Antibiotics	IC_{20}: 33 mg/L; IC_{50}: 334 mg/L; IC_{80}: >400 mg/L		
Diclofenac sodium	Antibiotics	IC_{20}: 33 mg/L; IC_{50}: 334 mg/L; IC_{80}: >400 mg/L		
Carbamazepine	Antibiotics	IC_{20}: 33 mg/L; IC_{50}: 334 mg/L; IC_{80}: >400 mg/L		
Sulfamethoxazole	Antibiotics	IC_{20}: 33 mg/L; IC_{50}: 334 mg/L; IC_{80}: >400 mg/L		
Clofibric acid	Antibiotics	IC_{20}: 33 mg/L; IC_{50}: 334 mg/L; IC_{80}: >400 mg/L		
Sulfanol NP-1	Antibiotics	IC_{20}: 33 mg/L; IC_{50}: 334 mg/L; IC_{80}: >400 mg/L		
Ofloxacin	Antibiotics	IC_{20}: 33 mg/L; IC_{50}: 334 mg/L; IC_{80}: >400 mg/L		Fountoulakis et al. (2004)
Diclofenac	Anti-inflammatory drug	2.5–25 mg/kg TSS	Increase VFAs production from 599 to 1113 mg COD/L	Hu et al. (2018b)
Diclofenac sodium	Anti-inflammatory drug	IC_{20}: 80 mg/L; IC_{50}: 120 mg/L; IC_{80}: 296 mg/L		Fountoulakis et al. (2004)
Triclocarban	Antibacterial agent	26.7 −520.5 mg/kg TSS	Increase VFAs production from 32.6 to 228.2 ± 3.6 mg COD/g VSS	Wang et al. (2017)
Para-chloro-meta-xylenol	Antibacterial agent	10–40 mg/g TSS	Increase VFAs production by 1.2–3.0 folds	Du et al. (2022)
Para-chloro-meta-xylenol	Antibacterial agent	100 mg/g TSS	Decrease VFAs production by 10%	Du et al. (2022)
Polyhexamethylene guanidine	Antibacterial agent	20–100 mg/g TSS	Increase VFAs production by 1.2–2.3 folds	Wang et al. (2022)
Polyhexamethylene guanidine	Antibacterial agent	400 mg/g TSS	Decrease VFAs production by 60%	Wang et al. (2022)
Triclosan	Antibacterial agent	IC_{50}: 35 mg/L		Symsaris et al. (2015)
Benzalkonium chloride	Biocides	0.02–100 mg/L	From no apparent effect to almost complete inhibition	(Flores et al., 2015)
Proxel LV	Biocides	IC_{20}: 976.1 mg/L; IC_{50}: 1003 mg/L		
Chlorhexidine	Biocides	50 mg/kg TSS	Increase VFAs from 567.3 in control to 2144.5 mg COD/L	Luo et al. (2019)
HTAB	Biocides	50 mg/kg TSS	Increase VFAs from 567.3 in control to 3400.0 mg COD/L	

Continued

Table 7.2 Influences of xenobiotic organic pollutants on the WAS anaerobic digestion.—cont'd

Names	Classifications	Concentration	Influences on WAS digestion	References
Propranolol hydrochloride	Phamaceuticals	IC_{20}: 7.6 mg/L; IC_{50}: 30 mg/L; IC_{80}: 67.5 mg/L		Fountoulakis et al. (2004)
Carbamazepine	Phamaceuticals	IC_{20}: 41 mg/L; IC_{50}: 220 mg/L		
Clofibric acid	Phamaceuticals	IC_{20} : >400 mg/L		
Sulfamethoxazole	Phamaceuticals	IC_{20} : >400 mg/L		
Nonylphenol	POPs	5–8 mg/L	12.7%–19.5% decrease in relative abundance of *Methanosaeta*	Kara Murdoch et al. (2018)
Nonylphenol	POPs	200 mg/kg TSS	Increase VFAs from 2856 in control to 5620 mg COD/L at pH 10 after long-time exposure	Duan et al. (2016)
Phenanthrene (PAH)	POPs	100 mg/kg TSS	Increase the acetic acid production by 80% at pH 10 after long-time exposure	Luo et al. (2016)
Di-ethylhexyl phthalate	POPs	120–160 mg/L	Decrease CH_4 production by 50% –60% after long-time exposure	Gavala et al. (2003)
Di-ethylhexyl phthalate	POPs	25–200 mg C/L	Gradual increase of di-ethylhexyl phthalate cause 6%–48% decrease CH_4 production over a period of 60 days	Battersby and Wilson (1988)
PFOA	POPs	3–60 μg/g TS	11.1%–19.2% decrease in methane production than the control	Wang et al. (2021)
BPA	POPs	50 mg/kg TSS	Increase VFAs production by 30% after long-time exposure	Jiang et al. (2021)
DOWFAX	Surfactant	IC_{10}: 95.3 mg/L; IC_{20}: 147.6 mg/L; IC_{50}: 311.5 mg/L;		Flores et al. (2015)
Triton X-100	Surfactant	IC_{10}: 16.91 mg/L; IC_{20}: 19.32 mg/L; IC_{50}: 24.26 mg/L;		Flores et al. (2015)
SDBS	Surfactant	Dependent on the length of alkyl chain in SDBS: inhibit the methanogenic activity totally with 100 mg/L C_{10}–C_{12} SDBS, but non-effect of C_{13}–C_{14} SDBS at 200 mg/L		Garcia et al. (2006)
SDBS	Surfactant	0.02 g/gTSS	Increase VFAs from 339.1 in control to 2559.1 mg COD/L, but inhibit methanogenesis	Jiang et al. (2007b)
SDBS	Surfactant	20–50 mg/L	Decreased maximal rate by 50%	Khalil et al. (1988)
SDBS	Surfactant	100 mg/L	Increase VFAs from 101.0 in control to 1005.0 mg COD/L	Luo et al. (2022)
SDS	Surfactant	0.1 g/gTSS	Increase VFAs from 191.1 in control to 243.04 mg COD/L, but inhibit methanogenesis	Jiang et al. (2007a)

Table 7.2 Influences of xenobiotic organic pollutants on the WAS anaerobic digestion.—cont'd

Names	Classifications	Concentration	Influences on WAS digestion	References
Rhamnolipid	Surfactant	0.04 g/gTSS	Increase VFAs production of 4.24-fold; CH_4 production reduce from 58.8 mL/g VSS to 2.0 mL notably	Zhou et al. (2013)
CTAB	Surfactant	100 mg/L	Increase VFAs from 101.5 in control to 816 mg COD/L	Fang et al. (2022)
SDBS	Surfactant	100 mg/L	Increase VFAs from 101.5 in control to 1102 mg COD/L	Fang et al. (2022)
HTAB	Surfactant	100 mg/L	Increase VFAs from 101.0 in control to 768.0 mg COD/L	Luo et al. (2022)
Saponin	Surfactant	0.10 g/gTSS	Increase VFAs production of 4-fold	Huang et al. (2015)
Surfactin	Surfactant	0.05 g/gTSS	Increase VFAs production of 4-fold	Huang et al. (2015)
Sophorolipid	Surfactant	0–0.1 g/gTSS	Increase VFAs production from 50.5 to 246.2 mg COD/g VSS	Xu et al. (2019)

[a]IC: Inhibitory concentrations, IC_{20}, IC_{50} and IC_{80} values measure the effectiveness of compound causing 20%, 50% and 80% inhibition toward biological or biochemical utility, respectively.

Some studies have reported the chronic influence of POPs on anaerobic digestion. Perfluorooctanoic acid (PFOA), which has received significant attention due to its widespread detection and toxicity, has inhibitory effects on methanogenesis. The cumulative methane production was 11.1%–19.2% lower in the presence of PFOA, ranging from 3 to 60 μg/g TSS (Wang et al., 2021). Similarly, Gavala et al. reported a decrease in methane production in the presence of 120–160 mg/L of DEHP after long-term exposure in anaerobic digesters. The methane decrease is attributed to the accumulation of DEHP or its metabolite, 2-ethyl hexanol, which was released from DEHP hydrolysis (Gavala et al., 2003).

It is intriguing to observe that PAHs and nonyphenol exposure had promoting effects on the production of VFAs during WAS alkaline fermentation (Duan et al., 2016; Luo et al., 2016). The average VFAs increased from 2236 to 3103 mg COD/L when the concentration of PAHs rose from 0 to 100 mg/kg TSS (Luo et al., 2016). The enhancement of VFAs was primarily attributed to an increase in acetic acid. Mechanism exploration revealed that PAHs and NP may improve microbial activities in extremely alkaline conditions (pH 10) and contribute to the acidification process that leads to acetic acid production. This was supported by the improved activities of phosphotransacetylase and AK, enzymes responsible for acetic acid production, and an increase in the quantity of corresponding gene copies. The variations in adenosine 5′-triphosphate content and membrane potential also indicated higher viabilities of acetogenic bacteria in PAH- or NP-fed anaerobic systems. Luo et al. proposed the assumption that these toxic POPs might activate the "self-protection" mechanism of acidogenic bacteria to produce more

EPS to protect themselves against the toxic compounds and extremely alkaline conditions. By increasing the thickness of cell walls, the mass transfer of toxicants into the inner cells could be reduced. The content of EPS produced by *Proteiniphilum acetatigenes*, a typical fermentative bacterium in digesters, was found to be enhanced by 32% in the PAHs reactors compared to the control (without PAHs addition). This suggests that acetogenic bacteria show stronger activity and viability, leading to increased acetic acid production (Luo et al., 2016).

Additionally, the presence of POPs has been observed to alter the distribution of microbial populations within anaerobic digesters. A study by Kara Murdoch et al. found that the abundance of the phylum Firmicutes was reduced from 74.7% in control reactors to 65.5% in the presence of PAHs, while the phyla *Proteobacteria* and *Bacteroidetes* exhibited an increased abundance. The addition of nonylphenol diethoxylate and NP was also found to significantly enhance the abundance of *Methanosarcina* by 200% while conversely decreasing the abundance of *Methanosaeta* by 12.7%–19.5%, which are the two dominant genera of acetoclastic methanogens in anaerobic digesters (Kara Murdoch et al., 2018). However, the efficiency of methane production and organic substrate degardation was not influenced by the presence of these POPs.

2.2.3 Influences of surfactants on the WAS anaerobic digestion

The widespread use of surfactants as detergents in households and industries results in their discharge into sewer systems and subsequent accumulation in WAS due to their low biodegradability, particularly chemical surfactants such as sodium dodecylbenzene sulfonate (SDBS), sodium dodecyl sulfate (SDS), and Triton X-100. One of the most commonly utilized commercial surfactants, SDBS, has been found in WAS at concentrations ranging from 4480 to 9233 mg/kg dry weight in sludge (Mungray & Kumar, 2009).

The impact of surfactants on the digestion of WAS has been extensively documented in the literature. Garcia et al. noted that low levels of SDBS may enhance methane production, but high doses of SDBS may exhibit partial or complete inhibition of methanogenic activity. This inhibitory effect was dependent on the characteristics of SDBS homologs, with longer alkyl chains generally showing lower toxicity. For instance, the methanogenic activity was completely inhibited with 100 mg/L of C10–C12 SDBS, while no complete inhibition was observed for the C13–C14 LAS (the most hydrophobic homologs) even at a concentration of 200 mg/L (Garcia et al., 2006). Similarly, Fang et al. demonstrated that the production of VFAs was enhanced by 11.0 times at 100 mg/L of LAS (Fang et al., 2022). Recently, the utilization of biosurfactants, such as alkyl polyglucose, rhamnolipid, saponin, and surfactin, has increased due to their lower toxicity, greater efficiency, and biodegradability at harsh temperatures or pHs when compared to their chemical counterparts. These biosurfactants were also found to have positive effects on the accumulation of VFAs (He et al., 2019; Luo et al., 2015; Xu et al., 2019a). However, the methanogenic activities were significantly depressed with high dosages. For example, the total methane production from glucose was reduced to 50% of its maximum rate at 20–50 mg/L of SDBS during anaerobic digestion, with SDBS being considered one of the most hazardous surfactants for acetoclastic methanogenesis (Khalil et al., 1988). The suppression ratio of SDS on methane generation was also reported to increase from 3% to 100% in the range of 0.02–0.3 g SDS/g TSS (Yoo et al., 2005).

The varied impact of surfactants on WAS digestion can be summarized as follows (shown in Fig. 7.3). On one hand, surfactants are recognized for their ability to enhance the apparent solubility of organic substrates due to the amphipathic nature of their molecules (Garcia et al., 2006; Luo et al., 2015). As a result, the organic substrates in WAS, particularly the EPS, are readily disrupted and released in the presence of surfactants (Lee et al., 2014). Furthermore, the reduction in surface tension at the solid-liquid interface facilitated by surfactants enhances the transport of organics to functional bacteria in the digester. Additionally, the occurrence of electrostatic interaction between extracellular enzymes and surfactants stabilizes the enzymes in fermentation liquids, leading to an increase in the activity of hydrolytic/fermentative enzymes such as protease and alkaline phosphatase (Jiang et al., 2007a,b). Biosurfactants, in particular, have been shown to enhance the carbon-to-nitrogen ratio in WAS digestion systems, which is often a drawback in conventional WAS digesters, and contribute to the production of VFAs (He et al., 2019; Luo et al., 2015). These findings suggest that the improvement of WAS digestion at low surfactant concentrations is mainly attributed to the increased bioavailability and subsequent biodegradation of organic substrates, particularly those refractory and hydrophobic ones (Luo et al., 2015, 2018a,b). For instance, Shao et al. demonstrated that the addition of SDBS could further enhance the digestion performance of digested WAS, which is rich in refractory organic matter, for methane production with less energy input through ultrasonication (Shao et al., 2018).

On the other hand, the potential negative impacts of surfactants on WAS digestion must also be considered. High concentrations of surfactants can lead to a loss of the integrity and function of cellular membranes in anaerobic microorganisms, thereby impairing their activities. This is particularly evident when severe EPS disintegration is caused by high surfactant levels, which are regarded as protective barriers for active microorganisms. The amphipathic structures of surfactants allow them to interact with hydrophilic proteins and disrupt hydrophobic lipids in the membranes of anaerobic microbes, leading to liquefaction and a loss of barrier properties. This would result in significant reductions in microbial activities, particularly the methanogens, which are known to be highly vulnerable in such environments (He et al., 2019; Luo et al., 2015). For instance, high doses of RL were shown to have negative impacts on the activities of dehydrogenases and AKs, with a 40% decrease in coenzyme F420 activity observed (Huang et al., 2015). Similarly, high concentrations of Triton X-100 (up to 160 mg/L) resulted in a decrease in methane accumulation by up to 85% due to the damaging interactions of Triton X-100 metabolites (mainly octylphenol and short-chained ethoxylates) with the membranes of methanogenic archaea (Flores et al., 2015; Wyrwas et al., 2013).

2.2.4 Mitigation methods for xenobiotic organic pollutants

Membrane technology is a physical treatment mechanism that utilizes membranes to filter out xenobiotic organic pollutants based on their size and properties (Rathi et al., 2021). The primary force driving filtration through a membrane is hydrostatic pressure. To date, high-pressure membrane processes such as nanofiltration and reverse osmosis have been widely adopted to eliminate xenobiotic organic pollutants (Egea-Corbacho Lopera et al., 2019), particularly in drinking water production from polluted surface water and recycling of drinking water. Therefore, it would prevent these toxic chemicals from entering the digester and affecting methane production. However, there are other membrane processes (such as forward osmosis, membrane distillation, and electrodialysis) that have the potential to mitigate ECs but have yet to be implemented on a large scale due to their high energy consumption and operating costs.

Ozone (O_3) is a promising chemical oxidation process that can be used to significantly reduce the xenobiotic organic pollutant load in WWTPs (Xiang et al., 2021). Ozone reacts directly with pollutants or indirectly via a secondary oxidant, hydroxyl radicals (HO·), that are formed through the side reaction of ozone with specific groups of effluents, such as phenols and amines (Rizzo et al., 2019). The oxidation efficiency of ozonation can be enhanced by combining ozone with H_2O_2 and UV radiation. Ozone is selective in nature and preferentially attacks electron-rich pollutants, such as sulfamethoxazoles, and contaminants with deprotonated amine groups, such as trimethoprim. Previous studies have reported a high removal efficiency (>95%) of diclofenac, carbamazepine, sulpiride, indomethacin, and trimethoprim at an ozone dose of 5 mg/L (Gogoi et al., 2018). Meanwhile, ozone was observed to improve methane production toward antibiotic-contaminated sludge digestion by 22%–32% (Wu et al., 2022). Further analysis found that ozone could not only enhance WAS solubilization but also accelerate antibiotic degradation (Fig. 7.4). Moreover, the combination of ozone and ultrasonication also exhibited considerable mitigation efficiency (approximately 30%) for sludge digestion under levofloxacin exposure (Zhao et al., 2020). Ultrasonication could help ozone to produce more hydroxyl radicals, which would be beneficial to promote the ring-opening of antibiotics and subsequently enhance antibiotic degradation. Despite its numerous advantages, ozonation does have one important downside: when insufficient ozone doses are applied, it can cause the formation of oxidation by-products instead of complete mineralization (de Oliveira et al., 2020). Furthermore, the introduction of additional chemicals to boost the decomposition of ozone and HO· generation may lead to secondary pollution, which is a cause for major concern (Menacherry et al., 2022; Yusuf et al., 2022).

2.3 Microplastics

In recent years, the presence of MPs in WWTPs and WAS has become increasingly prevalent due to the widespread usage of plastic products and their poor biodegradability (Li et al., 2018; Ziajahromi et al., 2016). The concentration of MPs in WAS has been

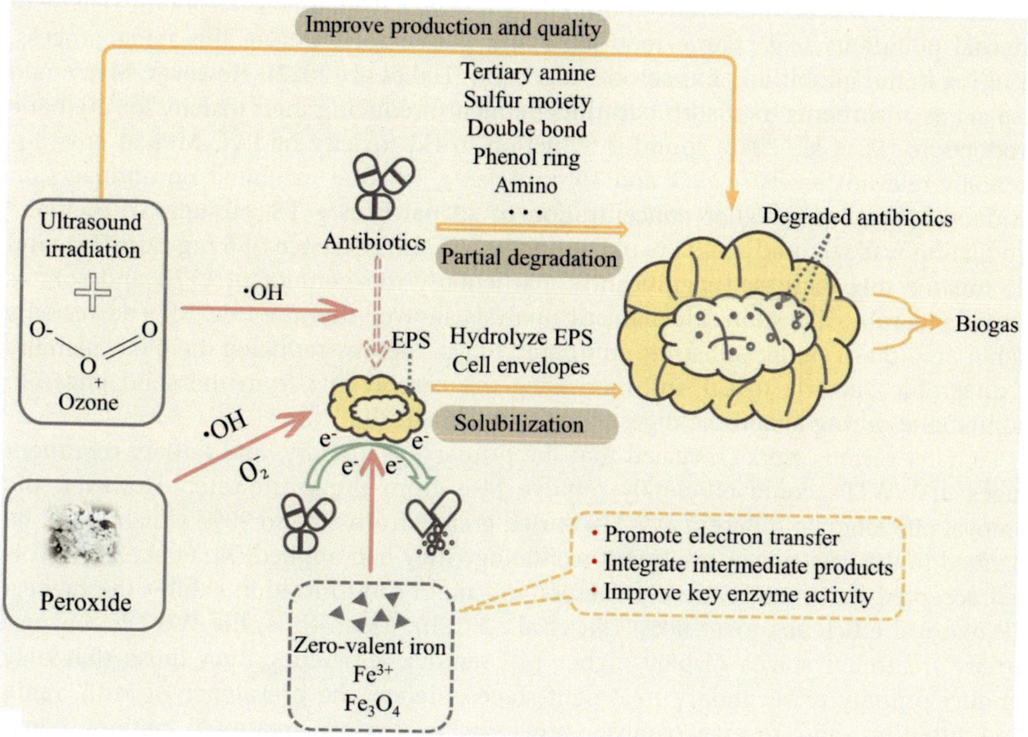

FIGURE 7.4 Proposed mitigation strategies to promote the performance of antibiotic-inhibited anaerobic digestion (Wu et al., 2022).

found to range from 1.60 to 56.4 × 10^3 particles/kg dry sludge, and polyethene (PE), polypropylene, polyamide, polyester, polystyrene, and polyethylene terephthalate are the most abundant MPs in the WWTPs (Li et al., 2018; Mahon et al., 2017). The impact of PVC MPs on WAS digestion was first investigated by Wei et al. At low levels (10 particles/g total suspended solids), PVC MPs slightly increased methane production by 5.9 ± 0.1%. But it exhibited a remarkable inhibition at higher levels (20–60 particles/g TSS). The study revealed that low levels of PVC MPs enhanced the solubilization of WAS, resulting in increased COD due to the release of lipids and nucleic acids. However, the effects of PVC on hydrolysis, acidification, and methanogenesis were different and dependent on the specific dose. Low levels of PVC MPs had no impact on WAS hydrolysis but improved the production of VFAs by 4.9%, resulting from increased soluble COD in solubilization. This further contributed to the improved methane production from WAS. In contrast, high levels of PVC MPs were observed to inhibit WAS hydrolysis, resulting in a reduction of VFA production by 6.9%–16.8% compared with the blank group. Also, the microbial community composition was significantly altered by the high levels of PVC MPs, reducing the abundance of fermentative microbes and the activities of hydrolytic and acidifying enzymes (Wei et al., 2019).

The toxicity was not only derived from the substance itself but also from the release of internal pollutants (e.g., flame retardants and plasticizers) during the aging process, resulting in the inhibition of anaerobic digestion (Dai et al., 2022). However, MPs could also act as adsorbents to adsorb harmful chemicals, reducing their toxicity for methane production. (Li et al., 2018) found a reduction in Cd toxicity by PVC-MPs at environmentally relevant levels (e.g., 1 and 10 particles/g TS) that exhibited non-impacts on production. Yet, at a higher concentration of 30 particles/g TS, cumulative methane production was restored to 89.7% of the control (in the presence of 5 mg Cd/g TS), and the organic flux increased significantly, particularly the content of VFAs and solutes during anaerobic digestion. Mechanistic analysis showed that the PVC-MPs adsorbed a higher amount of sludge substrate compared to Cd, thereby reducing the bioavailability of anaerobic bacteria to Cd and increasing the carbon flux from the solid phase to biomethane during anaerobic digestion (Liu et al., 2023b).

To date, various works revealed that the primary, secondary, and tertiary treatment stages in WWTPs could efficiently remove MPs from the wastewater. However, the removal efficiency in different WWTPs varied greatly from 50% to 99%, which might be ascribed to the wastewater treatment technology they had applied (Xu et al., 2021). It is well accepted that membrane bioreactors and rapid sand filtration exhibit the highest MP removal efficiency (over 90%) (Xu et al., 2019b). Meanwhile, the WWTPs that use tertiary treatment stages display higher MP removal efficiency than those that only conduct primary or secondary treatment stages. Hence, the prevalence of MBR, rapid sand filtration, and reverse osmosis processes as tertiary treatment options could effectively limit the MP release into subsequent treatment stages and natural water. Nevertheless, these methods only result in the phase transfer of MPs, not their complete removal or mineralization. Therefore, the MPs-targeted treatment technology should be further developed to mitigate its potential effects on anaerobic digestion and environmental risks.

2.4 Combined effects of mixed pollutants

It is widely recognized that WWTPs process a broad spectrum of wastewater, including municipal and industrial effluent, that contains a diverse range of pollutants. These pollutants co-occur in WAS with varying concentrations, and their combined presence can exacerbate the deleterious impacts of individual contaminants (Choi et al., 2018; Ekpeghere et al., 2017; Zhang and Li, 2018). For instance, the complexes of Ciprofloxacin and trimethoprim with Ag NPs were found to enhance the antimicrobial activity against *Bacillus* spp. by factors ranging from 0.1 to 5.3 and 3.0 to 6.1, respectively. This increase in antimicrobial activity was attributed to the chelating interactions between the hydroxyl and amino groups in antibiotics and Ag NPs (Naqvi et al., 2013). Nevertheless, limited information is available regarding the combined effects of multiple pollutants on the digestion of WAS (Table 7.3).

Table 7.3 Influences of mixed pollutants on the WAS anaerobic digestion.

Mixed pollutants	Objective	Influences on WAS digestion	References
CuO NPs and MWCNTs	*Propionibacterium acidipropionici*	Propionate production was increased by 41.4% compared with the sole presence of CuO NPs	Dong et al. (2021)
CuO NPs and MWCNTs	*Saccharofermentans acetigenes*	Acetate production was increased by 68.4% compared with the sole presence of CuO NPs	Hu et al. (2021)
PS MPs and PFOS	Sludge	H_2 production was increased by 48.4% compared with the sole presence of PS MPs	Chen et al. (2020a,b)
CeO_2 NPs and SDBS	Sludge	VFAs production was decreased by 33.9% compared with the sole presence of SDBS	Luo et al. (2022)
CeO_2 NPs and HTAB	Sludge	VFAs production was decreased by 33.9% compared with the sole presence of HTAB	Luo et al. (2022)
Fe_2O_3 NPs and SDBS	Sludge	VFAs production was decreased by 40.1% compared with the sole presence of SDBS	Fang et al. (2022)
Fe_2O_3 NPs and CTAB	Sludge	VFAs production was decreased by 20.3% compared with the sole presence of CTAB	Fang et al. (2022)
ZnO NPs and SDS	Sludge	CH_4 production efficiency was reduced by 82.7% compared with the sole presence of SDS	Zhu et al. (2020)
ZnO NPs and ciprofloxacin	Sludge	CH_4 production inhibition was reduced from 49.5% (individual presence of ZnO NPs) to 89.7% (coexistence of ZnO NPs and ciprofloxacin)	Zhao et al. (2018)

Regarding anaerobic digestion, a previous study found that norfloxacin and sulphamidine (500 mg/kg) inhibited the rate of methane production but not methane production (Aziz et al., 2022). ZnO NPs with antibiotics inhibited hydrolysis, fermentation, and methanogenesis at different digestion times. More influential on methane production than ZnO alone was the composite contamination, which had an acute synergistic toxic effect on methanogenesis over a short period of time. Complex contamination also has a differential effect on bacterial and archaeal communities during digestion. Meanwhile, the presence of surfactants was observed to augment the toxicity of Fe_2O_3 NPs in the biosynthesis of VFAs (Fang et al., 2022). The production of VFAs exhibited a marked reduction of 40.1% and 20.3% in the presence of surfactants, as compared to single-reactor systems containing SDBS and CTAB, respectively. Mechanistic investigations revealed that the dissolution, hydrolysis, and acidification steps of WAS were inhibited during the production of VFAs. The concurrent presence of Fe_2O_3 NPs and surfactants was found to inhibit acid-forming bacteria, such as *Saprospiraceae_uncultured* and *Bacteroidetes vadinHA17_norank*, as well as homologous metabolic functions associated with substrate hydrolysis (*malZ* and *pepD*), intracellular metabolism (*pfkA* and *gdhA*), the biosynthesis of VFAs (*bkdA* and *ackA*), and ATP production (*atpE* and *atpH*), resulting in a reduced output of VFAs. Moreover, surfactants facilitated the dispersion of coexisting Fe_2O_3 NPs, thereby maintaining their high reactivity to exacerbate toxicity.

Similar results were reported in the work of Luo et al. with respect to the impact of CeO_2 NPs and surfactants on the production of VFAs during anaerobic fermentation (Luo et al., 2022). It was found that CeO_2 NPs alone had no significant effect on VFA production; however, the presence of HTAB and SDBS induced improved dissolution and hydrolysis of the WAS and consequently increased the production of VFAs. On the other hand, the co-existence of CeO_2 NPs and surfactants led to a significant reduction in VFA production. The highest decrease was observed in the case of the combination of SDBS and CeO_2 NPs, with a 33.9% reduction compared to the SDBS reactor and a 10.7% decrease for the combination of HTAB and CeO_2 NPs compared to the HTAB reactor. The authors attributed the reduction in VFA production to the inhibitory effect on key anaerobic fermentation processes such as solubilization, hydrolysis, and acidification, as well as on the functional microorganisms involved in VFA biosynthesis and related gene expression. Further investigations revealed that the presence of surfactants improved the stability and dispersion of CeO_2 NPs, thereby promoting solubilization of Ce^{4+} and causing significant toxicity to functional anaerobes, and also increased the chances of interaction between microorganisms and CeO_2 NPs, thereby exacerbating toxicity.

Generally, the co-occurrence of ECs might aggravate their biotoxicity to WAS digestion. Therefore, the joint effects and potential mitigating strategies for these ECs are also critical to maintaining the high operational performance of WWTPs. Source control and segregation of ECs seem to be the most direct and effective approaches, but this is too ideal for a real application based on the imperfect discharge and collection systems. Also, this strategy requires cooperation between different administrative departments with strict legislation and mature management systems. Alternatively, some enhanced measures in coupling with conventional biological treatment systems have been adopted to remove undesirable ECs. Engineering bacterial consortia to target specific ECs is an environmentally friendly approach. However, the stable and efficient cultivation of an engineered consortium for treating diverse ECs is complicated. Meanwhile, these consortia should be compatible with those indigenous bacteria responsible for typical pollutant removal (e.g., COD, N, and P). Multiomics technologies provide abundant information on functional microorganisms with certain genetic traits, which would contribute to the cultivation of an engineered consortium (Bernard et al., 2022). In addition, advanced oxidation processes (AOPs) have proven to be a fast and effective method to simultaneously remove diverse ECs. However, the introduction of new chemicals and the generation of toxic by-products derived from AOPs might cause secondary pollution, which is a great concern. Moreover, the presence of huge amounts of typical pollutants will interfere with the removal process and decrease the selectivity and efficiency of removing target ECs (Hodges et al., 2018). The synchronous removal of different ECs via process upgrading and optimization is an intriguing option. For example, the effective removal of MPs from wastewater would undoubtedly benefit the decrease of adsorbed ECs on MPs. However, this is highly dependent on the disclosure of underlying mechanisms for the mutual interaction between ECs. Given the complexity of wastewater and identifying difficulties associated with the interaction between ECs, the incorporation of powerful computational tools (e.g., machine learning) into the development of new techniques for the control of ECs should be emphasized. Furthermore, the application of machine learning also provides the ability to predict the fates of specific ECs. Of note, the effectiveness, convenience, and environmental benefits and risks should all be comprehensively evaluated before selecting optimized EC control strategies.

3. Conclusion and prospects

In this review, the effects of various typical ECs on the anaerobic digestion of wastewater from WWTPs were analyzed. The findings indicate that different ECs can have varying influences on the processes of sludge hydrolysis, acidification, and methanogenesis, with susceptibility varying based on the specific species and concentration levels. While low doses of many pollutants may aid in anaerobic digestion, high levels of contaminants can lead to negative effects. The review highlights that methanogens are particularly vulnerable to toxic pollutants and that the methanogenic process is more easily impacted compared to hydrolysis and acidification.

In terms of future investigations, the review calls for a closer examination of the combined effects of mixed ECs on the anaerobic digestion of WWTPs, considering that many pollutants co-exist in WAS and may exhibit synergistic or antagonist effects. Additionally, research is needed to develop effective strategies for mitigating the negative effects of typical pollutants, such as PPCPs, given their high concentrations and strong toxicity, in order to ensure efficient anaerobic digestion.

References

Aziz, A., Sengar, A., Basheer, F., Farooqi, I.H., Isa, M.H., 2022. Anaerobic digestion in the elimination of antibiotics and antibiotic-resistant genes from the environment – a comprehensive review. Journal of Environmental Chemical Engineering 10 (1), 106423.

Baniamerian, H., Ghofrani-Isfahani, P., Tsapekos, P., Alvarado-Morales, M., Shahrokhi, M., Angelidaki, I., 2021. Multicomponent nanoparticles as means to improve anaerobic digestion performance. Chemosphere 283, 131277.

Battersby, N.S., Wilson, V., 1988. Evaluation of a serum bottle technique for assessing the anaerobic biodegradability of organic chemicals under methanogenic conditions. Chemosphere 17 (12), 2441–2460.

Bernard, C., Locard-Paulet, M., Noël, C., Duchateau, M., Giai Gianetto, Q., Moumen, B., Rattei, T., Hechard, Y., Jensen, L.J., Matondo, M., Samba-Louaka, A., 2022. A time-resolved multi-omics atlas of Acanthamoeba castellanii encystment. Nature Communications 13 (1), 4104.

Bueno-Lopez, J.I., Rangel-Mendez, J.R., Alatriste-Mondragon, F., Perez-Rodriguez, F., Hernandez-Montoya, V., Cervantes, F.J., 2018. Graphene oxide triggers mass transfer limitations on the methanogenic activity of an anaerobic consortium with a particulate substrate. Chemosphere 211, 709–716.

Cai, Q.-Y., Mo, C.-H., Wu, Q.-T., Zeng, Q.-Y., Katsoyiannis, A., Férard, J.-F., 2007. Bioremediation of polycyclic aromatic hydrocarbons (PAHs)-contaminated sewage sludge by different composting processes. Journal of Hazardous Materials 142 (1), 535–542.

Cervantes-Avilés, P., Ida, J., Toda, T., Cuevas-Rodríguez, G., 2018. Effects and fate of TiO2 nanoparticles in the anaerobic treatment of wastewater and waste sludge. Journal of Environmental Management 222, 227–233.

Chen, H., Chen, Y.G., Zheng, X., Li, X., Luo, J.Y., 2014. How does the entering of copper nanoparticles into biological wastewater treatment system affect sludge treatment for VFA production. Water Research 63, 125–134.

Chen, H., Zeng, X., Zhou, Y., Yang, X., Lam, S.S., Wang, D., 2020a. Influence of roxithromycin as antibiotic residue on volatile fatty acids recovery in anaerobic fermentation of waste activated sludge. Journal of Hazardous Materials 394, 122570.

Chen, W., Yuan, D., Shan, M., Yang, Z., Liu, C., 2020b. Single and combined effects of amino polystyrene and perfluorooctane sulfonate on hydrogen-producing thermophilic bacteria and the interaction mechanisms. Science of the Total Environment 703, 135015.

Choi, S., Johnston, M., Wang, G.S., Huang, C.P., 2018. A seasonal observation on the distribution of engineered nanoparticles in municipal wastewater treatment systems exemplified by TiO2 and ZnO. Science of the Total Environment 625, 1321–1329.

Dai, H., Gao, J., Li, D., Wang, Z., Zhao, Y., Cui, Y., 2022. Polyvinyl chloride microplastics changed risks of antibiotic resistance genes propagation by enhancing the removal of triclosan in partial denitrification systems with different carbon source. Chemical Engineering Journal 429, 132465.

de Oliveira, M., Frihling, B.E.F., Velasques, J., Filho, F.J.C.M., Cavalheri, P.S., Migliolo, L., 2020. Pharmaceuticals residues and xenobiotics contaminants: occurrence, analytical techniques and sustainable alternatives for wastewater treatment. Science of the Total Environment 705, 135568.

Dong, B., Xia, Z.H., Sun, J., Dai, X.H., Chen, X.M., Ni, B.J., 2019. The inhibitory impacts of nano-graphene oxide on methane production from waste activated sludge in anaerobic digestion. Science of the Total Environment 646, 1376–1384.

Dong, L., Wu, Y., Bian, Y., Zheng, X., Chen, L., Chen, Y., Zhang, X., 2021. Carbon nanotubes mitigate copper-oxide nanoparticles-induced inhibition to acidogenic metabolism of Propionibacterium acidipropionici by regulating carbon source utilization. Bioresource Technology 330, 125003.

Du, W., Wang, F., Fang, S., Huang, W., Cheng, X., Cao, J., Fang, F., Wu, Y., Luo, J., 2022. Antimicrobial PCMX facilitates the volatile fatty acids production during sludge anaerobic fermentation: insights of the interactive principles, microbial metabolic profiles and adaptation mechanisms. Chemical Engineering Journal 446, 137339.

Duan, X., Wang, X., Xie, J., Feng, L.Y., Yan, Y.Y., Zhou, Q., 2016. Effect of nonylphenol on volatile fatty acids accumulation during anaerobic fermentation of waste activated sludge. Water Research 105, 209–217.

Durham, J., Young, J.C., 2009. Toxic impact of commercial biocides on industrial wastewater treatment systems. Proceedings of the Water Environment Federation 2009 (10), 5670–5689.

Egea-Corbacho Lopera, A., Gutiérrez Ruiz, S., Quiroga Alonso, J.M., 2019. Removal of emerging contaminants from wastewater using reverse osmosis for its subsequent reuse: pilot plant. Journal of Water Process Engineering 29, 100800.

Ekpeghere, K.I., Lee, J.W., Kim, H.Y., Shin, S.K., Oh, J.E., 2017. Determination and characterization of pharmaceuticals in sludge from municipal and livestock wastewater treatment plants. Chemosphere 168, 1211–1221.

Fang, S., Guo, W., Feng, Q., Cao, W., Dou, Y., Huang, W., Wang, F., Cheng, X., Cao, J., Wu, Y., Luo, J., 2022. Surfactants aggravated the biotoxicity of Fe2O3 nanoparticles in the volatile fatty acids' biosynthesis during sludge anaerobic fermentation. ACS ES&T Water 2 (12), 2686–2697.

Flores, G.A.E., Fotidis, I.A., Karakashev, D.B., Kjellberg, K., Angelidaki, I., 2015. Effects of benzalkonium chloride, Proxel LV, P3 hypochloran, Triton X-100 and DOWFAX 63N10 on anaerobic digestion processes. Bioresource Technology 193, 393–400.

Fountoulakis, M., Drillia, P., Stamatelatou, K., Lyberatos, G., 2004. Toxic effect of pharmaceuticals on methanogenesis. Water Science and Technology 50 (5), 335–340.

Garcia, M.T., Campos, E., Sanchez-Leal, J., Ribosa, I., 2006. Effect of linear alkylbenzene sulphonates (LAS) on the anaerobic digestion of sewage sludge. Water Research 40 (15), 2958–2964.

García, A., Delgado, L., Torà, J.A., Casals, E., González, E., Puntes, V., Font, X., Carrera, J., Sánchez, A., 2012. Effect of cerium dioxide, titanium dioxide, silver, and gold nanoparticles on the activity of microbial communities intended in wastewater treatment. Journal of Hazardous Materials 199–200, 64–72.

Gavala, H.N., Alatriste-Mondragon, F., Iranpour, R., Ahring, B.K., 2003. Biodegradation of phthalate esters during the mesophilic anaerobic digestion of sludge. Chemosphere 52 (4), 673–682.

Gicheva, G., Yordanov, G., 2013. Removal of citrate-coated silver nanoparticles from aqueous dispersions by using activated carbon. Colloids and Surfaces A. Physicochemical and Engineering Aspects 431, 51–59.

Gogoi, A., Mazumder, P., Tyagi, V.K., Tushara Chaminda, G.G., An, A.K., Kumar, M., 2018. Occurrence and fate of emerging contaminants in water environment: a review. Groundwater for Sustainable Development 6, 169–180.

He, D., Xiao, J., Wang, D., Liu, X., Li, Y., Fu, Q., Li, C., Yang, Q., Liu, Y., Ni, B.-J., 2021. Understanding and regulating the impact of tetracycline to the anaerobic fermentation of waste activated sludge. Journal of Cleaner Production 313.

He, Q., Xu, P., Zhang, C., Zeng, G., Liu, Z., Wang, D., Tang, W., Dong, H., Tan, X., Duan, A., 2019. Influence of surfactants on anaerobic digestion of waste activated sludge: acid and methane production and pollution removal. Critical Reviews in Biotechnology 39 (5), 746–757.

Hodges, B.C., Cates, E.L., Kim, J.-H., 2018. Challenges and prospects of advanced oxidation water treatment processes using catalytic nanomaterials. Nature Nanotechnology 13 (8), 642–650.

Hu, J., Xu, Q., Li, X., Wang, D., Zhong, Y., Zhao, J., Zhang, D., Yang, Q., Zeng, G., 2018. Sulfamethazine (SMZ) affects fermentative short-chain fatty acids production from waste activated sludge. Science of the Total Environment 639.

Hu, J., Zhao, J., Wang, D., Li, X., Zhang, D., Xu, Q., Peng, L., Yang, Q., Zeng, G., 2018b. Effect of diclofenac on the production of volatile fatty acids from anaerobic fermentation of waste activated sludge. Bioresource Technology 254, 7–15.

Hu, W., Wu, Y., Bian, Y., Zheng, X., Chen, Y., Dong, L., Chen, Y., 2021. Joint effects of carbon nanotubes and copper oxide nanoparticles on fermentation metabolism towards Saccharofermentans acetigenes: enhancing environmental adaptability and transcriptional expression. Bioresource Technology 336, 125318.

Hua, L., Wu, W.X., Liu, Y.X., Tientchen, C.M., Cen, Y.X., 2008. Heavy metals and PAHs in sewage sludge from twelve wastewater treatment plants in Zhejiang province. Biomedical and Environmental Sciences 21 (4), 345–352.

Huang, X.F., Shen, C.M., Liu, J., Lu, L.J., 2015. Improved volatile fatty acid production during waste activated sludge anaerobic fermentation by different bio-surfactants. Chemical Engineering Journal 264, 280–290.

Huang, J., Cao, C., Liu, J., Yan, C., Xiao, J., 2019. The response of nitrogen removal and related bacteria within constructed wetlands after long-term treating wastewater containing environmental concentrations of silver nanoparticles. Science of the Total Environment 667, 522–531.

Jiang, S., Chen, Y., Zhou, Q., 2007a. Effect of sodium dodecyl sulfate on waste activated sludge hydrolysis and acidification. Chemical Engineering Journal 132 (1), 311–317.

Jiang, S., Chen, Y., Zhou, Q., Gu, G., 2007b. Biological short-chain fatty acids (SCFAs) production from waste-activated sludge affected by surfactant. Water Research 41 (14), 3112–3120.

Jiang, X., Yan, Y., Feng, L., Wang, F., Guo, Y., Zhang, X., Zhang, Z., 2021. Bisphenol A alters volatile fatty acids accumulation during sludge anaerobic fermentation by affecting amino acid metabolism, material transport and carbohydrate-active enzymes. Bioresource Technology 323, 124588.

Kara Murdoch, F., Murdoch, R.W., Gürakan, G.C., Sanin, F.D., 2018. Change of microbial community composition in anaerobic digesters during the degradation of nonylphenol diethoxylate. International Biodeterioration & Biodegradation 135, 1–8.

Kato, S., Hashimoto, K., Watanabe, K., 2012. Methanogenesis facilitated by electric syntrophy via (semi) conductive iron-oxide minerals. Environmental Microbiology 14 (7), 1646–1654.

Khalil, E.F., Whitmore, T.N., Gamal-El-Din, H., El-Bassel, A., Lloyd, D., 1988. The effects of detergents on anaerobic digestion. Applied Microbiology and Biotechnology 29 (5), 517–522.

Kökdemir Ünşar, E., Perendeci, N.A., 2018. What kind of effects do Fe2O3 and Al2O3 nanoparticles have on anaerobic digestion, inhibition or enhancement? Chemosphere 211, 726–735.

Lee, W.S., Chua, A.S.M., Yeoh, H.K., Ngoh, G.C., 2014. A review of the production and applications of waste-derived volatile fatty acids. Chemical Engineering Journal 235, 83–99.

Li, X.Q., Brown, D.G., Zhang, W.X., 2007. Stabilization of biosolids with nanoscale zero-valent iron (nZVI). Journal of Nanoparticle Research 9 (2), 233–243.

Li, N., Liu, H.J., Xue, Y.G., Wang, H.Y., Dai, X.H., 2017. Partition and fate analysis of fluoroquinolones in sewage sludge during anaerobic digestion with thermal hydrolysis pretreatment. Science of the Total Environment 581, 715–721.

Li, X., Chen, L., Mei, Q., Dong, B., Dai, X., Ding, G., Zeng, E.Y., 2018. Microplastics in sewage sludge from the wastewater treatment plants in China. Water Research 142, 75–85.

Liu, D., Gu, W., Zhou, L., Lei, J., Wang, L., Zhang, J., Liu, Y., 2023a. From biochar to functions: lignin induced formation of Fe3C in carbon/Fe composites for efficient adsorption of tetracycline from wastewater. Separation and Purification Technology 304, 122217.

Liu, X., Deng, Q., Du, M., Lu, Q., Zhou, W., Wang, D., 2023b. Microplastics decrease the toxicity of cadmium to methane production from anaerobic digestion of sewage sludge. Science of the Total Environment 869, 161780.

Luna-delRisco, M., Orupold, K., Dubourguier, H.C., 2011. Particle-size effect of CuO and ZnO on biogas and methane production during anaerobic digestion. Journal of Hazardous Materials 189 (1–2), 603–608.

Luo, J., Feng, L., Chen, Y., Li, X., Chen, H., Xiao, N., Wang, D., 2014. Stimulating short-chain fatty acids production from waste activated sludge by nano zero-valent iron. Journal of Biotechnology 187, 98–105.

Luo, J.Y., Feng, L.Y., Chen, Y.G., Sun, H., Shen, Q.T., Li, X., Chen, H., 2015. Alkyl polyglucose enhancing propionic acid enriched short-chain fatty acids production during anaerobic treatment of waste activated sludge and mechanisms. Water Research 73, 332–341.

Luo, J.Y., Chen, Y.G., Feng, L.Y., 2016. Polycyclic aromatic hydrocarbon affects acetic acid production during anaerobic fermentation of waste activated sludge by altering activity and viability of acetogen. Environmental Science & Technology 50 (13), 6921–6929.

Luo, J.Y., Wu, J., Zhang, Q., Feng, Q., Wu, L.J., Cao, J.S., Li, C., Fang, F., 2018a. Efficient production of short-chain fatty acids from anaerobic fermentation of liquor wastewater and waste activated sludge by breaking the restrictions of low bioavailable substrates and microbial activity. Bioresource Technology 268, 549–557.

Luo, J.Y., Zhang, Q., Wu, L.J., Feng, Q., Fang, F., Xue, Z.X., Li, C., Cao, J.S., 2018b. Improving anaerobic fermentation of waste activated sludge using iron activated persulfate treatment. Bioresource Technology 268, 68–76.

Luo, J.Y., Wu, L.J., Zhang, Q., Fang, F., Feng, Q., Xue, Z.X., Cao, M., Peng, Z.Q., Li, C., Cao, J.S., 2019. How do biocides that occur in waste activated sludge affect the resource recovery for short-chain fatty acids production. ACS Sustainable Chemistry & Engineering 7 (1), 1648–1657.

Luo, J., Zhang, Q., Zhao, J., Wu, Y., Wu, L., Li, H., Tang, M., Sun, Y., Guo, W., Feng, Q., Cao, J., Wang, D., 2020. Potential influences of exogenous pollutants occurred in waste activated sludge on anaerobic digestion: a review. Journal of Hazardous Materials 383, 121176.

Luo, J., Cao, W., Guo, W., Fang, S., Huang, W., Wang, F., Cheng, X., Du, W., Cao, J., Feng, Q., Wu, Y., 2022. Antagonistic effects of surfactants and CeO2 nanoparticles co-occurrence on the sludge fermentation process: novel insights of interaction mechanisms and microbial networks. Journal of Hazardous Materials 438, 129556.

Luo, J., Li, Y., Huang, W., Wang, F., Fang, S., Cheng, X., Feng, Q., Fang, F., Cao, J., Wu, Y., 2023. Dissimilarity of different cephalosporins on volatile fatty acids production and antibiotic resistance genes fates during sludge fermentation and underlying mechanisms. Chinese Chemical Letters 34 (4), 107661.

Ma, J., Quan, X., Si, X., Wu, Y., 2013. Responses of anaerobic granule and flocculent sludge to ceria nanoparticles and toxic mechanisms. Bioresource Technology 149, 346–352.

Mahon, A.M., Connell, B.O., Healy, M.G., Connor, I.O., Officer, R., Nash, R., Morrison, L., 2017. Microplastics in sewage sludge: effects of treatment. Environmental Science & Technology 51 (2), 810.

Mai, D.T., Stuckey, D.C., Oh, S., 2018. Effect of ciprofloxacin on methane production and anaerobic microbial community. Bioresource Technology 261, 240–248.

Mailler, R., Gasperi, J., Patureau, D., Vulliet, E., Delgenes, N., Danel, A., Deshayes, S., Eudes, V., Guerin, S., Moilleron, R., Chebbo, G., Rocher, V., 2017. Fate of emerging and priority micropollutants during the sewage sludge treatment: case study of Paris conurbation. Part 1: contamination of the different types of sewage sludge. Waste Management 59, 379–393.

Menacherry, S.P.M., Aravind, U.K., Aravindakumar, C.T., 2022. Critical review on the role of mass spectrometry in the AOP based degradation of contaminants of emerging concern (CECs) in water. Journal of Environmental Chemical Engineering 10 (4), 108155.

Montes-Grajales, D., Fennix-Agudelo, M., Miranda-Castro, W., 2017. Occurrence of personal care products as emerging chemicals of concern in water resources: a review. Science of the Total Environment 595, 601–614.

Mu, H., Chen, Y.G., 2011. Long-term effect of ZnO nanoparticles on waste activated sludge anaerobic digestion. Water Research 45 (17), 5612–5620.

Mu, H., Chen, Y.G., Xiao, N.D., 2011. Effects of metal oxide nanoparticles (TiO2, Al2O3, SiO2 and ZnO) on waste activated sludge anaerobic digestion. Bioresource Technology 102 (22), 10305–10311.

Mungray, A.K., Kumar, P., 2009. Fate of linear alkylbenzene sulfonates in the environment: a review. International Biodeterioration & Biodegradation 63 (8), 981–987.

Nanninga, N., 2009. Cell structure, organization, bacteria and archaea. Encyclopedia of Microbiology 331 (1), 357–374.

Naqvi, S.Z.H., Kiran, U., Ali, M.I., Jamal, A., Hameed, A., Ahmed, S., Ali, N., 2013. Combined efficacy of biologically synthesized silver nanoparticles and different antibiotics against multidrug-resistant bacteria. International Journal of Nanomedicine 8.

Östman, M., Lindberg, R.H., Fick, J., Björn, E., Tysklind, M., 2017. Screening of biocides, metals and antibiotics in Swedish sewage sludge and wastewater. Water Research 115, 318–328.

Park, C.M., Chu, K.H., Her, N., Jang, M., Baalousha, M., Heo, J., Yoon, Y., 2017. Occurrence and removal of engineered nanoparticles in drinking water treatment and wastewater treatment processes. Separation and Purification Reviews 46 (3), 255–272.

Rathi, B.S., Kumar, P.S., Show, P.-L., 2021. A review on effective removal of emerging contaminants from aquatic systems: current trends and scope for further research. Journal of Hazardous Materials 409, 124413.

Richmond, E.K., Rosi, E.J., Walters, D.M., Fick, J., Hamilton, S.K., Brodin, T., Sundelin, A., Grace, M.R., 2018. A diverse suite of pharmaceuticals contaminates stream and riparian food webs. Nature Communications 9 (1), 4491.

Rizzo, L., Malato, S., Antakyali, D., Beretsou, V.G., Đolić, M.B., Gernjak, W., Heath, E., Ivancev-Tumbas, I., Karaolia, P., Lado Ribeiro, A.R., Mascolo, G., McArdell, C.S., Schaar, H., Silva, A.M.T., Fatta-Kassinos, D., 2019. Consolidated vs new advanced treatment methods for the removal of contaminants of emerging concern from urban wastewater. Science of the Total Environment 655, 986–1008.

Salih, H.H., Sorial, G.A., Patterson, C.L., Sinha, R., Krishnan, E.R., 2012. Removal of trichloroethylene by activated carbon in the presence and absence of TiO2 nanoparticles. Water, Air, & Soil Pollution 223 (5), 2837–2847.

Sanz, J.L., Rodríguez, N., Amils, R., 1996. The action of antibiotics on the anaerobic digestion process. Applied Microbiology and Biotechnology 46 (5), 587–592.

Shao, L.M., Li, W.P., Lu, F., He, P.J., 2018. Anaerobic re-digestion of digested sludge post-treated by ultrasound: effect of the adding linear alkylbenzene sulfonate. Journal of Chemical Technology and Biotechnology 93 (8), 2464–2470.

Shi, J., Dang, Q., Zhang, C., Zhao, X., 2022. Insight into effects of polyethylene microplastics in anaerobic digestion systems of waste activated sludge: interactions of digestion performance, microbial communities and antibiotic resistance genes. Environmental Pollution 310, 119859.

Symsaris, E.C., Fotidis, I.A., Stasinakis, A.S., Angelidaki, I., 2015. Effects of triclosan, diclofenac, and nonylphenol on mesophilic and thermophilic methanogenic activity and on the methanogenic communities. Journal of Hazardous Materials 291, 45–51.

Tang, M., Wu, Y., Zeng, X., Yang, X., Wang, D., Chen, H., 2021. Unveiling the different faces of chlortetracycline in fermentative volatile fatty acid production from waste activated sludge. Bioresource Technology 329, 124875.

Tezel, U., 2009. Fate and Effect of Quaternary Ammonium Compounds in Biological Systems. Georgia Institute of Technology.

Tian, T., Qiao, S., Li, X., Zhang, M.J., Zhou, J.T., 2017. Nano-graphene induced positive effects on methanogenesis in anaerobic digestion. Bioresource Technology 224, 41–47.

Troester, M., Brauch, H.-J., Hofmann, T., 2016. Vulnerability of drinking water supplies to engineered nanoparticles. Water Research 96, 255–279.

Walser, T., Limbach, L.K., Brogioli, R., Erismann, E., Flamigni, L., Hattendorf, B., Juchli, M., Krumeich, F., Ludwig, C., Prikopsky, K., Rossier, M., Saner, D., Sigg, A., Hellweg, S., Gunther, D., Stark, W.J., 2012. Persistence of engineered nanoparticles in a municipal solid-waste incineration plant. Nature Nanotechnology 7 (8), 520–524.

Wang, D.B., Chen, Y.G., 2016. Critical review of the influences of nanoparticles on biological wastewater treatment and sludge digestion. Critical Reviews in Biotechnology 36 (5), 816–828.

Wang, C., Jeong, M., 2018. What makes you choose Airbnb again? An examination of users' perceptions toward the website and their stay. International Journal of Hospitality Management 74, 162–170.

Wang, T., Zhang, D., Dai, L., Chen, Y., Dai, X., 2016. Effects of metal nanoparticles on methane production from waste-activated sludge and microorganism community Shift in anaerobic granular sludge. Scientific Reports 6, 25857.

Wang, Y., Wang, D., Liu, Y., Wang, Q., Chen, F., Yang, Q., Li, X., Zeng, G., Li, H., 2017. Triclocarban enhances short-chain fatty acids production from anaerobic fermentation of waste activated sludge. Water Research 127, 150–161.

Wang, C., Wu, L., Zhang, Y.-T., Wei, W., Ni, B.-J., 2021. Unravelling the impacts of perfluorooctanoic acid on anaerobic sludge digestion process. Science of the Total Environment 796, 149057.

Wang, F., Wu, Y., Du, W., Shao, Q., Huang, W., Fang, S., Cheng, X., Cao, J., Luo, J., 2022. How does the polyhexamethylene guanidine interact with waste activated sludge and affect the metabolic functions in anaerobic fermentation for volatile fatty acids production. Science of the Total Environment 839, 156329.

Wei, W., Huang, Q.-S., Sun, J., Wang, J.-Y., Wu, S.-L., Ni, B.-J., 2019. Polyvinyl chloride microplastics affect methane production from the anaerobic digestion of waste activated sludge through leaching toxic bisphenol-A. Environmental Science & Technology 53 (5), 2509–2517.

Wei, Y., Zhou, A., Duan, Y., Liu, Z., He, Z., Zhang, J., Liang, B., Yue, X., 2022. Unraveling the behaviors of sulfonamide antibiotics on the production of short-chain fatty acids by anaerobic fermentation from waste activated sludge and the microbial ecological mechanism. Chemosphere 296, 133903.

Wieck, S., Olsson, O., Kümmerer, K., 2018. Not only biocidal products: washing and cleaning agents and personal care products can act as further sources of biocidal active substances in wastewater. Environment International 115 (1), 247–256.

Wu, X., Narsimhan, G., 2008. Characterization of secondary and tertiary conformational changes of beta-lactoglobulin adsorbed on silica nanoparticle surfaces. Langmuir 24 (9), 4989–4998.

Wu, Q., Zou, D., Zheng, X., Liu, F., Li, L., Xiao, Z., 2022. Effects of antibiotics on anaerobic digestion of sewage sludge: performance of anaerobic digestion and structure of the microbial community. Science of the Total Environment 845, 157384.

Wu, Y., Hu, W., Zhu, Z., Zheng, X., Chen, Y., Chen, Y., 2023. Enhanced volatile fatty acid production from food waste fermentation via enzymatic pretreatment: new insights into the depolymerization and microbial traits. ACS ES&T Engineering 3 (1), 26–35.

Wyrwas, B., Dymaczewski, Z., Zgola-Grzeskowiak, A., Szymanski, A., Franska, M., Kruszelnicka, I., Ginter-Kramarczyk, D., Cyplik, P., Lawniczak, L., Chrzanowski, L., 2013. Biodegradation of Triton X-100 and its primary metabolites by a bacterial community isolated from activated sludge. Journal of Environmental Management 128, 292–299.

Xiang, L., Xie, Z., Guo, H., Song, J., Li, D., Wang, Y., Pan, S., Lin, S., Li, Z., Han, J., Qiao, W., 2021. Efficient removal of emerging contaminant sulfamethoxazole in water by ozone coupled with calcium peroxide: mechanism and toxicity assessment. Chemosphere 283, 131156.

Xie, J., Duan, X., Feng, L.Y., Yan, Y.Y., Wang, F., Dong, H.Q., Jia, R.Y., Zhou, Q., 2019. Influence of sulfadiazine on anaerobic fermentation of waste activated sludge for volatile fatty acids production: focusing on microbial responses. Chemosphere 219, 305–312.

Xiu, Z.M., Jin, Z.H., Li, T.L., Mahendra, S., Lowry, G.V., Alvarez, P.J.J., 2010. Effects of nano-scale zero-valent iron particles on a mixed culture dechlorinating trichloroethylene. Bioresource Technology 101 (4), 1141–1146.

Xu, Q.X., Li, X.M., Ding, R.R., Wang, D.B., Liu, Y.W., Wang, Q.L., Zhao, J.W., Chen, F., Zeng, G.M., Yang, Q., Li, H.L., 2017. Understanding and mitigating the toxicity of cadmium to the anaerobic fermentation of waste activated sludge. Water Research 124, 269–279.

Xu, Q.X., Liu, X.R., Wang, D.B., Liu, Y.W., Wang, Q.L., Ni, B.J., Li, X.M., Yang, Q., Li, H.L., 2019a. Enhanced short-chain fatty acids production from waste activated sludge by sophorolipid: performance, mechanism, and implication. Bioresource Technology 284, 456–465.

Xu, X., Jian, Y., Xue, Y., Hou, Q., Wang, L., 2019b. Microplastics in the wastewater treatment plants (WWTPs): occurrence and removal. Chemosphere 235, 1089–1096.

Xu, Z., Bai, X., Ye, Z., 2021. Removal and generation of microplastics in wastewater treatment plants: a review. Journal of Cleaner Production 291, 125982.

Yang, Y., Chen, Q., Wall, J.D., Hu, Z.Q., 2012. Potential nanosilver impact on anaerobic digestion at moderate silver concentrations. Water Research 46 (4), 1176–1184.

Yang, Y., Gajaraj, S., Wall, J.D., Hu, Z.Q., 2013a. A comparison of nanosilver and silver ion effects on bioreactor landfill operations and methanogenic population dynamics. Water Research 47 (10), 3422–3430.

Yang, Y., Guo, J.L., Hu, Z.Q., 2013b. Impact of nano zero valent iron (NZVI) on methanogenic activity and population dynamics in anaerobic digestion. Water Research 47 (17), 6790–6800.

Yoo, D.S., Lee, B.S., Kim, E.K., 2005. Characteristics of microbial biosurfactant as an antifungal agent against plant pathogenic fungus. Journal of Microbiology and Biotechnology 15 (6), 1164–1169.

Yusuf, A., Giwa, A., Eniola, J.O., Amusa, H.K., Bilad, M.R., 2022. Recent advances in catalytic sulfate radical-based approach for removal of emerging contaminants. Journal of Hazardous Materials Advances 7, 100108.

Zaidi, A.A., RuiZhe, F., Malik, A., Khan, S.Z., Bhutta, A.J., Shi, Y., Mushtaq, K., 2019. Conjoint effect of microwave irradiation and metal nanoparticles on biogas augmentation from anaerobic digestion of green algae. International Journal of Hydrogen Energy 44 (29), 14661–14670.

Zandvoort, M.H., van Hullebusch, E.D., Fermoso, F.G., Lens, P.N.L., 2006. Trace metals in anaerobic granular sludge reactors: bioavailability and dosing strategies. Engineering in Life Sciences 6 (3), 293–301.

Zhang, X.Y., Li, R.Y., 2018. Variation of antibiotics in sludge pretreatment and anaerobic digestion processes: degradation and solid-liquid distribution. Bioresource Technology 255, 266–272.

Zhang, Z., Guo, L., Wang, Y., Zhao, Y., She, Z., Gao, M., Guo, Y., 2020. Application of iron oxide (Fe3O4) nanoparticles during the two-stage anaerobic digestion with waste sludge: impact on the biogas production and the substrate metabolism. Renewable Energy 146, 2724–2735.

Zhang, Z., Liu, H., Wen, H., Gao, L., Gong, Y., Guo, W., Wang, Z., Li, X., Wang, Q., 2021. Microplastics deteriorate the removal efficiency of antibiotic resistance genes during aerobic sludge digestion. Science of the Total Environment 798, 149344.

Zhao, Y.X., Chen, Y.G., 2011. Nano-TiO2 enhanced photofermentative hydrogen produced from the dark fermentation liquid of waste activated sludge. Environmental Science & Technology 45 (19), 8589–8595.

Zhao, L., Ji, Y., Sun, P., Li, R., Xiang, F., Wang, H., Ruiz-Martinez, J., Yang, Y., 2018. Effects of individual and complex ciprofloxacin, fullerene C60, and ZnO nanoparticles on sludge digestion: methane production, metabolism, and microbial community. Bioresource Technology 267, 46–53.

Zhao, L., Ji, Y., Sun, P.Z., Deng, J.H., Wang, H.Y., Yang, Y.K., 2019a. Effects of individual and combined zinc oxide nanoparticle, norfloxacin, and sulfamethazine contamination on sludge anaerobic digestion. Bioresource Technology 273, 454–461.

Zhao, M., Gong, H., Ma, M., Dong, L., Huang, M., Wan, R., Gu, H., Kang, Y., Li, D., 2019b. A comparative antibacterial activity and cytocompatibility for different top layers of TiN, Ag or TiN-Ag on nanoscale TiN/Ag multilayers. Applied Surface Science 473, 334–342.

Zhao, Q., Li, M., Zhang, K., Wang, N., Wang, K., Wang, H., Meng, S., Mu, R., 2020. Effect of ultrasound irradiation combined with ozone pretreatment on the anaerobic digestion for the biosludge exposed to trace-level levofloxacin: degradation, microbial community and ARGs analysis. Journal of Environmental Management 262, 110356.

Zhou, A.J., Yang, C.X., Guo, Z.C., Hou, Y.A., Liu, W.Z., Wang, A.J., 2013. Volatile fatty acids accumulation and rhamnolipid generation in situ from waste activated sludge fermentation stimulated by external rhamnolipid addition. Biochemical Engineering Journal 77, 240–245.

Zhu, K., Zhang, L., Mu, L., Ma, J., Wang, X., Li, C., Cui, Y., Li, A., 2020. Antagonistic effect of zinc oxide nanoparticle and surfactant on anaerobic digestion: focusing on the microbial community changes and interactive mechanism. Bioresource Technology 297, 122382.

Ziajahromi, S., Neale, P.A., Leusch, F.D., 2016. Wastewater treatment plant effluent as a source of microplastics: review of the fate, chemical interactions and potential risks to aquatic organisms. Water Science and Technology 74 (10), 2253.

PART III

Control strategies of emerging contaminants in organic wastes treatment

8

Role of biochar in removal of contaminants from organic wastes: A special insights to eco-restoration and bio-economy

Ram Kumar Ganguly and Susanta Kumar Chakraborty
DEPARTMENT OF ZOOLOGY, VIDYASAGAR UNIVERSITY, MIDNAPORE, WEST BENGAL, INDIA

1. Introduction

The term "biochar" describes a substance rich in carbon. Biochar is typically produced by thermally processing biomass, such as wood, dung, or leaves, in an atmosphere devoid of oxygen. Owing to its unique physicochemical properties and many potential uses in areas including energy production, agriculture, ecological restoration, and climate change mitigation measures, biochar has attracted significant interest. The production of biochar ranges from small-scale to large-scale, utilizing the pyrolysis of organic biomass. Pyrolysis is a sort of thermochemical reaction that transforms biomass into biochar, bio-oil, and syngas at temperatures ranging from 350 to 700°C in an oxygen-free environment (Verma and Singh, 2019) (Table 8.1). However, both gasification and pyrolysis are thermal reactions that generate biochar in its rigid solid state (Lehmann et al., 2015). Pyrolysis is classified into two distinct categories: slow pyrolysis and fast pyrolysis; consequently, they differ in terms of residence duration and heating rate. Slow pyrolysis yields more syngas, whereas quick pyrolysis releases more oils and solvents. Similarly, slow pyrolysis generates more biochar (36%) than quick pyrolysis (17%) or gasification (12%) (Uchimiya et al., 2011). Many different carbonaceous feedstocks (mainly organic wastes) can be used to create biochar, which implicitly assists with waste disposal. Owing to its inexpensive manufacturing and feasibility, biochar has been used in water and soil remediation as an economical alternative in lieu of activated carbon. Such properties help to get rid of an array of pollutants such as organic volatile compounds, metallic ions, chemical pesticides,

Occurrence and Behavior of Emerging Contaminants in Organic Wastes and Their Control Strategies
https://doi.org/10.1016/B978-0-443-13585-9.00009-4

Table 8.1 Elemental composition of biochar under different pyrolysis temperature.

Biomass	Pyrolysis temperature (°C)	Carbon (%)	Hydrogen (%)	Oxygen (%)	References
Rice straw	300	64	4.8	30	Zhang et al. (2019)
	600	85	2.57	12	Zhang et al. (2019)
Cotton stem	550	72	2.8	13	Shaheen et al. (2019)
Mulberry	350	68	4.6	25	Zama et al. (2017)
	650	80	1.63	16	Zama et al. (2017)
Cassava waste	350	59	3.6	23	Li et al. (2016)
	550	69	2.46	10	Li et al. (2016)
Pine wood	700	79	3	15	Jiang et al. (2016)
Wood chip	620	82	1.5	6	Cely et al. (2014)
Peanut	300	68	3.9	26	Ahmad et al. (2012)
	400	60	35	11	Liang et al. (2019)
	700	84	1.8	13	Ahmad et al. (2012)

Biomass represent the feed stock used in production of biochar.

pharmaceutical products, dyes, and polycyclic aromatic hydrocarbons (PAH) (El-Naggar et al., 2021; Zhao et al., 2021).

An increasing number of soil ecological degradation events such as soil nutrient loss, soil deterioration, and contamination of the soil have been observed over time (Ganguly and Chakraborty, 2020; Ganguly et al., 2021). Several additional environmental problems are caused by these processes, which include fluctuations in temperature, shortages of water, declining productivity of land, and food availability (Yuan et al., 2019; Zou et al., 2022). The problem of soil pollution is widespread throughout the world, as organic and inorganic contaminants have profound adverse impacts on the natural world. Consequently, to solve these environmental challenges, sustainable methods for eco-restoration are required. According to Cheng et al. (2020), biochar exhibits potential as an appropriate amendment because it has an array of advantages, like restoration of soil pH and organic matter (OM) content, increasing the soil's ability to retain water, reducing the concentration of pollutants, increasing crop yields, and preventing the accumulation of pollutants (Cheng et al., 2020). The interactions involving metal ions and biochar principally entail redox reactions, noncovalent interactions, complexation, precipitation, and cation exchange in the presence of appropriate pH, dissolved organic carbon content, ash content, etc. Earlier investigations have demonstrated that hydrophobic, pi-pi bond, and hydrogen bond interactions play significant roles in the interactions of biochar with organic contaminants.

The present research review has investigated all the potential applications of biochar in mitigating environmental pollutants alongside putting an insight toward the framework of the circular bio-economy.

2. Adsorption capacity of biochar

It is widely accepted that organic contaminants and heavy metal ions interact with terrestrial and aquatic ecosystems. The radioactive substances and ions of heavy metals that are emitted by mining, farming, and other activities are found to contaminate the environment ubiquitously. However, unlike organic contaminants, the ions of heavy metals are incapable of being transformed into other forms, whereas there is a high risk of toxicity for radionuclides even at low concentrations. Thus, in order to lessen the extent of their environmental impact and risk to human health, it is essential to remove such ions from the environment effectively. Biochar and its modifications play an important role in eliminating such inorganic and organic contaminants from the environment (Fig. 8.1).

Magnetic biochar (MBC) and nonmagnetic biochar (NBC) were manufactured by wet-fast pyrolysis and implemented for the elimination of Pb(II) and Cd(II) ions from sewage water. The Langmuir model of sorption capacities at pH 5.0 and 45°C has been found to be ~40 mg/g for Pb(II) and ~16 mg/g for Cd(II) on NBC, which were more substantial than those of MBC [~27 mg/g for Pb(II) and ~11 mg/g for Cd(II)] (Hu et al.,

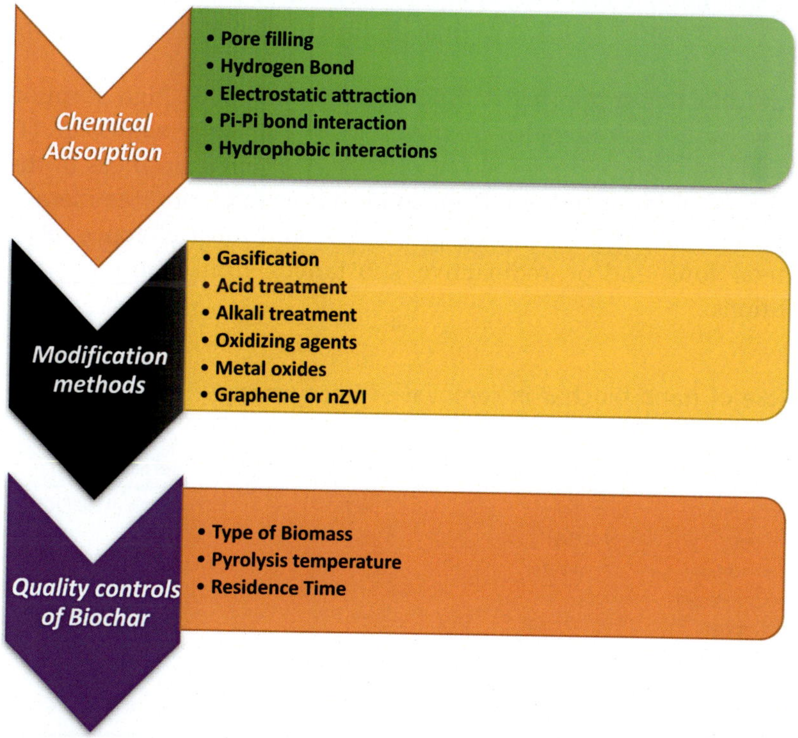

FIGURE 8.1 Different aspects of biochar in removal of environmental contaminants.

2020). The adsorption sequence of Pb(II), Cu(II), and Cd(II) on MgFe2O4/biochar magnetic composites was Pb(II) > Cu(II) > Cd(II), which corresponded to the order of occurrence of their covalent indices (Jung et al., 2018). Liu et al. (2020a) utilized silicon on cornstalk biochar (SiBC) to remove Cu(II) ions from liquids. Chemical adsorption, interface complexation, and precipitation all played a role in the exceptionally high and rapid sorption efficacy of Cu (II) on SiBC.

Li et al. (2015) have demonstrated an increased adsorption efficiency of Cu(II) in the presence of tylosin and/or sulfamethoxazole (Li et al., 2015). The competitive adhesion of tylosin, sulfamethoxazole, and Cu (II) on nano-hydroxyapatite-modified biochar recommended that the electrostatic interaction, π-π interaction, and H-bond are the primary interactions for the adsorption of organic and inorganic contaminants.

The biochar-supported nano iron sulfide (nFeS) aggregates included carboxymethyl cellulose (CMC) with a ratio of FeS:CMC:Biochar as 1:1:1 has demonstrated a Cr(VI) sorption capacity of 130.5 mg/g at pH 5 in comparison to FeS (38.6 mg/g) and biochar (25.4 mg/g) (Lyu et al., 2017). The high adsorption was attributed to surface complexation as well as chemical adsorption, with approximately half of the adsorbent Cr(VI) being converted to Cr(III) through the formation of surface precipitation. The addition of nano zero-valent iron (nZVI) or the incorporation of nitrogen- or sulfur-based functional groups together could preferentially adsorb U(VI) and decrease high-valent U(VI) to low-valent U(IV) thereby immobilizing U(VI) in the surrounding environment.

Biochar and biochar-supported composites demonstrated high sorption capacity for metal ions and radionuclides removal from aqueous solutions (Table 8.2). However, metal ion adsorption should be addressed if circumstances in the environment alter and metal ions are freed from solid matter to create free metal ions afterward. Adsorption is well known as one of the best and most effective techniques for removing metal ions and/or radioactive substances from enormous amounts of aqueous solutions.

Table 8.2 Role of nano biochar in removal of heavy metal.

Biomass	Heavy metals	Environment	Adsorption capacity (mg/g)	References
Corn straw	Cadmium	Soil	18.7	Zhang et al. (2022)
	Lead	Soil	126	
Dendro	Cadmium	Water	922	Ramanayaka et al. (2020)
	Chromium		7.4	
Wood	Copper	Water	22	Safari et al. (2019)
Bark	Copper	Soil	121.5	Arabyarmohammadi et al. (2018)
	Lead		336	
	Zinc		135	

Biomass represent the feed stock used in production of biochar.

3. Potential role of microbial augmentation in biochar: A synergism

Biochar supports microorganisms with complementary interactions by functioning as a source of vital nutrients, boosting microbial colonization, supplying microbial habitat, and diminishing the harmful effects of pollutants from the surroundings (Table 8.3). As the application of biochar involves the immobilization of essential nutrients, the potential of biochar to boost the microbial degradation of organic pollutants would thus be determined by the overall impact of the foregoing synergistic and antagonistic effects on the growth of microbes like *Sphingomonas, Enterobacter, Bacillus*, etc. (Dai et al., 2021; Yaashikaa et al., 2021). Biochar's permeable composition may aid microorganisms in fostering the development of microbial communities and safeguarding advantageous microbes from predators and severe conditions such as exposure to chemicals and poor growing circumstances.

The immobilization of microbial cells (1–10 μm) within the crevices and on the outermost layer of biochar is contingent upon changes in pH, electrochemical potential, structural arrangement, as well as the habitat. Earlier investigations have demonstrated improvements in soil health upon biochar amendment in various attributes like extracellular soil enzyme activity (Cui et al., 2013), microbial communities, microbial respiration (Wei et al., 2021), etc. Biochar is recognized for serving as a favorable microbiome due to its robust porous nature, i.e., nano (<10 Å), micro (<20 Å), meso (20–500 Å), and macro (>500 Å) pores accomplished in varied pyrolysis environments (Zhao et al., 2021).

Table 8.3 Role of microbes-biochar synergism in removal of environmental contaminants.

Biomass	Microbes	Contaminants	Removal (%)	Mechanism	References
Rice straw	*Bacillus* sp.	Cadmium (II)	95%	Electrostatic attraction	
Cow manure	*Acidothermus*	Tetracycline	30%	pH mediated removal	Yue et al. (2021)
Pig manure	*Geobacter*	Chromium (VI)	46%–74%	Fe-dissimilarity reduction	Xu et al. (2019)
Casuarina seed	*Alcaligenes faecalis*	Methylene blue	89%	Electrostatic attraction	Bharti et al. (2019)
Bamboo	*Pseudomonas*	Toluene	99%	Adsorption	Lin et al. (2019)
Peanut	*Pseudomonas*	Phenol	45%	Pi-pi, london dispersion force	Zhang et al. (2018)
Rice straw	*Mycobacterium*	PAH	63%	Hydrophobic interaction	Xiong et al. (2017)
Bamboo	*Streptomyces*	Quinolone	100%	Fe2O3-biochar protected bacteria	Zhuang et al. (2015)

PAH represents poly aromatic hydrocarbons. II and VI represent the valency state of metal ions. Biomass represent the feed stock used in production of biochar.

The pores in biochar produced under low pyrolysis temperatures (about 350°C) offered an optimal environment for microbial colonization.

Biochar has been found to contribute accessible forms of humic acid-like compounds as well as to promote denitrifying microbial development in wetland environments (Zhou et al., 2019). Microscopic pores of biochar may create an ideal anoxic condition for the denitrification process, which was prevalent for biochar developed at elevated temperatures (550–800°C) (Zhang et al., 2021). OM is one of the most critical features of biochar that can impact the growth of microbial cells. According to the pH of the substrate, the OM might remain adhered on the exterior surface of biochar and regulate the population of microbes (Bai et al., 2022). Biochar is an environment-friendly substance that boosts the microbial population and activity. From the standpoint of engineering practice and the concept of microbial inoculation, it can be assumed that biochar carrying the microbial agent is advantageous for preservation, transport, and utilization, which proves to be advantageous for its substantial value in compost manufacturing. In an effort to enhance the composting of pig manure, Tu et al. (2019) employed commercial bacterium-loaded bamboo biochar (Tu et al., 2019). The application of biochar as an exogenous feedstock source for microbial-driven bioremediation has demonstrated significant improvements in microbial proliferation, stability, biodegradation, release of humic acid-like substances, and elimination of pollutants (Zhao et al., 2020). Earlier research investigations have revealed that the application of biochar could assist in eliminating contaminants more synergistically by accelerating microbial degradation and sorption of pollutants (Lou et al., 2019). QY1, a consortium of microbes consisting of *Methylobacterium* sp., *Burkholderia* sp., and *Stenotrophomonas* sp., has been employed in conjunction with biochar manufactured from water hyacinth to remove PAHs from polluted environments containing heavy metals such as Cu(II), Cd (II), and Cr(VI). Such an experiment has demonstrated biodegradation of phenanthrene (94.5%) and 17.8% of 10 mg/L pyrene after 7 days in a single system (Li et al., 2021). The use of biochar additionally boosted the degradation efficacy of sulfamethoxazole by *Pseudomonas stutzeri* and *Shewanella putrefaciens* to 62 (61.8%) and 68.7%, respectively, and for chloramphenicol, it was 85.8% and 85.7%, respectively (Yang et al., 2021). Although prior research investigations have indicated that biochar amendment reduces protist and bacterial prey-predator interactions, which can confer an adverse effect on trophic levels (Liu et al., 2020; Asiloglu et al., 2021b). It has been widely hypothesized that biochar shields bacteria from predation by letting them wander in the micropore habitat, which provides inadequate space for predatory organisms. This theory has been reinforced by an overall reduction in the abundance of bacterivorous nematodes, amoeba, and comparatively big predators in soils with biochar supplements (Kamau et al., 2019). Microbiological investigations have revealed that the charosphere, the immediate vicinity of soil encompassing biochar particles, provides an appropriate habitat for microorganisms, which boosts the growth of microbial communities (Palansooriya et al., 2019). The enhanced activities of bacterial and fungal populations, along with nutritional enrichment, attract the predatory protists in the charosphere zone of biochar.

In summary, biochar has been credited with at least three aspects: (i) Improvements to physicochemical attributes like nutrient immobilization (Na+, K+), enhanced soil pH, electrical conductivity, good C:N ratio, (ii) accessibility to donors of electrons and energy resources, and (iii) promoting and safeguarding the microbiome.

4. Application of biochar: Environmental perspectives

Biochar is a budget-friendly carbon-based substance. It has a substantial surface area, microporosity, and ion exchange capability, implying that biochar has a wide range of environmental applications (Fig. 8.2).

4.1 Soil remediation

Biochar has been utilized for clearing up organic contaminants and metals of terrestrial environment. Adsorption is the primary mechanism through which biochar improves soil health. Surface complexation coupled with hydrogen binding, electrostatic attractions, acid-base interactions, and pi-pi interactions are all part of the biochar adsorption action. High-molecular-weight PAHs (poly aromatic hydrocarbons) were more resistant

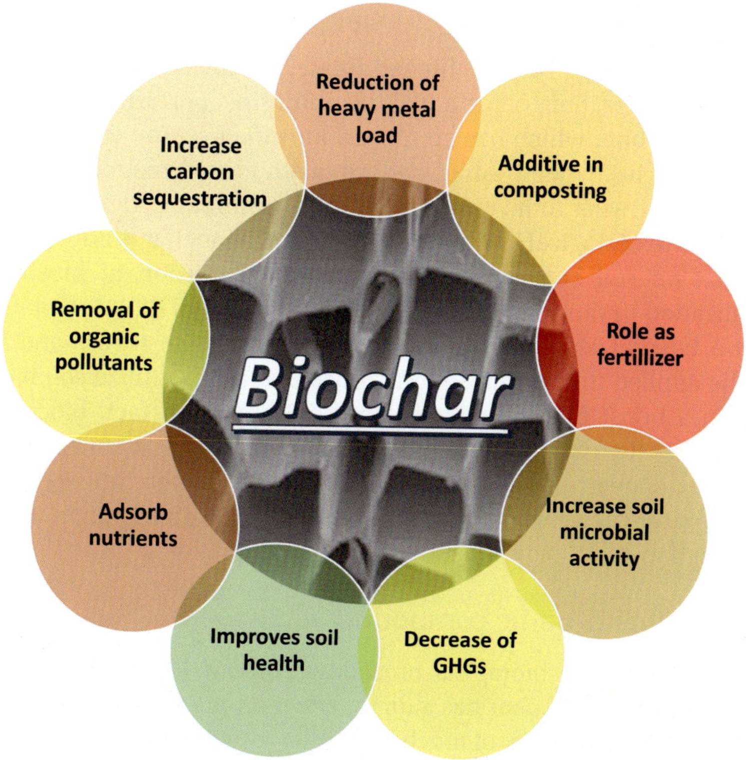

FIGURE 8.2 Different possible applications of biochar in environment.

to decomposition than lower-molecular-weight PAHs, like three- to four-ring PAHs. The inclusion of biochar in soil can boost the activity of microbes. The adsorption capability of the biochar to the metal gets varied. Among the metals, biochar had the most Pb absorption. Furthermore, when the metals existed alongside one another, Cd adsorption on the charcoal was rapidly replaced by other metal ions. Freshwater hyacinth-derived biochar can successfully remove around 90% of As(V) (Zhang et al., 2016), although rice straw-derived biochar exhibited the greatest degree of Zn^2 adsorption (Lu et al., 2017).

The sulfur-altered rice husk charcoal increased Hg absorption by creating insoluble HgS (O'Connor et al., 2018a). Biochar blended with nZVI improved Cu and As adsorption, resulting in the maximum environmental recuperation of the soil microbial population.

In a nutshell, the incorporation of biochar into soil boosts its fertility for many different reasons, such as (1) greater water holding capacity; (2) increased soil assembly stability; (3) diminished soil compaction; and (4) lowered bulk density of the soil and boosted porosity Nelissen et al. (2015); Peake et al. (2014). Furthermore, aged biochar had more distinctive characteristics than freshly prepared biochar. The aged biochar mineralized at a faster pace than the new biochar.

4.2 Sequestration of carbon

The threat of climate change has heightened interest in lowering carbon dioxide emissions into the atmosphere. Soil, being a major carbon sink, performs an essential part in the global cycle of carbon, which has an immediate effect on the environment. Carbon capture and storage have been advocated as methods of reducing carbon dioxide emissions in the soil. Owing to its extremely compact aromatic framework, biochar has excellent resistance to biological degradation. As a result, biochar is anticipated to have a beneficial impact on soil carbon sequestration. The inclusion of biochar enhanced carbon mineralization in the soil. According to Cross and Sohi (2011), carbon in biochar could be of two types: Liable carbon and recalcitrant carbon (Cross and Sohi, 2011). When biochar is applied to the soil, soil microbes may easily utilize soluble, liable carbon, resulting in a boost in the mineralization of carbon. In fact, biochar comprises significantly more recalcitrant carbon than labile carbon. Recalcitrant carbon can persist in the soil for an extended period of time. As a result, the carbon inflow triggered by the introduction of biochar is greater than the emissions of carbon generated by relevant carbon mineralization.

4.3 Composting of organic compound

Composting can involve the biological breakdown of OM (Ganguly and Chakraborty, 2018; Ganguly et al., 2022). Biochar has a direct effect on microbes, which in turn affects composting. The following effects of biochar on microorganisms throughout the organic solid waste composting process are: (1) To provide a habitat for microorganisms; (2) to provide favorable population growth for microorganisms; and (3) to enrich microbial diversity. It has been found that the inclusion of biochar expedites the breakdown of organic solid waste. Several investigations have demonstrated that biochar may eliminate contaminants, such as inorganic and organic contaminants, from water and sewage water through sorption. Antibiotics are increasingly common organic contaminants in the natural environment (Wang et al., 2019a,b). It was shown that sludge-derived biochar is an affordable and recyclable adsorbent for antimicrobial drug recovery.

The process of adsorption of contaminants by biochar in water is determined by the physical and chemical characteristics of the relevant contaminants as well as the nature of biochar. For example, sawdust-derived biochar can totally remove 20.3 mg/L of sulfamethoxazole, but wood-derived biochar has a substantially lower effectiveness for the removal of sulfamethoxazole (Reguyal et al., 2017). The elimination effectiveness of sulfamethoxazole was lowest for biochar obtained from organic farms (6%) (Lin et al., 2017a). The difference in pyrolysis temperature has a profound role in the elimination of tetracycline by biochar generated from rice husk. As a result, it is inferred that the temperature of pyrolysis had a significant impact on the potential for adsorption of biochar. In conjunction with pyrolysis temperature, other parameters like pyrolysis time may modify the physicochemical features of biochar, regulating its adsorption ability. The contamination of heavy metals is an important challenge in the restoration of the environment. Adsorption is an efficient method for removing heavy metals from aquatic environments. It should be highlighted that the presence of functional groups undoubtedly affects the potential for adsorption. For example, amino-modified biochar clearly improved Cu(II) adsorption via robust surface complexation (Yang and Jiang, 2014). According to Zhang et al. (2019), biochar produced at a high temperature performed well in removing Cr(VI). Besides adsorption, biochar can harbor microorganisms, which improve the reduction of organic load (Zhang et al., 2019). Luo et al. (2015) observed that the number of colonies of Archaea was significantly greater due to the inclusion of fruitwood-derived biochar, which reduced the adverse effects of ammonia and acids on the microbes and augmented the degree of microbial activity (Luo et al., 2015).

4.4 Organic catalyst and activator

Biochar can also act as a sort of catalyst. It was employed as a catalyst in the manufacturing of biodiesel. As an example, after sulfonation and smudging, hardwood-derived biochar has shown significant activity for the esterification process of free fatty acids. According to Zhong et al. (2019), the hydrophobicity and strong acidity of sulfonic acid-functionalized mesoporous biochar increased the catalytic efficiency and durability in the alkylation reaction involving 2-methylfuran with cyclopentanone (Zhong et al., 2019). Biochar, as one of the carbon compounds, includes a lot of chemical groups that can activate persulfate. However, a few investigations have been undertaken to explore persulfate activation by biochar for organic contamination remediation. Wang et al.

(2017) discovered that sludge-derived biochar may successfully stimulate persulfate for 4-chlorophenol breakdown (Wang et al., 2017). Within 100 min, the elimination efficiency of 4-chlorophenol was 92.3%. Biochar is capable of activating persulfate; however, the generation of reactive compounds can be accomplished in varying amounts.

Furthermore, metal-ion-loaded biochar may improve the decomposition of polychlorinated biphenyl (Dong et al., 2017). However, the durability of the metal-biochar composites must be considered, and the impact of biochar on persulfate activation ought to be addressed. Metal ions, for example, are abundant in sewage waste. As a result, biochar generated from sewage waste has a higher degree of persulfate activation. Furthermore, biochar can be used to rid of sewage waste. However, several detailed research investigations are required to understand the biochar-persulfate activation.

5. Role of nanobiochar in removal of environmental contaminants

Nanobiochar's structure and chemical makeup have an immense effect on its capacity to absorb contaminants. For instance, nanobiochar with a greater cation content could boost its ion exchange with heavy metals, securing higher concentrations of metal ions. An enhanced aromatic nanobiochar framework may result in more π-π interactions among nanobiochar and organic contaminants. Consequentially, when using nanobiochar as a supplement for ecological remediation, its structure and chemical makeup must also be considered. The agglomeration ability, colloidal strength, and electrokinetic characteristics of nanobiochar can influence its ability to adsorb contaminants, which could be measured using the zeta potential (Filipinas et al., 2021). The nature of biomass, pyrolysis temperature, and surface charges play an important role in influencing the zeta potential of nanobiochar. Owing to the substantial variances in the characteristics of the contaminants, the adsorption activity of nanobiochar for various contaminants may vary considerably. In the case of contaminants such as heavy metals, features like valence electrons, electronegativity, hydration radius, and hydrolysis constant constitute the primary parameters that regulate the nanobiochar restoration approach.

In general, nanobiochar can be manufactured using two methods. The first is the synthetic generation of biochar, and the second is the physical degradation of biochar in the natural environment caused by aging (Huang et al., 2021). Nanobiochar has a substantially more effective surface area than bulk biochar, with a lesser hydrodynamic radius, a higher negative zeta potential, and additional oxygen-containing functional moieties.

It has been found that feedstock with a substantial amount of hemicellulose typically yields nanobiochar with a low percentage of carbon along with a significant amount of oxygen (Weber and Quicker 2018), whereas materials with a substantial amount of lignin typically yield nanobiochar with a significant agglomeration potential (Föhr et al., 2017). Some methods for manufacturing nanobiochar are as follows:

- Ball milling is a prominent method for manufacturing nanobiochar because it leads to the effective breakdown of the biochar into particles of nanoscale size. The ball milling technique is being shown to be a relatively inexpensive and reproducible green strategy.
- The double disc milling technique was employed to create polygonal platelet-shaped aggregates with dimensions ranging from 50 to 150 nm.
- In the sonication procedure, the probe needle is immobilized underneath the suspension's surface, wherein biochar molecules are fragmented and the graphite-like framework of the biomass is stripped away by ripples of waves. The primary method comprises the utilization of ultrasound-induced microjets and shocks, which generate additional pores, open blocked pores, and alter the biochar framework, resulting in increased microporosity.

Several other technologies, like acid digestion, centrifugation technology, and microwave pyrolysis, along with several modification techniques, are employed to increase the potential of nanobiochar.

- Instances for heavy metal removal and organic pollutants

Nanobiochar has been shown to be a useful tool for facilitating heavy metal immobilization in the natural environment. Previous investigation has demonstrated that nanobiochar can effectively absorb Cd, Pb, or Cu in water-based environments, and the heavy metal removal rate steadily rises with the rise in concentration of nanobiochar. The surge in the elimination rate of heavy metals could be attributed to adsorbent exchangeable sites and the higher specific surface area of nanobiochar. According to Lyu et al. the degree of nickel (Ni) sorption improved with the supplementation of low concentrations of nanobiochar but diminished with higher concentations. Such facts demonstrated a heterogeneous agglomeration of nanobiochar with minerals from the soil, which can be attributed to electrostatic interactions among positively charged soil minerals and oppositely charged nanobiochar substrates (Zhang et al., 2022). Lyu et al. used ball milling to generate thiol-coated nanobiochar, which imparts the adsorption potential of positively charged mercury (Hg) to 320 mg/g.

Ma et al. (2019) have demonstrated the maximal adsorption capacity of diethyl phthalate using maize straw nanobiochar (34 mg/g) and rice husk straw nanobiochar (28 mg/g) (Ma et al., 2019). For tetracycline, the adsorption capacity of nanobiochar pyrolyzed at 700°C was 268.3 mg/g (Li et al., 2020). The administration of modified nanobiochar for the elimination of metformin hydrochloride (antidiabetic medication) (10 mg L 1) from drinking water, sewage water, and seawater has been permitted by 87.0%, 97.0%, and 92.0%, respectively.

Nanobiochar has the ability to greatly stabilize or eliminate hazardous chemicals from soil, potentially providing an optimal environment for the development of plants.

Fan et al. (2020) hypothesized that the interactions among negatively charged layers bearing oxygen-containing functional molecules (serving as donors of electrons) coupled with positively charged minerals are the main driver for heavy metal sorption by nanobiochar. Wang et al. (2020) demonstrated that in aqueous settings, the formation of complexes between positively charged −OH groups of nanobiochar with the oppositely charged Cr(VI) was the most significant component that governs the adsorption mechanism. When nanobiochar is introduced into the natural environment, it has the capability to stabilize a significant number of contaminants. This will significantly restrict the mobility of pollutants and their bioavailability, minimizing the hazard to ecosystems and human well-being.

6. Bio-economy with biochar

According to Bugge et al. (2016), the term "bio-economy" refers to the research and utilization of bio-resources, which includes the implementation of biotechnology to produce novel bio-products with commercial value (Bugge et al., 2016). The feedstock (organic biomass) and the biochar that is produced from it serve as essential elements of the bioeconomy. In simple terms, the long-term viability of the bio-economy is contingent upon manufacturing, advertising, public awareness campaigns, and commercialization. All of these bio-economic activities provide chances for both immediate and long-term employment. The bio-economy is also significantly impacted by the bio-product's reliability, safety, and availability (Fuentes-Saguar et al., 2017). With the rise of the global population, there is an increase in demand for energy in different sectors. Fossil fuels continue to be the main energy source. It can be turned into energy by mechanical, biological, physical, and thermochemical processes, making biomass an appealing source of sustainable energy. Due to its advantages in terms of sustainability, economics, and rising demand in the sectors of energy and the environment, biomass has recently been transformed into biochar (Fig. 8.3). Biochar is a commercially viable and lucrative bio-product as it can be used in different industrial sectors. Biochar generated from biomass and its applications as organic soil improvements have the potential to affect the global economy, particularly in the farming and forest domains (Fig. 8.4). It has an impact on allied industries, such as those involved in the supply of biochar machinery. According to some experts, carbon dioxide retention costs could be critical to biochar viability. Biochar can be encouraged by enacting a CO_2 sequester payment scheme. Meanwhile, cultivators may find it profitable to implement biochar for the growth of high-yielding cash crops. Harsono et al. (2013) estimated the financial assumptions for biochar manufacturing in Selangor to be 533 US dollars per year, with an overall profit from biochar sales of 8012 US dollars per year (Harsano et al., 2013). Thus, the net future value for biochar manufacturing, which was estimated by the total amount of money invested and the revenue from net sales, pointed out that biochar was economically feasible. According

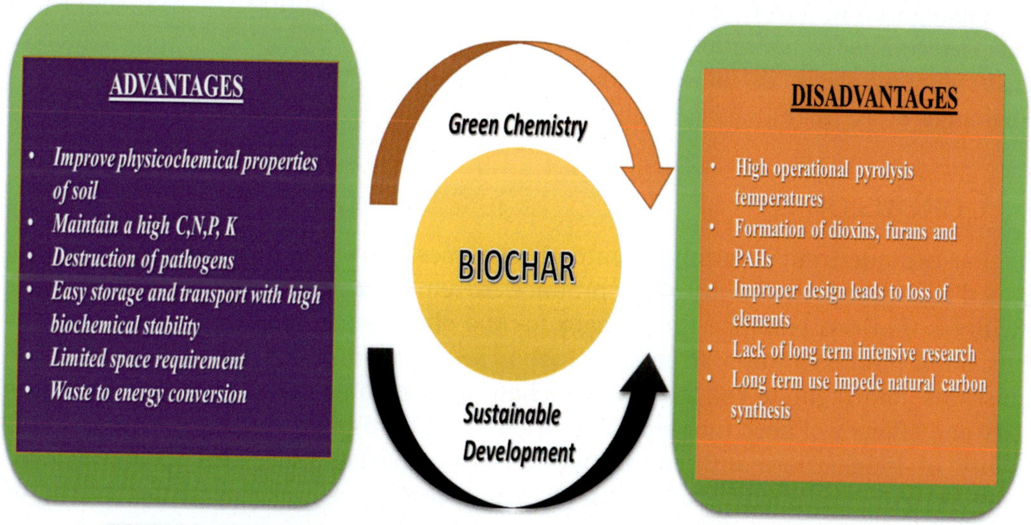

FIGURE 8.3 Advantages and disadvantages of using biochar for sustainable development.

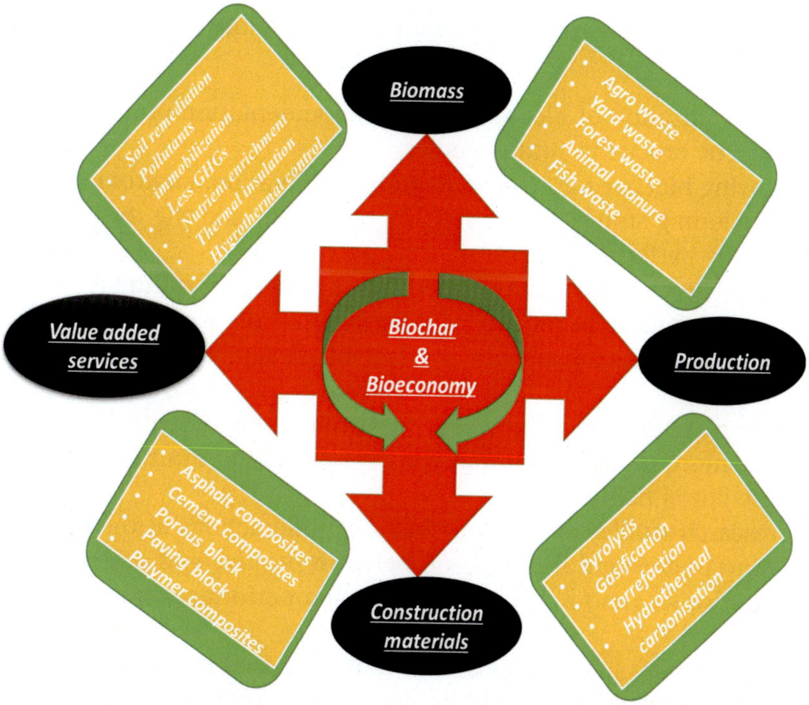

FIGURE 8.4 Role of biochar in supporting circular bioeconomy.

to Shabangu et al. (2014), the economic viability for the manufacture of biochar has been determined by its retail value, resulting in a break-even point of approximately \$220/t for pyrolysis at 300°C and approximately \$280/t for pyrolysis at 450°C (Shabangu et al., 2014).

7. Conclusion

Biochar is made from a variety of natural resources known as biomass that are combusted under various conditions to produce extremely diverse chemical and physical attributes that can enhance the efficacy for the elimination of toxic substances from soils, boost photosynthesis, strengthen carbon sequestration, minimize greenhouse gas emissions, and regulate soil erosion. Biochar has the potential to reduce the accessibility and efficacy of both inorganic and organic contaminants in soil. The amount and quality of biochar are influenced by the raw material and pyrolysis conditions. However, prior to its applications to soil, it is crucial to understand the degree of biochar's potential to retain pollutants, as it might rise, fall, or change over time as its adsorption pockets become saturated with OM and contaminants. Biochar plays an important role in the removal of toxic components, the mitigation of several environmental challenges, and supporting the bio-economy. However, more intensive research works on the following points is required for its long-term implementation in the environment:

- The industrial utilization of biochar and the implementation of biochar quality guidelines will be crucial in future studies.
- When employing biochar for environmental rehabilitation, environmental risk assessments, including studies of the viability of biochar and the toxicity of contaminants, should be performed.
- The deliberate use of biochar may impede the synthesis of naturally occurring organic carbon, resulting in reduced soil fertility and the worsening of climate change. Hence, both the advantages and lacunas should be considered prior to its use.
- Since biochar plays an important role in controlling the dynamics of soil microbial communities, more research investigations are needed to examine its long-term effects on agricultural fields.
- Extensive research studies on biochar-microbial interactions at the molecular level are required to understand the trade-off between positive and detrimental effects on microorganisms so as to optimize remediation performance at the real-time scale.
- Research studies over the application of biochar based on techno-economic-environmental perspectives should be done for its long-term use in the environment.

References

Ahmad, M., Lee, S.S., Dou, X., Mohan, D., Sung, J.K., Yang, J.E., Ok, Y.S., 2012. Effects of pyrolysis temperature on soybean stover-and peanut shell-derived biochar properties and TCE adsorption in water. Bioresource Technology 118, 536–544.

Arabyarmohammadi, H., Darban, A.K., Abdollahy, M., Yong, R., Ayati, B., Zirakjou, A., van der Zee, S.E., 2018. Utilization of a novel chitosan/clay/biochar nanobiocomposite for immobilization of heavy metals in acid soil environment. Journal of Polymers and the Environment 26, 2107–2119.

Asiloglu, R., Sevilir, B., Samuel, S.O., Aycan, M., Akca, M.O., Suzuki, K., Harada, N., 2021. Effect of protists on rhizobacterial community composition and rice plant growth in a biochar amended soil. Biology and Fertility of Soils 57, 293–304.

Bai, S.H., Omidvar, N., Gallart, M., Kämper, W., Tahmasbian, I., Farrar, M.B., van Zwieten, L., 2022. Combined effects of biochar and fertilizer applications on yield: a review and meta-analysis. Science of the Total Environment 808, 152073.

Bharti, V., Vikrant, K., Goswami, M., Tiwari, H., Sonwani, R.K., Lee, J., Singh, R.S., 2019. Biodegradation of methylene blue dye in a batch and continuous mode using biochar as packing media. Environmental Research 171, 356–364.

Bugge, M.M., Hansen, T., Klitkou, A., 2016. What is the bioeconomy? A review of the literature. Sustainability 8 (7), 691.

Cely, P., Tarquis, A.M., Paz-Ferreiro, J., Mendez, A., Gasco, G., 2014. Factors driving the carbon mineralization priming effect in a sandy loam soil amended with different types of biochar. Solid Earth 5 (1), 585–594.

Cheng, D., Ngo, H.H., Guo, W., Chang, S.W., Nguyen, D.D., Zhang, X., Liu, Y., 2020. Feasibility study on a new pomelo peel derived biochar for tetracycline antibiotics removal in swine wastewater. Science of the Total Environment 720, 137662.

Cross, A., Sohi, S.P., 2011. The priming potential of biochar products in relation to labile carbon contents and soil organic matter status. Soil Biology and Biochemistry 43 (10), 2127–2134.

Cui, L., Yan, J., Yang, Y., Li, L., Quan, G., Ding, C., Chang, A., 2013. Influence of biochar on microbial activities of heavy metals contaminated paddy fields. Bioresources 8 (4).

Dai, Z., Xiong, X., Zhu, H., Xu, H., Leng, P., Li, J., Xu, J., 2021. Association of biochar properties with changes in soil bacterial, fungal and fauna communities and nutrient cycling processes. Biochar 3, 239–254.

Dong, C.D., Chen, C.W., Hung, C.M., 2017. Synthesis of magnetic biochar from bamboo biomass to activate persulfate for the removal of polycyclic aromatic hydrocarbons in marine sediments. Bioresource Technology 245, 188–195.

El-Naggar, A., Chang, S.X., Cai, Y., Lee, Y.H., Wang, J., Wang, S.L., Ok, Y.S., 2021. Mechanistic insights into the (im) mobilization of arsenic, cadmium, lead, and zinc in a multi-contaminated soil treated with different biochars. Environment International 156, 106638.

Fan, Z., Zhang, Q., Li, M., Sang, W., Qiu, Y., Wei, X., Hao, H., 2020. Removal behavior and mechanisms of Cd (II) by a novel MnS loaded functional biochar: influence of oxygenation. Journal of Cleaner Production 256, 120672.

Filipinas, J.Q., Rivera, K.K.P., Ong, D.C., Pingul-Ong, S.M.B., Abarca, R.R.M., de Luna, M.D.G., 2021. Removal of sodium diclofenac from aqueous solutions by rice hull biochar. Biochar 3, 189–200.

Föhr, J., Ranta, T., Suikki, J., Soininen, H., 2017. Manufacturing of torrefied pellets without a binder from different raw wood materials in the pilot plant. Wood Research 62 (3), 481–494.

Fuentes-Saguar, P.D., Mainar-Causapé, A.J., Ferrari, E., 2017. The role of bioeconomy sectors and natural resources in EU economies: a social accounting matrix-based analysis approach. Sustainability 9 (12), 2383. https://doi.org/10.3390/su9122383.

Ganguly, R.K., Chakraborty, S.K., 2018. Assessment of microbial roles in the bioconversion of paper mill sludge through vermicomposting. Journal of Environmental Health Science and Engineering 16, 205–212.

Ganguly, R.K., Chakraborty, S.K., 2020. Eco-management of industrial organic wastes through the modified innovative vermicomposting process: a sustainable approach in tropical countries. Earthworm Assisted Remediation of Effluents and Wastes 161–177.

Ganguly, R.K., Al-Helal, M.A., Chakraborty, S.K., 2022. Management of invasive weed Chromolaena odorata (Siam weed) through vermicomposting: an eco-approach utilizing organic biomass valorization. Environmental Technology & Innovation 28, 102952.

Ganguly, R.K., Mukherjee, A., Chakraborty, S.K., Verma, J.P., 2021. Impact of agrochemical application in sustainable agriculture. In: New and Future Developments in Microbial Biotechnology and Bioengineering. Elsevier, pp. 15–24.

Harsono, S.S., Grundman, P., Lau, L.H., Hansen, A., Salleh, M.A.M., Meyer-Aurich, A., Ghazi, T.I.M., 2013. Energy balances, greenhouse gas emissions and economics of biochar production from palm oil empty fruit bunches. Resources, Conservation and Recycling 77, 108–115.

Hu, B., Ai, Y., Jin, J., Hayat, T., Alsaedi, A., Zhuang, L., Wang, X., 2020. Efficient elimination of organic and inorganic pollutants by biochar and biochar-based materials. Biochar 2, 47–64.

Huang, X., Zhu, S., Zhang, H., Huang, Y., Wang, X., Wang, Y., Chen, D., 2021. Biochar nanoparticles induced distinct biological effects on freshwater algae via oxidative stress, membrane damage, and nutrient depletion. ACS Sustainable Chemistry & Engineering 9 (32), 10761–10770.

Jiang, S., Huang, L., Nguyen, T.A., Ok, Y.S., Rudolph, V., Yang, H., Zhang, D., 2016. Copper and zinc adsorption by softwood and hardwood biochars under elevated sulphate-induced salinity and acidic pH conditions. Chemosphere 142, 64–71.

Jung, K.W., Lee, S.Y., Lee, Y.J., 2018. Hydrothermal synthesis of hierarchically structured birnessite-type MnO2/biochar composites for the adsorptive removal of Cu (II) from aqueous media. Bioresource Technology 260, 204–212.

Kamau, S., Karanja, N.K., Ayuke, F.O., Lehmann, J., 2019. Short-term influence of biochar and fertilizer-biochar blends on soil nutrients, fauna and maize growth. Biology and Fertility of Soils 55, 661–673.

Lehmann, J., Kuzyakov, Y., Pan, G., Ok, Y.S., 2015. Biochars and the plant-soil interface. Plant and Soil 395, 1–5.

Li, T., Han, X., Liang, C., Shohag, M.J.I., Yang, X., 2015. Sorption of sulphamethoxazole by the biochars derived from rice straw and alligator flag. Environmental Technology 36 (2), 245–253.

Li, F., Feng, D., Deng, H., Yu, H., Ge, C., 2016. Effects of biochars prepared from cassava dregs on sorption behavior of ciprofloxacin. Procedia Environmental Sciences 31, 795–803.

Li, R., Zhang, Y., Deng, H., Zhang, Z., Wang, J.J., Shaheen, S.M., Du, J., 2020. Removing tetracycline and Hg (II) with ball-milled magnetic nanobiochar and its potential on polluted irrigation water reclamation. Journal of Hazardous Materials 384, 121095.

Li, M., Yin, H., Zhu, M., Yu, Y., Lu, G., Dang, Z., 2021. Co-metabolic and biochar-promoted biodegradation of mixed PAHs by highly efficient microbial consortium QY1. Journal of Environmental Sciences 107, 65–76.

Liang, J., Fang, Y., Luo, Y., Zeng, G., Deng, J., Tan, X., Ye, S., 2019. Magnetic nanoferromanganese oxides modified biochar derived from pine sawdust for adsorption of tetracycline hydrochloride. Environmental Science and Pollution Research 26, 5892–5903.

Lin, L., Jiang, W., Xu, P., 2017. Comparative study on pharmaceuticals adsorption in reclaimed water desalination concentrate using biochar: impact of salts and organic matter. Science of the Total Environment 601, 857–864.

Lin, C.W., Tsai, S.L., Lai, C.Y., Liu, S.H., Wu, C.H., 2019. Biodegradation kinetics and microbial dynamics of toluene removal in a two-stage cell-biochar-filled biotrickling filter. Journal of Cleaner Production 238, 117940.

Liu, W., Li, Y., Feng, Y., Qiao, J., Zhao, H., Xie, J., Liang, S., 2020. The effectiveness of nanobiochar for reducing phytotoxicity and improving soil remediation in cadmium-contaminated soil. Scientific Reports 10 (1), 858.

Lou, L., Huang, Q., Lou, Y., Lu, J., Hu, B., Lin, Q., 2019. Adsorption and degradation in the removal of nonylphenol from water by cells immobilized on biochar. Chemosphere 228, 676–684.

Luo, C., Lü, F., Shao, L., He, P., 2015. Application of eco-compatible biochar in anaerobic digestion to relieve acid stress and promote the selective colonization of functional microbes. Water Research 68, 710–718.

Lyu, H., Tang, J., Huang, Y., Gai, L., Zeng, E.Y., Liber, K., Gong, Y., 2017. Removal of hexavalent chromium from aqueous solutions by a novel biochar supported nanoscale iron sulfide composite. Chemical Engineering Journal 322, 516–524.

Ma, S., Jing, F., Sohi, S.P., Chen, J., 2019. New insights into contrasting mechanisms for PAE adsorption on millimeter, micron-and nano-scale biochar. Environmental Science and Pollution Research 26, 18636–18650.

Nelissen, V., Ruysschaert, G., Manka'Abusi, D., D'Hose, T., De Beuf, K., Al-Barri, B., Boeckx, P., 2015. Impact of a woody biochar on properties of a sandy loam soil and spring barley during a two-year field experiment. European Journal of Agronomy 62, 65–78.

O'Connor, D., Peng, T., Li, G., Wang, S., Duan, L., Mulder, J., Hou, D., 2018. Sulfur-modified rice husk biochar: a green method for the remediation of mercury contaminated soil. Science of the Total Environment 621, 819–826.

Palansooriya, K.N., Wong, J.T.F., Hashimoto, Y., Huang, L., Rinklebe, J., Chang, S.X., Ok, Y.S., 2019. Response of microbial communities to biochar-amended soils: a critical review. Biochar 1, 3–22.

Peake, L.R., Reid, B.J., Tang, X., 2014. Quantifying the influence of biochar on the physical and hydrological properties of dissimilar soils. Geoderma 235, 182–190.

Ramanayaka, S., Tsang, D.C., Hou, D., Ok, Y.S., Vithanage, M., 2020. Green synthesis of graphitic nanobiochar for the removal of emerging contaminants in aqueous media. Science of the Total Environment 706, 135725.

Reguyal, F., Sarmah, A.K., Gao, W., 2017. Synthesis of magnetic biochar from pine sawdust via oxidative hydrolysis of FeCl2 for the removal sulfamethoxazole from aqueous solution. Journal of Hazardous Materials 321, 868–878.

Safari, S., von Gunten, K., Alam, M.S., Hubmann, M., Blewett, T.A., Chi, Z., Alessi, D.S., 2019. Biochar colloids and their use in contaminants removal. Biochar 1 (2), 151–162.

Shabangu, S., Woolf, D., Fisher, E.M., Angenent, L.T., Lehmann, J., 2014. Techno-economic assessment of biomass slow pyrolysis into different biochar and methanol concepts. Fuel 117, 742–748.

Shaheen, S.M., Niazi, N.K., Hassan, N.E., Bibi, I., Wang, H., Tsang, D.C., Rinklebe, J., 2019. Wood-based biochar for the removal of potentially toxic elements in water and wastewater: a critical review. International Materials Reviews 64 (4), 216–247.

Tu, Z., Ren, X., Zhao, J., Awasthi, S.K., Wang, Q., Awasthi, M.K., Li, R., 2019. Synergistic effects of biochar/microbial inoculation on the enhancement of pig manure composting. Biochar 1, 127–137.

Uchimiya, M., Wartelle, L.H., Klasson, K.T., Fortier, C.A., Lima, I.M., 2011. Influence of pyrolysis temperature on biochar property and function as a heavy metal sorbent in soil. Journal of Agricultural and Food Chemistry 59 (6), 2501–2510.

Verma, L., Singh, J., 2019. Synthesis of novel biochar from waste plant litter biomass for the removal of Arsenic (III and V) from aqueous solution: a mechanism characterization, kinetics and thermodynamics. Journal of Environmental Management 248, 109235.

Wang, J., Liao, Z., Ifthikar, J., Shi, L., Du, Y., Zhu, J., Chen, Z., 2017. Treatment of refractory contaminants by sludge-derived biochar/persulfate system via both adsorption and advanced oxidation process. Chemosphere 185, 754–763.

Wang, C., Gu, L., Ge, S., Liu, X., Zhang, X., Chen, X., 2019a. Remediation potential of immobilized bacterial consortium with biochar as carrier in pyrene-Cr (VI) co-contaminated soil. Environmental Technology 40 (18), 2345–2353.

Wang, J., Zhuan, R., Chu, L., 2019b. The occurrence, distribution and degradation of antibiotics by ionizing radiation: an overview. Science of the Total Environment 646, 1385–1397.

Wang, K., Sun, Y., Tang, J., He, J., Sun, H., 2020. Aqueous Cr (VI) removal by a novel ball milled Fe0-biochar composite: role of biochar electron transfer capacity under high pyrolysis temperature. Chemosphere 241, 125044.

Weber, K., Quicker, P., 2018. Properties of biochar. Fuel 217, 240–261.

Wei, Z., Wang, J.J., Fultz, L.M., White, P., Jeong, C., 2021. Application of biochar in estrogen hormone-contaminated and manure-affected soils: impact on soil respiration, microbial community and enzyme activity. Chemosphere 270, 128625.

Xiong, B., Zhang, Y., Hou, Y., Arp, H.P.H., Reid, B.J., Cai, C., 2017. Enhanced biodegradation of PAHs in historically contaminated soil by M. ágilvum inoculated biochar. Chemosphere 182, 316–324.

Xu, X., Huang, H., Zhang, Y., Xu, Z., Cao, X., 2019. Biochar as both electron donor and electron shuttle for the reduction transformation of Cr (VI) during its sorption. Environmental Pollution 244, 423–430.

Yaashikaa, P.R., Kumar, P.S., Saravanan, A., Vo, D.V.N., 2021. Advances in biosorbents for removal of environmental pollutants: a review on pre-treatment, removal mechanism and future outlook. Journal of Hazardous Materials 420, 126596.

Yang, G.X., Jiang, H., 2014. Amino modification of biochar for enhanced adsorption of copper ions from synthetic wastewater. Water Research 48, 396–405.

Yang, F., Jian, H., Wang, C., Wang, Y., Li, E., Sun, H., 2021. Effects of biochar on biodegradation of sulfamethoxazole and chloramphenicol by Pseudomonas stutzeri and Shewanella putrefaciens: microbial growth, fatty acids, and the expression quantity of genes. Journal of Hazardous Materials 406, 124311.

Yuan, P., Wang, J., Pan, Y., Shen, B., Wu, C., 2019. Review of biochar for the management of contaminated soil: preparation, application and prospect. Science of the Total Environment 659, 473–490.

Yue, Y., Liu, Y.J., Wang, J., Vukanti, R., Ge, Y., 2021. Enrichment of potential degrading bacteria accelerates removal of tetracyclines and their epimers from cow manure biochar amended soil. Chemosphere 278, 130358.

Zama, E.F., Zhu, Y.G., Reid, B.J., Sun, G.X., 2017. The role of biochar properties in influencing the sorption and desorption of Pb (II), Cd (II) and As (III) in aqueous solution. Journal of Cleaner Production 148, 127–136.

Zhao, L., Xiao, D., Liu, Y., Xu, H., Nan, H., Li, D., Cao, X., 2020. Biochar as simultaneous shelter, adsorbent, pH buffer, and substrate of Pseudomonas citronellolis to promote biodegradation of high concentrations of phenol in wastewater. Water Research 172, 115494.

Zhao, D., Yan, B., Liu, C., Yao, B., Luo, L., Yang, Y., Zhou, Y., 2021. Mitigation of acidogenic product inhibition and elevated mass transfer by biochar during anaerobic digestion of food waste. Bioresource Technology 338, 125531.

Zhang, F., Wang, X., Xionghui, J., Ma, L., 2016. Efficient arsenate removal by magnetite-modified water hyacinth biochar. Environmental Pollution 216, 575–583.

Zhang, G., Guo, X., Zhu, Y., Liu, X., Han, Z., Sun, K., Han, L., 2018. The effects of different biochars on microbial quantity, microbial community shift, enzyme activity, and biodegradation of polycyclic aromatic hydrocarbons in soil. Geoderma 328, 100–108.

Zhang, K., Sun, P., Zhang, Y., 2019. Decontamination of Cr (VI) facilitated formation of persistent free radicals on rice husk derived biochar. Frontiers of Environmental Science & Engineering 13, 1–9.

Zhang, X., Wells, M., Niazi, N.K., Bolan, N., Shaheen, S., Hou, D., Wang, Z., 2022. Nanobiochar-rhizosphere interactions: implications for the remediation of heavy-metal contaminated soils. Environmental Pollution 299, 118810.

Zhang, Y., Wang, J., Feng, Y., 2021. The effects of biochar addition on soil physicochemical properties: a review. Catena 202, 105284.

Zhong, Y., Deng, Q., Zhang, P., Wang, J., Wang, R., Zeng, Z., Deng, S., 2019. Sulfonic acid functionalized hydrophobic mesoporous biochar: design, preparation and acid-catalytic properties. Fuel 240, 270–277.

Zhou, X., Wang, R., Liu, H., Wu, S., Wu, H., 2019. Nitrogen removal responses to biochar addition in intermittent-aerated subsurface flow constructed wetland microcosms: enhancing role and mechanism. Ecological Engineering 128, 57–65.

Zhuang, H., Han, H., Xu, P., Hou, B., Jia, S., Wang, D., Li, K., 2015. Biodegradation of quinoline by Streptomyces sp. N01 immobilized on bamboo carbon supported Fe3O4 nanoparticles. Biochemical Engineering Journal 99, 44–47.

Zou, R., Qian, M., Wang, C., Mateo, W., Wang, Y., Dai, L., Lei, H., 2022. Biochar: from by-products of agro-industrial lignocellulosic waste to tailored carbon-based catalysts for biomass thermochemical conversions. Chemical Engineering Journal 135972.

9

Effect of additives on the reduction of antibiotic resistance genes during composting of dewatered sludge

Jiwei Shi[1], Bangchi Wang[1], Jiachen Xie[1] and Kui Huang[1,2]

[1]SCHOOL OF ENVIRONMENTAL AND MUNICIPAL ENGINEERING, LANZHOU JIAOTONG UNIVERSITY, LANZHOU, CHINA; [2]KEY LABORATORY OF YELLOW RIVER WATER ENVIRONMENT IN GANSU PROVINCE, LANZHOU, CHINA

1. Reduction technologies for sludge antibiotic resistance genes

Antibiotics are a specific class of secondary metabolites that are produced by microorganisms or higher plants and animals and can inhibit pathogenic bacteria or other organisms and interfere with the normal growth, metabolism, and development of living cells (Finlay et al., 1950). Antibiotic use levels are high globally, with particular regions such as China exhibiting extremely high levels of use in medicine and agriculture, resulting in an increase in the abundance and diversity of antibiotic resistance genes (ARGs) and their transmission throughout the environment. As a relatively novel type of environmental pollutant, ARGs are characterized as being highly persistent, difficult to degrade, and capable of rapid transmission, potentially having a more detrimental effect on the environment than the antibiotic residues themselves (Kummerer et al., 2004). Therefore, for the protection of human and environmental health, it is essential that methods be developed for the effective removal of antibiotics and ARGs from wastewater and wastewater sludge.

Wastewater treatment plants are known to be important reservoirs of harmful contaminants such as antibiotics and ARGs, as they generally receive wastewater through complex pipe networks from various sources, typically including a combination of domestic, industrial, and medical sewage. The mixing, transfer, and transmission of ARGs in wastewater have resulted in the presence of large quantities of ARGs in wastewater and sludge (Rizzo et al., 2013; Su et al., 2014), which may enter the receiving environment through the discharge of wastewater effluent or the reuse of sludge resources (Brjesson et al., 2009). Wastewater treatment plants mainly control and abate ARGs in sludge through disinfection processes (such as UV or liquid chlorine disinfection),

Occurrence and Behavior of Emerging Contaminants in Organic Wastes and Their Control Strategies
https://doi.org/10.1016/B978-0-443-13585-9.00004-5

Table 9.1 Common antibiotics and ARGs reduction techniques in sludge with their merits and demerits.

Method	Reduction technique	Merits	Demerits
Conventional disinfection	UV disinfection	No disinfection by-products formed, can effectively and rapidly inactivate resistant bacteria.	No continuous disinfection capability, photo-reanimation phenomenon.
	Chlorine disinfection	Low cost, simple operation.	Chlorination by-products formed.
Advanced oxidation	Ozone oxidation	Fast reaction speed, complete sterilization.	High cost, performance affected by temperature and pH.
	Photocatalytic oxidation	Nontoxic, odorless, complete disinfection.	Catalyst may be inactivated, low light source utilization.
Biological treatment	Aerobic composting	Effectively kills pathogenic bacteria in sludge and produces an organic fertilizer product.	Safety has not been confirmed experimentally, long fermentation cycle times.
	Phytoremediation	Safe and environmentally friendly, can be used on a large scale.	Performance influenced by plant species selection.
	Anaerobic digestion	Can achieve some abatement of antibiotics and ARGs, low energy consumption.	Long cycle times, low abatement effect.
	Membrane bioreactor	High retention rate, low carbon footprint.	Antibiotics and ARGs may contaminate the environment with sludge reuse.
	Artificial wetland treatment	Low operational costs, simple processes.	Large footprint, complex mechanism of action.

aerobic composting, anaerobic digestion, ozone oxidation, or other treatment processes (Brjesson et al., 2009). The commonly applied ARG abatement techniques and their advantages and disadvantages are shown in Table 9.1.

Dewatered sludge is a final product of wastewater treatment, and large amounts of antibiotics and ARGs in wastewater are transferred to the sludge in all stages of the wastewater treatment process. A diverse range of ARGs are known to be present in dewatered sludge, often in high abundance, such as tetracyclines (TCs), quinolones, and sulfonamides (SAs). The reuse of sludge as an agricultural resource can lead to the spread of ARGs throughout the environment, posing a high risk to both human and environmental health. Consequently, the control and reduction of antibiotics and ARGs during sludge treatment prior to reuse are urgently required.

2. Effect of aerobic composting on antibiotic resistance genes

Aerobic composting is extensively used as a biological treatment method prior to the utilization of solid waste resources, as it can effectively reduce the harmful components of municipal sludge, improving its value and safety as a resource. Aerobic composting not only kills pathogenic bacteria in sludge but also greatly reduces the volume of sludge after composting (Xie et al., 2019). Previous studies have confirmed that composting has a positive effect on the removal of antibiotics and ARGs from sludge (Xie et al., 2019). However, the efficiency of the reduction of antibiotics and ARGs during composting can be affected by the raw material being composted, the bacterial colony structure, the physicochemical properties and composition of sludge (e.g., heavy metals), and the use of additives, as shown in Fig. 9.1.

The presence of antibiotics in compost materials does not usually affect the composting process, although they may influence the microbial community structure and result in an increase in the abundance and diversity of ARGs. It has previously been reported that aerobic composting can reduce the concentration of penicillin and other pharmaceuticals in waste sludge to below detectable limits, meeting the requirements of the composting standard, but may significantly increase the abundance of ARGs (Yang et al., 2016). Furthermore, the addition of corn stalks containing antibiotic-resistance bacteria (ARB) to compost has been shown to increase the detection rate of ARB in the final compost product (Zachery et al., 2020).

Microbes are the main carriers of ARGs, and microbial community structure is a major factor affecting the reduction of antibiotics and ARGs during composting (Nao et al., 2020; Huang et al., 2021). Firmicutes, Proteobacteria, Actinobacteria, and Bacteroidetes are the major host phyla associated with ARGs. In the composting process, the distribution of potential host bacteria varies according to the raw material, composting time, and compost pile depth (Devin et al., 2017). The composition of bacterial communities is closely related to the abundance of ARGs, with changes in ARGs significantly influenced by microbial evolution. Therefore, microbial community succession may be the main driving force behind ARG dynamics in mixed composting conditions.

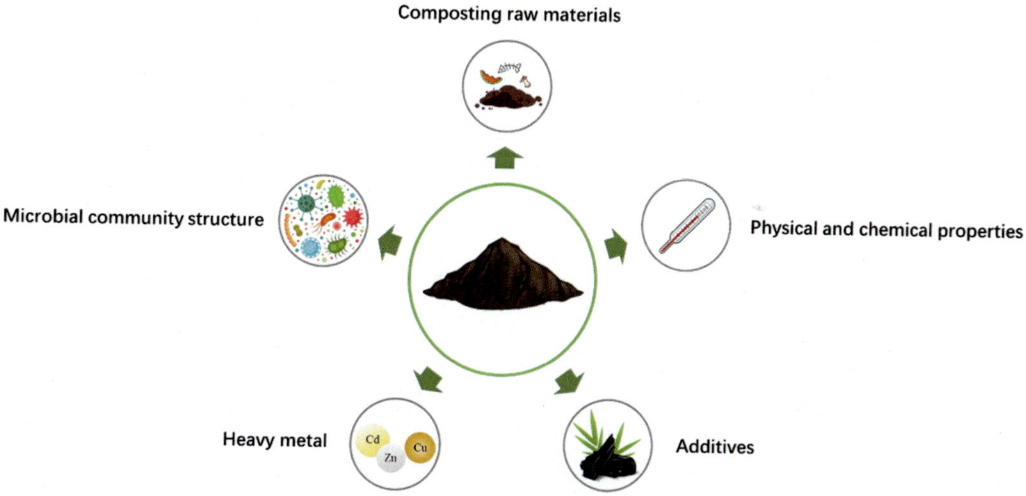

FIGURE 9.1 Factors influencing the reduction efficiency of antibiotics and resistance genes during composting.

It has been established that pH affects the antibiotic degradation process during composting, with penicillin more easily degraded under alkaline composting conditions (Ning et al., 2018). The high temperatures generated during composting can significantly reduce ARGs in sludge (Wu et al., 2011), and studies have found that temperature can affect the relative abundance of ARGs remaining after sludge composting. A comparison of the treatment of sludge by ultrahigh-temperature composting and ordinary composting showed that microbial diversity was significantly reduced by ultrahigh-temperature composting, with the abundance of ARGs and mobile genetic elements in sludge also significantly reduced, exhibiting a good overall performance. However, temperature indirectly affects antibiotic degradation by enhancing microbial activity and altering the microbial community structure (Nao et al., 2020).

Some studies have found that heavy metals affect the ARGs, metal resistance genes, and microbial community composition in composting. For example, high copper concentrations inhibit the degradation of SAs (Liu et al., 2015a,b) and tetracycline antibiotics (Gao et al., 2018), while high zinc concentrations can lead to a decrease or delay in the compost pile temperature peak (Duan et al., 2016). Copper, zinc, and other heavy metals have been found to be significantly positively correlated with the abundance of ARGs. In addition, a significant difference has been observed in the abundance of ARGs under different copper concentrations, indicating that copper increases the ecological risk posed by ARGs. The evolution of ARGs is greatly affected by their host organism, with the presence of copper, zinc, and other heavy metals driving the selection of these resistance genes in bacterial hosts to a certain extent, which may also inhibit the adaptation of hosts to extreme environments. Therefore, due to the long-term presence of coselective pressures (Li et al., 2015), the risk of transmission of ARGs may increase.

Additives have been shown to have a significant impact on the reduction of ARGs during composting. Therefore, additives have been widely explored to determine their impact during the composting of ARGs. Research has shown that the addition of rice straw and sawdust to compost can enhance the removal of sulfonamide antibiotics, increase the efficiency of antibiotic degradation, and inhibit the spread of ARGs (Qiu et al., 2012). Furthermore, various studies have shown that some additives can significantly reduce the relative abundance of ARGs following sludge composting, reducing the potential environmental risks and improving compost quality.

3. Common antibiotic resistance genes in sludge

3.1 Tetracyclines

TCs are a class of broad-spectrum antibiotics that consist of a four-ring structure. Due to their good antibacterial properties and minimal side effects, they are widely used in the treatment and control of bacterial infections and as feed additives to promote animal growth. TCs can form salts with bases, exist in a dissolved form under alkaline conditions, and are easily degraded while being more stable under acidic conditions (Michael et al., 2001).

After entering the human or animal body, only a portion of TCs can be metabolized and absorbed, with the remaining unabsorbed TCs and their metabolites excreted in urine and feces, eventually entering the wastewater treatment system (Chung and Fok, 2010). In addition, pharmaceutical and medical wastewater contains high concentrations of antibiotics, which are also discharged into the wastewater treatment network. However, common wastewater treatment technologies cannot remove high concentrations of TCs, allowing them to enter the receiving water body via the effluent. TCs can also adsorb and accumulate in activated sludge, and if not properly treated, TCs will ultimately enter the soil with sludge disposal, further increasing the risk of soil and groundwater pollution (Chung and Fok, 2010).

Research has shown that after sludge composting, the residual concentrations of tetracycline, doxycycline, and oxytetracycline all significantly decreased, with an overall TC reduction efficiency of >85%, with degradation mainly occurring during the heating and high-temperature stages (Heuer et al., 2011).

3.2 Sulfonamides

SAs are a class of artificially synthesized antibacterial drugs that are characterized by the presence of an organosulfur group. Due to their low cost, broad antibacterial spectrum, high stability, and strong hydrophilicity, SAs can interfere with and inhibit the growth and reproduction of bacteria. SAs are widely used in medicine, agriculture, animal husbandry, and aquaculture (Abdallah et al., 2016). The prolonged accumulation of SAs in water can induce the production of sulfonamide-resistant bacteria, presenting a significant risk to the environment, particularly when these bacteria reach large abundances.

In recent years, sulfonamide ARGs have been detected in various environmental media, with the detection rates of sul1 and *sul2* in particular being relatively high in aquatic environments, sediments, and sewage treatment plants (Pallares-Vega et al., 2019). *Sul1* is mainly proliferated and disseminated in the environment through horizontal transfer of the integron *IntI1* (Heuer et al., 2011). Studies have shown that the abundance of *sul1* and *sul2* can be reduced using artificial wetland systems and other deep treatment units, which is beneficial to reducing the risk of ARG transmission. Research on changes in the abundance of sulfonamide resistance genes sul1 and *sul2* during the process of sewage sludge composting showed that the relative abundance of both genes decreased during the heating stage, with the abundance of *sul1* reducing to its minimum level in the high-temperature treatment stage. However, the relative abundance of both *sul1* and *sul2* increased significantly during the cooling and maturation stages. These findings indicate that temperature may affect the relative abundance of sulfonamide resistance genes, while horizontal gene transfer may be an important pathway for the dissemination of these resistance genes.

3.3 Macrolides

Macrolide antibiotics are a universally utilized class of antibiotics, the most commonly used being erythromycin, clarithromycin, roxithromycin, and azithromycin. Erythromycin was first discovered and is the most widely used macrolide antibiotic. Currently, macrolide antibiotics are mainly produced by fermentation, resulting in a large amount of antibiotic-containing broth and waste, which poses an environmental risk and has limited the development and potential use of macrolide antibiotics in many regions worldwide. For example, the annual production capacity of erythromycin in China is over 10,000 tons, although actual annual production is around 6000 tons due to the environmental issues associated with production (Brjesson et al., 2009).

Approximately 30%–90% of antibiotics are excreted into the environment via feces and urine in the form of the original compound or its metabolites, which can be toxic to the receiving ecosystem and change the original bacterial community structure. This ultimately leads to the production of ARB and the spread of ARGs (Kummerer, 2004). In the macrolide production process, a large amount of solid and liquid antibiotic-containing waste is produced. Previous literature has reported that the erythromycin esterases encoded by genes *ere*A, *ere*B, *ere*C, and *ere*D can hydrolyze the lactone bond of macrolide antibiotics, thereby rendering them inactive (Xie et al., 2019; Xiao et al., 2018; Yang et al., 2016; Zachery et al., 2020).

3.4 Integrons

Strokes and Hall (Nao et al., 2020) first proposed the concept of integrons in 1989, which are natural cloning and expression elements containing site-specific recombination systems and gene cassettes, which can be classified based on the different integrases they encode. Currently, many types of integrase genes have been discovered, with the encoded integrons playing important roles in mediating and disseminating bacterial resistance and bacterial genomic evolution (Huang et al., 2021; Devin et al., 2017). Integrons can also be divided into six classes based on the different conservational regions that encode integrase genes. Five of these integron classes contain antibiotic resistance gene cassettes (Ning et al., 2018), with the integrons most closely related to pathological bacterial resistance being mainly classes 1, 2, and 3. The class 2 integron is located on the transposon *Tn*7 and its derivatives, with the integrase gene *intI*2 being a defective form of *intI*1. The 3′ conserved region of the class 2 integron contains 5 *tns* genes that can assist in transposon movement, with only a few gene cassettes having been identified on this type of integron (Wu et al., 2011).

At present, integrons in bacteria are mainly detected using one of two molecular biology techniques: (1) Nucleic acid hybridization: Probes are designed based on the conserved sequences of both ends of the class 1 integron to detect integrons by nucleic acid hybridization of plasmid and chromosomal DNA. In addition, gene cassette probes can be used to determine the types of gene cassettes present; (2) PCR detection: This is the most widely used detection method, in which primers are designed based on the conserved sequences on both ends of the class 1 integron and the resistance gene sequences, to determine integrons and gene cassettes by PCR amplification followed by sequence analysis. Numerous studies have shown that integrons are widely present in various clinical ARBs, and the antimicrobial resistance gene products encoded by the gene cassettes they carry have been shown to tolerate almost all clinical antibiotics. Integrons are often found on movable genetic elements such as transposons and plasmids, accelerating the dissemination of ARGs among bacteria (Duan et al., 2016). Integrons play an important role in the transfer of resistance genes, as they have the ability to capture and store exogenous genes. Multidrug-resistant integrons can rapidly increase the number of resistant bacterial strains and the variety of antibiotics to which they are resistant (Yin et al., 2016).

4. Common additives and their effect on antibiotic resistance gene removal

4.1 Porous additives

Porous additives mainly improve the permeability of oxygen and affect the succession of microbial communities by changing the pore structure of the compost media, thereby increasing the rate of ARG removal (He et al., 2022). Some porous expansion agents, such as biochar, can directly adsorb antibiotics due to their particular chemical functional groups (Sun et al., 2018), which reduces the secretion pressure of antibiotics on ARGs and thereby reduces the risk of ARG transmission. However, different additives have different effects on the compost media structure, particularly in terms of porosity, with the shape of the additive material and the uniformity of the pores playing an important role in the removal of ARGs during the aerobic composting process.

Diatomaceous earth is a highly porous silicate material that is often used in the composting process to reduce greenhouse gas emissions, while its porosity is beneficial for solid waste decomposition and compost maturation (Ren et al., 2019). Bentonite is a nonmetallic mineral composed of montmorillonite, which has a porous structure and a large specific surface area, as well as a high cation exchange capacity and water absorption capacity. Bentonite modification can significantly change the physicochemical properties of compost, improving enzyme activities and promoting the formation of high-quality compost products (Fernández-Calviño et al., 2015). In addition, previous studies have shown that bentonite can significantly reduce the abundance of ARGs in plant roots and contaminated soil (Gao et al., 2018).

Fly ash is a fine solid ash present in the flue gas discharged following coal combustion, containing various components such as silica and alumina. Fly ash has the characteristics of low bulk density, high porosity, and a large specific surface area, resulting in good permeability in liquids and air with high adsorption capability (Du et al., 2014). A key benefit of fly ash as a composting additive is that the final compost product is rich in many of the minerals necessary for plant growth (Zhang et al., 2009; Chee-Sanford et al.,

2009; Michael et al., 2001), improving the quality and value of compost products. In addition, the low bulk density and high porosity of fly ash help maintain air circulation within the compost heap (Qiu et al., 2012; Michael et al., 2001; Bent et al., 2003). When composting with fly ash as an additive, the pH of the compost remains high throughout the entire process, which is beneficial as the composting process is more stable under alkaline conditions (Abdallah et al., 2016). However, excessively high pH levels may affect the quality of the final compost product. Previous studies have shown that fly ash contains high levels of calcium, potassium, magnesium, and phosphorus, with high buffering capacity, water retention, and porosity, making it a suitable compost additive (Pallares-Vega et al., 2019).

4.2 Carbonaceous additives

The C/N ratio of compost is an important factor in ensuring normal aerobic composting performance. Typically, municipal sludge has a high nitrogen content with a low carbon content, and therefore, the C/N ratio of single-component sludge compost may not achieve satisfactory results. Agricultural waste materials such as straw, rice husks, and sawdust have the opposite properties to sludge, with a high carbon content and a low nitrogen content, resulting in a higher C/N ratio than sludge. Therefore, the inclusion of carbonaceous additives during sludge composting can effectively adjust the C/N ratio of the compost matrix, which benefits aerobic composting and promotes the utilization of sludge and agricultural solid waste as a valuable resource. Additionally, many previous studies have shown that the use of additives such as straw and rice husk material can enhance the reduction of antibiotics and ARGs during sludge composting.

Scholars have studied the effect of adding straw during the aerobic composting of sludge and found that after combining straw with dehydrated sludge and adjusting the C/N ratio to 30:1, the total gene copy number of target ARGs was reduced by 3.61-fold in the high temperature stage as compared to day 1. Furthermore, compared with the control group, the addition of straw caused the total gene copy number of target ARGs to decrease by 21.2% on the 32nd day. After aerobic composting, tetracycline and β-lactam resistance genes (*tet*M, *tet*Q, *bla*TEM, *bla*OXA, and the multidrug efflux pump *mex*F) and macrolides resistance genes (*erm*B and *mef*A) were significantly reduced by more than 90% and 70%, respectively.

4.3 Mechanism of antibiotic removal by additives during composting

The differences in additive pore structure, specific surface area, oxygen-containing functional groups, and other characteristics result in various adsorption mechanisms for organic pollutants such as antibiotics (Rosales et al., 2017). In addition, the chemical characteristics of different antibiotics can also influence the adsorption mechanism

(Jung et al., 2015). As shown in Fig. 9.2, the common adsorption mechanisms of biomass carbon additives for antibiotics include the following:

(1) π-πEDA interactions: The π-π electron donor-acceptor (π-πEDA) interaction is currently considered the main mechanism for the adsorption of antibiotics by carbonaceous materials. The process involves the aromatic ring and unsaturated structure (such as sulfonamide, amine, and ketone groups) on antibiotics serving as π electron acceptors, while the perpendicular π orbitals formed by each atom on the surface of the carbonaceous material act as electron donors, thereby creating a π-πEDA interaction (Finlay et al., 1950).

(2) Electrostatic interactions: Electrostatic interaction is one of the main mechanisms for the adsorption of ionized antibiotics by biochar (Kummerer, 2004). The surface of biochar is usually negatively charged, which generates an electrostatic attraction between biochar and positively charged organic compounds (Rizzo et al., 2013). When the solution pH is greater than the pH_{pzc}, the biochar surface becomes negatively charged, while a solution pH lower than the pH_{pzc} causes the biochar surface to be positively charged. Therefore, the initial pH of the solution plays an important role in the process of biochar adsorption (Su et al., 2014).

(3) Hydrophobic adsorption: Hydrophobic interaction is another mechanism for biochar to adsorb antibiotics. Generally, biochar with low surface oxidation exhibits hydrophobicity and can react with hydrophobic organic compounds through hydrophobic interactions, removing organic pollutants (Brjesson et al., 2009).

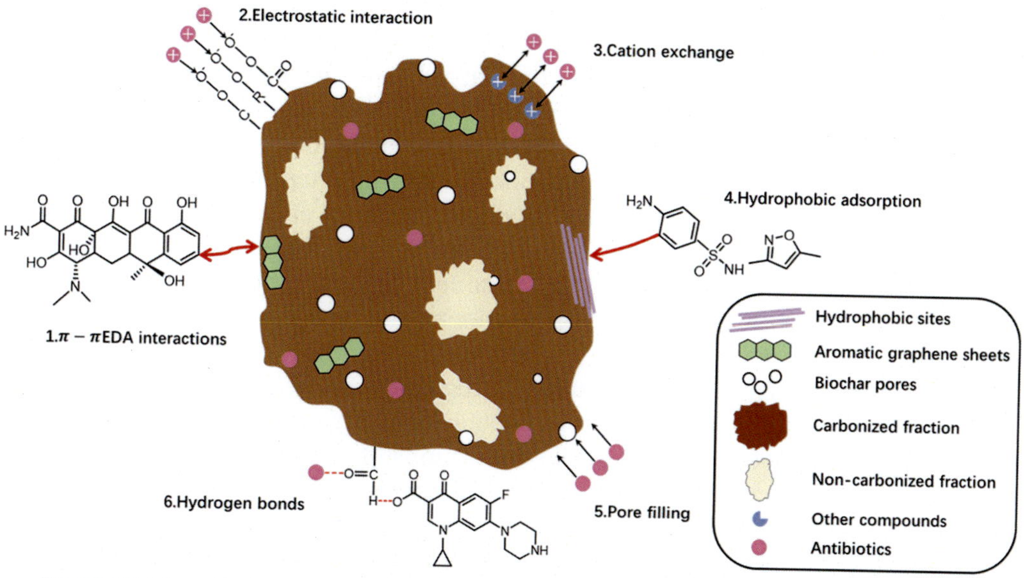

FIGURE 9.2 Schematic diagram of the mechanism of antibiotic removal by additives during composting.

(4) Pore filling: Pore filling is also considered to be an important mechanism for biochar to adsorb antibiotics. The adsorption of organic pollutants on biochar can be achieved by diffusion into biochar surface pores (adsorption pore filling) or the organic substrate of uncarbonized fractions (partitioning) (Brjesson et al., 2009).

(5) Hydrogen bonds: In addition to electrostatic interactions, strong hydrogen bonds and negative charge-assisted H-bond formation may be another important mechanism for biochar to adsorb antibiotics (Xie et al., 2019). The surface of carbon materials contains oxygen-containing functional groups such as carboxyl, lactone, and phenol hydroxyl groups, which can form hydrogen bonds with the benzene ring electron of antibiotics, thereby enhancing the adsorption capacity of carbon materials for antibiotics (Xiao et al., 2018). Similarly, the benzene ring on the carbon material surface can also form hydrogen bonds with oxygen-containing functional groups on antibiotics, serving as an electron donor. Studies have shown that the adsorption of sulfonamide drugs by natural soil, humus, and clay minerals occurs via the formation of hydrogen bonds between sulfonamide drugs and polar adsorbents, which has also been confirmed in the adsorption mechanism of sulfonamide drugs on carbon materials.

The removal of ARGs by composting additives usually involves a combination of multiple adsorption mechanisms, as shown in Fig. 9.2.

5. Conclusions

Urban sludge not only contains a large amount of organic matter, nitrogen, phosphorus, and other nutrients but also residual heavy metals and pollutants such as antibiotics and ARGs. If appropriate treatment is not performed prior to the utilization of sludge resources, the contaminants remaining within sludge pose a serious environmental pollution risk, particularly in terms of the antibiotic and ARG content, which cause incalculable harm to both human and environmental health. Aerobic composting is one of the main strategies used to reduce the risk posed by sludge and allow its reutilization as an agricultural resource. Aerobic composting not only transforms and degrades organic matter into more stable and low-environmental-impact substances but has also been shown to reduce the abundance of antibiotics and specific ARGs in sludge. The composition and succession of microbial communities play a crucial role in the reduction of antibiotics and ARGs in sludge during composting. The reduction of ARGs during composting is influenced by various factors, including the substrate composition, temperature, pH, and the characteristics of the compost materials. In order to achieve a better composting effect and further reduce the content of antibiotics and ARGs in sludge, additives with a high carbon content or porosity, such as straw, sawdust, and biochar, are usually added to the sludge prior to composting. These additives can significantly affect the physical and chemical properties of the compost environment, further reducing the abundance of pollutants in sludge such as antibiotics and ARGs through direct adsorption or by changing the structure and abundance of microbial communities during composting. Therefore, additives can reduce the transfer and dissemination of ARGs, minimizing the hazards posed by sludge to human and environmental health.

Credit authorship contribution statement

Jiwei Shi: Visualization, writing—original draft.

Bangchi Wang: Writing—original draft, formal analysis.

Jiachen Xie: Supervision, data curation.

Kui Huang: Conceptualization, methodology, writing, review and editing.

Acknowledgments

This work was supported by the Science and Technology Program Foundation of Gansu Province (22JR5RA335, 22JR9KA034) and the Natural Science Foundation of China (51868036, 52000095).

References

Abdallah, H., et al., 2016. Monitoring of twenty-two sulfonamides in edible tissues: investigation of new metabolites and their potential toxicity. Food Chemistry 192, 212–227. https://doi.org/10.1016/j.foodchem.2015.06.090.

Bent, H., et al., 2003. Characterisation of the abiotic degradation pathways of oxytetracyclines in soil interstitial water using LC-MS-MS. Chemosphere 50 (10), 1331–1342. https://doi.org/10.1016/S0045-6535(02)00660-9.

Brjesson, S., Melin, S., Matussek, A., et al., 2009. A seasonal study of the mecA gene and *Staphylococcus aureus* including methicillin-resistant *S. aureus* in a municipal wastewater treatment plant. Water Research 43 (15), 925–932. https://doi.org/10.1016/j.watres.2008.11.042.

Chee-Sanford Joanne, C., et al., 2009. Fate and transport of antibiotic residues and antibiotic resistance genes following land application of manure waste. Journal of Environmental Quality 38 (3), 1086–1108. https://doi.org/10.2134/jeq2008.0128.

Chung, R., Fok, K., 2010. Sorption and dissipation of tetracyclines in soils and compost. Pedosphere 20 (6), 807–816. https://doi.org/10.1016/S1002-0160(10)60072-8.

Devin, B., Hao, X., Edward, T., et al., 2017. Effect of co-composting cattle manure with construction and demolition waste on the archaeal, bacterial, and fungal microbiota, and on antimicrobial resistance determinants. PLoS One 11 (6), 0157539. https://doi.org/10.1371/journal.pone.0157539.

Du, J., Ren, H., Geng, J., et al., 2014. Occurrence and abundance of tetracycline, sulfonamide resistance genes, and class 1 integron in five wastewater treatment plants. Environmental Science and Pollution Research 21 (12), 7276–7284. https://doi.org/10.1007/s11356-014-2738-8.

Duan, M., et al., 2016. Effects of sulphamethazine and zinc on the functional diversity of microbial communities during composting. Environmental Technology 37 (11), 1357–1368. https://doi.org/10.1080/21622515.2015.1129325.

Fernández-Calviño, D., Rodríguez-Salgado, I., Pérez-Rodríguez, P, et al., 2015. Time evolution of the general characteristics and Cu retention capacity in an acid soil amended with a bentonite winery waste. Journal of Environmental Management 150, 435–443. https://doi.org/10.1016/j.jenvman.2014.12.024.

Finlay, A.C., Hobby, G.L., P'An, S.Y., et al., 1950. Terramycin, a new antibiotic. Science 111, 85. https://doi.org/10.1126/science.111.2877.85.

Gao, P., et al., 2018. Long-term impact of a tetracycline concentration gradient on the bacterial resistance in anaerobic-aerobic sequential bioreactors. Chemosphere 205, 308–316. https://doi.org/10.1016/j.chemosphere.2018.04.155.

He, X., Xiong, J., Yang, Z., et al., 2022. Exploring the impact of biochar on antibiotics and antibiotics resistance genes in pig manure aerobic composting through untargeted metabolomics and metagenomics. Bioresource Technology 352. https://doi.org/10.1016/j.biortech.2022.127118.

Heuer, H., Schmitt, H., Smalla, K., 2011. Antibiotic resistance gene spread due to manure application on agricultural fields. Current Opinion in Microbiology 14 (3), 236–243. https://doi.org/10.1016/j.mib.2011.04.009.

Huang, X., et al., 2021. Fitness reduction of antibiotic resistance by an extra carbon source during swine manure composting. Environmental Pollution 277, 116819. https://doi.org/10.1016/j.envpol.2021.116819.

Jung, C., Boateng, L., Flora, J., et al., 2015. Competitive adsorption of selected non-steroidal anti-inflammatory drugs on activated biochars: Experimental and molecular modeling study. Chemical Engineering Journal 264, 1–9. https://doi.org/10.1016/j.cej.2014.11.076.

Kummerer, K., 2004. Resistance in the environment. Journal of Antimicrobial Chemotherapy 54 (2), 311–320. https://doi.org/10.1093/jac/dkh325.

Li, Y., et al., 2015. Effects of Cu exposure on enzyme activities and selection for microbial tolerances during swine-manure composting. Journal of Hazardous Materials 283, 512–518. https://doi.org/10.1016/j.jhazmat.2014.09.071.

Liu, B., et al., 2015. Effects of composting process on the dissipation of extractable sulfonamides in swine manure. Bioresource Technology 175, 284–290. https://doi.org/10.1016/j.biortech.2014.10.101.

Liu, Y., Hu, X., Zhang, Q., et al., 2015. Application of biochar for the removal of pollutants from aqueous solutions. Chemosphere: Environmental Toxicology and Risk Assessment 125. https://doi.org/10.4172/2161-0525.C1.009.

Michael, K., et al., 2001. Formation of anhydrotetracycline during a high-temperature treatment of animal-derived feed contaminated with tetracycline. Food Chemistry 75 (4), 423–429. https://doi.org/10.1016/S0308-8146(01)00185-8.

Nao, K., et al., 2020. Tylosin degradation during manure composting and the effect of the degradation byproducts on the growth of green algae. Science of the Total Environment 718©, 137295. https://doi.org/10.1016/j.scitotenv.2020.137295.

Ning, L., et al., 2018. Variations in the fate and risk analysis of amoxicillin and its degradation products during pig manure aerobic composting. Journal of Hazardous Materials 346, 234–241. https://doi.org/10.1016/j.jhazmat.2017.12.034.

Pallares-Vega, R., Blaak, H., Plaats, R., et al., 2019. Determinants of presence and removal of antibiotic resistance genes during WWTP treatment: a cross-sectional study. Water Research 161, 319–328. https://doi.org/10.1016/j.watres.2019.05.049.

Qiu, J., et al., 2012. Effects of conditioners on sulfonamides degradation during the aerobic composting of animal manures. Procedia Environmental Sciences 16, 17–24. https://doi.org/10.1016/j.proenv.2012.10.004.

Ren, X., Wang, Q., Zhao, J., et al., 2019. Improvement of cleaner composting production by adding Diatomite: From the nitrogen conservation and greenhouse gas emission. Bioresource Technology 286. https://doi.org/10.1016/j.biortech.2019.121377.

Rizzo, L., et al., 2013. Urban wastewater treatment plants as hotspots for antibiotic resistant bacteria and genes spread into the environment: a review. Science of the Total Environment 447, 345–360. https://doi.org/10.1016/j.scitotenv.2012.12.013.

Rosales, E., Meijide, J., Pazos, M., et al., 2017. Challenges and recent advances in biochar as low-cost biosorbent: From batch assays to continuous-flow systems. Bioresource Technology 246, 176–192. https://doi.org/10.1016/j.biortech.2017.06.084.

Su, H., Ying, G., He, L., et al., 2014. Antibiotic resistance, plasmid-mediated quinolone resistance (PMQR) genes and amp C gene in two typical municipal wastewater treatment plants. Environmental Science: Processes and Impacts 16 (2), 324–332. https://doi.org/10.1039/C3EM00527C.

Sun, W., Gu, J., Wang, X., et al., 2018. Impacts of biochar on the environmental risk of antibiotic resistance genes and mobile genetic elements during anaerobic digestion of cattle farm wastewater. Bioresource Technology 256, 342–349. https://doi.org/10.1016/j.biortech.2018.02.052.

Wu, X., et al., 2011. The behavior of tetracyclines and their degradation products during swine manure composting. Bioresource Technology 102 (10), 5924–5931. https://doi.org/10.1016/j.biortech.2011.02.038.

Xiao, Z., Li, G., Chen, X., et al., 2018. Influence of carbon-rich amendments on penicillin degradation during pharmaceutical sewage sludge composting. Environmental Chemistry 37 (8), 1728–1737. https://doi.org/10.1071/en18182.

Xie, W., Zou, X., Liu, D., et al., 2019. Dynamics of metal(loid) resistance genes driven by succession of bacterial community during manure composting. Environmental Pollution 255. https://doi.org/10.1016/j.envpol.2019.07.082.

Yang, L., Zhang, S., Chen, Z., et al., 2016. Maturity and security assessment of pilot-scale aerobic co-composting of penicillin fermentation dregs(PFDs) with sewage sludge. Bioresource Technology 204, 185–191. https://doi.org/10.1016/j.biortech.2015.12.057.

Yin, Y., et al., 2016. Effects of copper on the abundance and diversity of ammonia oxidizers during dairy cattle manure composting. Bioresource Technology 221, 181–187. https://doi.org/10.1016/j.biortech.2016.09.031.

Zachery, R., et al., 2020. Corn stalk residue may add antibiotic-resistant bacteria to manure composting piles. Journal of Environmental Quality 49 (3), 745–753. https://doi.org/10.1002/jeq2.20138.

Zhang, T., Zhang, M., Zhang, X., et al., 2009. Tetracycline resistance genes and tetracycline resistant lactose-fermenting *Enterobacteriaceae* in activated sludge of sewage treatment plants. Environmental Science and Technology 43 (10), 3455–3460. https://doi.org/10.1021/es803565n.

10

Removal of antibiotic resistance genes in sewage sludge vermicomposting: a mini-review

Licheng Zhu[1], Zilong Wu[1], Jin Chen[1] and Kui Huang[1,2]

[1]SCHOOL OF ENVIRONMENTAL AND MUNICIPAL ENGINEERING, LANZHOU JIAOTONG UNIVERSITY, LANZHOU, CHINA; [2]KEY LABORATORY OF YELLOW RIVER WATER ENVIRONMENT IN GANSU PROVINCE, LANZHOU, CHINA

1. Introduction

Dewatered sludge, as a by-product of wastewater treatment, inevitably accumulates large amounts of organic, inorganic, and emerging biological pollutants, making it a worldwide problem of effective sludge treatment and disposal (Huang and Xia et al., 2018). For instance, unused antibiotics, primarily excreted by humans and animals through urine and feces, enter into wastewater treatment plants (WWTP) and build up antibiotic resistance genes (ARGs) in sewage sludge (Alonso and Tolmasky, 2020). The literature shows that the elimination rate of ARGs in sludge treatment was lower compared to sewage treatment (Yang et al., 2014).

By using the synergistic effects of earthworms and microorganisms, sludge vermicomposting (SV) can not only decompose and transform organic matter in sludge but also reduce and stabilize the sludge (Cui and Li et al., 2018). Compared to other sludge stabilization methods, the final product of sludge vermicompost is more stable, homogeneous, and contains higher levels of nutrients and probiotics, which are regarded as high-quality soil amendments (Xia and Chen et al., 2019). However, the quality of sludge vermicompost has been affected by contaminations such as heavy metals, antibiotics, and pathogens (Yang et al., 2014). Different from traditional biological pollutants, ARGs do not rely on toxicity but render host cells resistant to antibiotics through a unique transmission method (Ramos et al., 2022). These ARGs can spread via both vertical and horizontal gene transfer (HGT) and can enter the food chain and damage human health (Xia and Chen et al., 2019). Nonetheless, the SV can partially reduce the abundance of ARGs in the sludge. Some studies have demonstrated that the environmental risk of sludge vermicompost intensifies when used as a biofertilizer (Zhang et al., 2016; Cui and Fu et al., 2022). The different parameters, including the type of ARGs and microbial

Occurrence and Behavior of Emerging Contaminants in Organic Wastes and Their Control Strategies
https://doi.org/10.1016/B978-0-443-13585-9.00007-0

communities in the initial sludge, composting conditions such as composting temperature, earthworm inoculation density, and residence time of earthworms in the pile shape the reduction efficiency (Liu and Sun et al., 2019; Huang and Xia et al., 2020). Therefore, it is necessary to improve the ARG reduction efficiency of SV technology. However, up to now, relevant research indicates that the changes in experimental conditions still have selectivity on the reduction of ARGs in sludge (Zhang and Chen et al., 2016; Cui and Sartaj et al., 2019). Different additives, such as cow dung, biomass charcoal, bentonite, and other materials, have been reported to reduce the residues of antibiotics and ARGs in vermicomposting (Huang et al., 2020; Guan and Peng, 2021; Miao and Wang et al., 2023). Among different additives, biochar can accelerate the degradation of organic matter and passivate harmful substances in sludge. In addition, it could improve microbial activities and promote the proliferation of microbes to accelerate sludge stabilization (Zhang et al., 2014). Moreover, plant alkaloids contained in residues from traditional Chinese medicine, as macromolecular substances in plants, can interfere with bacterial DNA synthesis, which can inhibit bacterial activity (William Chin and Chen et al., 2010). This paper discussed the effect of additives on the removal of ARGs during SV and evaluated how the removal rate varies with different composting conditions.

2. Species and abundance of ARGs in vermicomposting products

Dewatered sludge, a byproduct of WWTP, contains a high abundance of antibiotic-resistant bacteria (ARB) and ARGs (Huang et al., 2020). Previous studies have shown that ARGs usually accumulate in sludge during sewage treatment (He and Zhou et al., 2019), with diverse characteristics in sewage and vermicompost (Yang et al., 2014). The typical ARGs found in sewage and vermicompost included tetracyclines (*tet*X, *tet*M, *tet*G), quinolones (*qnr*A, *qnr*S), macrolides (*erm*B, *erm*F), sulfonamides (*sul*1, *sul*2), the integron gene (*int*I1), and the transposon gene (tnpA-4), indicating horizontal transmission of ARGs among bacterial cells. The absolute abundance of different ARGs can be found in Fig. 10.1. The most abundant ARGs in sludge vermicompost were sul1, *sul*2, and *tet*X, with an absolute abundance of 9.0×10^8, 1.0×10^9–5.0×10^{12}, and 4.0×10^{12} copies/g, respectively. The high abundance of *sul*1 and *tet*X is likely due to the frequent use of sulfonamides and tetracyclines in clinical and livestock breeding. In addition, the absolute abundance of ARGs decreased with earthworm digestion in the sludge. The removal rate of *qnr*S, *tet*G, and *tet*X was >90%, while the removal rate of *qnr*A, *sul*1, and *tet*M ranged between 31% –73%. These results suggest that a portion of ARGs in excess sludge can be reduced during earthworm intestinal digestion (Xia and Chen et al., 2019).

3. Effects of additives on ARGs in vermicomposting

Prior studies have reported that the addition of biochar to SV can affect the ARGs (Huang et al., 2020). In addition to biochar application, herbal medicine residues (HMR) have also

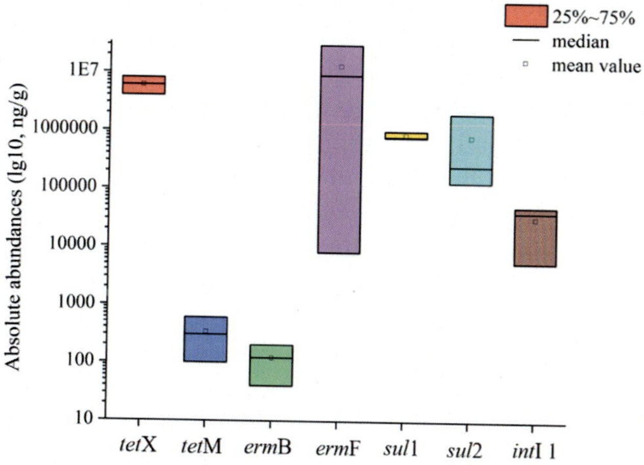

FIGURE 10.1 Absolute abundance of ARGs and *intI*1 gene in vermicompost.

been found to inhibit the HGT of ARGs, thereby reducing the risk of ARG transmission (Zhang et al., 2017). This chapter will mainly discuss the effects of additives such as biochar and HMR on the removal of ARGs in the vermicomposting of sewage sludge. In this review, the control group was represented by vermicomposting without any additives.

3.1 Effect of biochar on ARGs in sludge vermicomposting

The effectiveness of vermicomposting with different additives in removing ARGs can be found in Fig. 10.2. The results suggest that the treatments with corncob biochar and rice husk biochar significantly decreased ($P < 0.05$) compared to the control treatment, which received no additives. Meanwhile, the reduction of *erm*F was more obvious at the

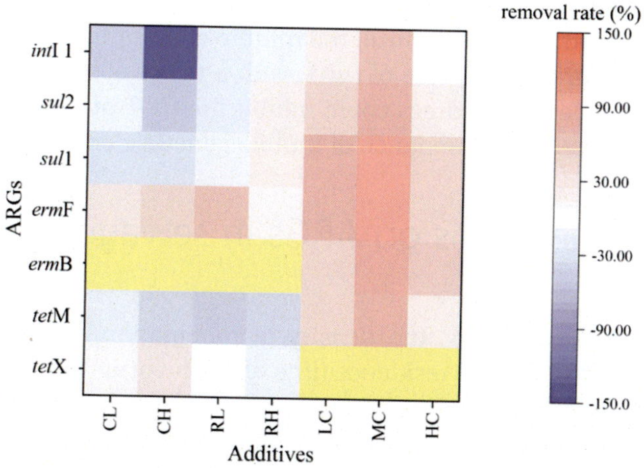

FIGURE 10.2 Removal of ARGs by vermicomposting with different additives. LC, MC, and HC represent herbal residue treatments spiked with low (10%), medium (30%), and high (50%) concentrations, respectively; CL and CH represent corn cob biochar treatments with low (1.25%) and high (5%) concentrations, respectively; RL and RH represent rice husk biochar treatments with low (1.25%) and high (5%) concentrations, respectively.

low content of rice husk biochar (1.25%). However, the treatments with biochar from corncob and rice husk showed a significantly higher abundance of *tet*X than *tet*X and *tet*M, while *tet*M showed little change. This difference in the effect of biochar on tetX and tetM is likely due to the enzyme modification gene and ribosomal protein protection gene, respectively (Zhang and Gu et al., 2018) and the distinct properties of the genes themselves. For sulfonamide ARGs, the total abundance of *sul*1 and *sul*2 in vermicomposting products increased at different levels with the addition of corncob biochar and rice hull biochar. Nevertheless, there was no significant difference in abundance between the corncob biochar and the rice hull biochar. The results also demonstrated that the abundance of ARGs (e.g., *int*I1) was the highest in the composting products of corncob biochar treatment. Prior studies have verified that the high abundance of the *int*I1 gene in excess sludge could increase the risk of HGT of ARGs, which in turn could increase the abundance of ARGs (Zhang and Li et al., 2020). In addition, Chao et al. (2019) found that the earthworm gut was a favorable environment for ARGs transmission, and the presence of corncob biochar increased the microbial diversity in the earthworm gut, which further explained the reduction of ARGs abundance by biochar.

3.2 Effect of herbal medicine residues on ARGs in sludge vermicomposting

The absolute abundance of *erm*B declined during the vermicomposting process. The treatment with medium content of dregs (30% dregs + 70% sludge) showed the best removal effect on *erm*B, with a removal rate of 94.8% higher than the original sludge. The abundance of *tet*M increased in the control and high content of dregs (50% dregs + 50% sludge). The *tet*M decreased significantly ($P < 0.05$) only with the medium content of HMR treatment, with a removal rate of 69.7%. Moreover, HMR treatments decreased the absolute abundance of *sul*1 and *sul*2 genes compared with control. This indicated that the addition of traditional HMR could effectively reduce the abundance of *sul*1 and *sul*2 in vermicomposting, with a removal rate of up to 90.1%. The abundance of *intI*1 in SV was reduced after adding HMR. Compared with the control, the middle content residue group had the highest removal effect on the abundance of *int*I1, with a removal rate of 71.4% ($P < 0.05$). This indicated that the decoction dregs could inhibit the HGT of ARGs and weaken the spread of ARGs by reducing the abundance of mobile genetic elements.

4. Effects of experimental conditions on ARGs in sewage sludge vermicomposting

Numerous factors, such as species of earthworms, the density of inoculation, composting temperature, composting humidity, and the residence time of earthworms in the composting substrate, shape the removal of ARGs during vermicomposting (Cui and Fan et al., 2020; Cui et al., 2022). Therefore, the effects of biological and physicochemical conditions on ARG removal were studied in this paper.

Table 10.1 Removal of ARGs by sludge vermicomposting at different temperatures (15°C, 20°C, and 25°C).

	ARGs/MGE						
Temperature/(°C)	***qnr*A**	***qnr*S**	***sul*1**	***tet*M**	***tet*G**	***tet*X**	***int*I1**
15	64.7	85.3	−1.5	77.5	8.9	33.0	−40.7
20	57.1	85.3	18.2	84.0	18.3	18.1	−9.7
25	9.5	−2.9	−3.0	11.6	−43.6	12.8	−15.4

4.1 Effect of composting temperature on ARGs in sludge vermicomposting

Among the different variables, the temperature is relatively easy to control. A suitable temperature can improve the stability of vermicomposting, specifically in terms of dehydrogenase activity and dissolved organic carbon (Zhang and Li et al., 2020). Given the close relationship between ARGs and bacteria during the biological treatment of organic waste (Su and Wei et al., 2015; Miller and Novak et al., 2016), it can be hypothesized that the fate of ARGs in the vermicomposting of excess sludge may vary at different temperatures due to changes in the bacterial population. For instance, Hou et al. (2021) found lower bacterial diversity and biostability of vermicompost at 25°C compared to 15°C and 20°C. However, the effect of temperature on ARGs during excess sludge composting has received little attention. Cui et al. (2022) conducted a series of vermicomposting experiments across a moderate temperature gradient (15°C–25°C) to study the mechanisms that regulate the impact of temperature change on ARGs. The results showed that vermicomposting had little impact on ARG removal at 25°C but was beneficial for the removal of *int*I1 compared with 15°C and 20°C. The abundance of *qnr*A and *qnr*S decreased significantly ($P < 0.05$) during vermicomposting, with the highest removal rates at 15°C and 20°C. However, the *sul* gene was not significantly affected by the vermicomposting and temperature conditions. The *tet*M gene was significantly reduced, with the optimum removal at 20°C, while the abundance of the *tet*G and *tet*X genes did not change significantly. The alterations in the qnrA, qnrS, and tetM genes under both 15°C and 20°C temperature conditions were observed to conform to the first-order kinetic model. Surprisingly, the expected positive correlation between temperature and the kinetic coefficients was not observed, presenting an intriguing finding worthy of further investigation. The removal effect of ARGs by vermicomposting at different temperatures can be found in Table 10.1. The results show the highest removal rate of ARGs and good biological stability of the final product at 20°C, which could be considered a relatively suitable temperature for vermicomposting.

4.2 Effect of earthworm biomass on ARGs in sludge vermicomposting

In the industrialization of vermicomposting, huge earthworm biomass is also a valuable by-product. Earthworms have the ability to rapidly reproduce, with a single potential producing up to 28 offspring every 6 months under optimal conditions of moisture, temperature, and food (Ramos et al., 2022). The rapid growth of earthworm biomass will further promote vermicomposting technology to degrade various harmful substances in excess sludge. Some studies (Cui and Sartaj et al., 2019) have suggested that the high inoculation density of earthworms can improve the removal efficiency of ARGs, which may be related to earthworm intestinal microorganisms. A series of vermicomposting systems with suitable earthworm inoculation density ranges (100, 180, and 360 earthworms/3 kg) were designed to investigate and clarify the influence mechanism of earthworm biomass on ARGs. The removal efficiency of ARGs obtained by different earthworm biomass in vermicomposting can be found in Table 10.2. The abundance of *sul*1, *erm*F, and *tet*M genes simultaneously decreased after composting. Moreover, the removal rates of the three ARGs at an inoculation density of 360 earthworms/3 kg were higher than those of the other two treatments, with the best removal rates above 80% compared with the control treatment. However, the abundance of the *sul*2 gene after vermicomposting showed the opposite trend. The abundance of the *sul*2 gene in the compost at the inoculation density of 360 earthworms/3 kg increased compared with the other two treatments. This suggests that while a higher inoculation density can effectively improve the removal rate of *sul*1, *erm*F, and *tet*M genes, an excessively high inoculation density might also increase the living space and food competition within the microbial population, which might influence the treatment efficiency of vermicomposting.

4.3 Effect of earthworm residence time on ARGs in vermicomposting

The retention time of earthworms in the compost can affect the maturity of the final products. Based on the retention time, the vermicomposting process could be divided into an initial stage and a final stage (Cui et al., 2020). Prior studies have found that the surface area of sludge particles is increased and microbial activities are stimulated due to the ingestion of earthworms, gizzard grinding, and substrate degradation, which makes the microorganisms have high activity in the initial stage of composting (Horn and Schramm et al., 2003). The final stage of vermicomposting is characterized by reduced earthworm and microbial activity, which is likely due to the low utilization of fresh residues and increased refractory organic matter (Che and Lin et al., 2020). Therefore,

Table 10.2 Removal of ARGs by different earthworm biomass in vermicomposting.

	ARGs/MGE				
Number (earthworms/3 kg)	***erm*F**	***tet*M**	***sul*1**	***sul*2**	***int*I1**
100	35.5	91.3	16.7	68	48.9
180	33.3	84.6	6.2	60	46.2
360	84.2	91	81.8	−25	−30.8

Table 10.3 Removal of ARGs in vermicomposting by earthworm residence time.

	ARGs/MGE							
Times (days)	***erm*B**	***erm*F**	***tet*G**	***tet*X**	***sul*1**	***qnr*S**	***qnr*A**	***intI*1**
10	77.8	0	−6	−4.2	−71.4	−135.3	−3.8	−43.8
20	79.6	22.2	−14	−25.5	−93	17.6	43.4	−162.5
30	84.3	65.3	28	34	31.8	76.5	45.3	−15.4

previous studies (Che et al., 2020; Kui et al., 2020) have utilized a set of continuous vermicomposting systems with earthworm retention times (10, 20, and 30 days) using 10 kg of excess sludge as composting substrates to clarify the influence of earthworm retention time on ARGs removal effect. The abundance of *erm*B and *qnr*A decreased by 79.6% and 43.4%, respectively, at the initial stage of 10–20 days, while there was almost no removal effect on other ARGs (Table 10.3). All the sludge in the compost was transformed into vermicompost after 30 days of vermicomposting. Additionally, with the increase in composting time, the removal efficiency of *erm*B, *erm*F, and *qnr*S genes was the most obvious, and the removal rate was 65.3%–84.3%. Overall, these results indicate that the removal efficiency of ARGs in vermicompost increased with the improvement of earthworm residence time.

5. Conclusion

The addition of biochar to SV could improve the removal efficiency of ARGs (i.e., *erm*F and *sul*1). On the other hand, the addition of herbal medicine significantly improved the removal efficiency of macrolides, tetracyclines, and sulfonamide ARGs ($P < 0.05$). Moreover, the HMR with 30% content was found to be the best to improve the removal of ARGs. In addition, optimum vermicomposting, with the highest removal rate of ARGs and good final product stability, was observed at 20°C. In addition, the high inoculation density of earthworms significantly improved the removal efficiency of *sul*1, *erm*F, and *tet*M genes ($P < 0.05$), while the removal effect on *erm*B, *erm*F, and *qnr*S genes increased with increase in an composting time.

Considering the widespread use of sludge vermicompost as a biofertilizer, it is imperative to comprehensively assess the implications of ARGs on WWTPs when utilizing sludge vermicompost as a product. Since the intestinal microorganisms and ARGs of earthworms are crucial in reducing ARGs during vermicomposting, the effect of additives on these factors should be a key focus of future research.

Acknowledgments

This work was supported by the Science and Technology Program Foundation of Gansu Province (22JR5RA335, 22JR9KA034) and the Natural Science Foundation of China (51868036, 52000095, 52160012).

Declarations of competing interest

The authors declare that they have no known competing financial interests or personal relationships that could have appeared to influence the work reported in this paper.

References

Alonso, J.C., Tolmasky, M.E. (Eds.), 2020. Plasmids: Biology and Impact in Biotechnology and Discovery. John Wiley and Sons.

Chao, H., Kong, L., Zhang, H., et al., 2019. Metaphire guillelmi gut as hospitable micro-environment for the potential transmission of antibiotic resistance genes. Science of the Total Environment 669, 353–361. https://doi.org/10.1016/j.scitotenv.2019.03.017.

Che, J., Lin, W., Ye, J., et al., 2020. Insights into compositional changes of dissolved organic matter during a full-scale vermicomposting of cow dung by combined spectroscopic and electrochemical techniques. Bioresource Technology 301, 122757. https://doi.org/10.1016/j.biortech.2020.122757.

Chin, L.W., Cheng, Y.W., Lin, S.S., et al., 2010. Anti-herpes simplex virus effects of berberine from Coptidis rhizoma, a major component of a Chinese herbal medicine, Ching-Wei-San. Archives of Virology 155, 1933–1941. https://doi.org/10.1007/s00705-010-0779-9.

Cui, G., Li, F., Li, S., et al., 2018. Changes of quinolone resistance genes and their relations with microbial profiles during vermicomposting of municipal excess sludge. Science of the Total Environment 644, 494–502. https://doi.org/10.1016/j.scitotenv.2018.07.015.

Cui, G., Bhat, S.A., Li, W., et al., 2019. Gut digestion of earthworms significantly attenuates cell-free and-associated antibiotic resistance genes in excess activated sludge by affecting bacterial profiles. Science of the Total Environment 691, 644–653. https://doi.org/10.1016/j.scitotenv.2019.07.177.

Cui, G., Lü, F., Zhang, H., et al., 2020. Critical insight into the fate of antibiotic resistance genes during biological treatment of typical biowastes. Bioresource Technology 317, 123974. https://doi.org/10.1016/j.biortech.2020.123974.

Cui, G., Fu, X., Bhat, S.A., et al., 2022. Temperature impacts fate of antibiotic resistance genes during vermicomposting of domestic excess activated sludge. Environmental Research 207, 112654. https://doi.org/10.1016/j.envres.2021.112654.

Guan, M., Peng, L., et al., 2021. Effect of corncob biochar on microbial population and ARGs in sludge vermicompost. China Environmental Science 41 (06), 2744–2751. https://doi.org/10.19674/j.cnki.issn1000-6923.20210223.014.

He, P., Zhou, Y., Shao, L., et al., 2019. The discrepant mobility of antibiotic resistant genes: evidence from their spatial distribution in sewage sludge flocs. Science of the Total Environment 697, 134176. https://doi.org/10.1016/j.scitotenv.2019.134176.

Horn, M.A., Schramm, A., Drake, H.L., 2003. The earthworm gut: an ideal habitat for ingested N2O-producing microorganisms. Applied and Environmental Microbiology 69 (3), 1662–1669. https://doi.org/10.1128/AEM.69.3.1662-1669.2003.

Hou, S., Xuyang, L., Hui, Z., et al., 2021. Analysis of the effect of temperature on vermicomposting of municipal sludge based on EEM and PCR-DGGE. Ecology and Environment 30 (5), 1060. https://doi.org/10.16258/j.cnki.1674-5906.2021.05.019.

Huang, K., Xia, H., Wu, Y., et al., 2018. Effects of earthworms on the fate of tetracycline and fluoroquinolone resistance genes of sewage sludge during vermicomposting. Bioresource Technology 259, 32–39. https://doi.org/10.1016/j.biortech.2018.03.021.

Huang, K., Xia, H., Zhang, Y., et al., 2020. Elimination of antibiotic resistance genes and human pathogenic bacteria by earthworms during vermicomposting of dewatered sludge by metagenomic analysis. Bioresource Technology 297, 122451. https://doi.org/10.1016/j.biortech.2018.03.021.

Kui, H., Jingyang, C., Mengxin, G., et al., 2020. Effects of biochars on the fate of antibiotics and their resistance genes during vermicomposting of dewatered sludge. Journal of Hazardous Materials 397, 122767. https://doi.org/10.1016/j.jhazmat.2020.122767.

Liu, K., Sun, M., Ye, M., et al., 2019. Coexistence and association between heavy metals, tetracycline and corresponding resistance genes in vermicomposts originating from different substrates. Environmental Pollution 244, 28–37. https://doi.org/10.1016/j.envpol.2018.10.022.

Miao, L., Wang, Y., Zhang, M., et al., 2023. Effects of hydrolyzed polymaleic anhydride addition combined with vermicomposting on maturity and bacterial diversity in the final vermicompost from the biochemical residue of kitchen waste. Environmental Science and Pollution Research 30 (4), 8998–9010. https://doi.org/10.1007/s11356-022-20795-w.

Miller, J.H., Novak, J.T., Knocke, W.R., et al., 2016. Survival of antibiotic resistant bacteria and horizontal gene transfer control antibiotic resistance gene content in anaerobic digesters. Frontiers in Microbiology 7, 263. https://doi.org/10.3389/fmicb.2016.00263.

Ramos, R.F., Santana, N.A., de Andrade, N., et al., 2022. Vermicomposting of cow manure: Effect of time on earthworm biomass and chemical, physical, and biological properties of vermicompost. Bioresource Technology 345, 126572. https://doi.org/10.1016/j.biortech.2021.126572.

Su, J.Q., Wei, B., Ou-Yang, W.Y., et al., 2015. Antibiotic resistome and its association with bacterial communities during sewage sludge composting. Environmental Science and Technology 49 (12), 7356–7363. https://doi.org/10.1021/acs.est.5b01012.

Xia, H., Chen, J., Chen, X., et al., 2019. Effects of tetracycline residuals on humification, microbial profile and antibiotic resistance genes during vermicomposting of dewatered sludge. Environmental Pollution 252, 1068–1077. https://doi.org/10.1016/j.envpol.2019.06.048.

Yang, Y., Li, B., Zou, S., et al., 2014. Fate of antibiotic resistance genes in sewage treatment plant revealed by metagenomic approach. Water research 62, 97–106. https://doi.org/10.1016/j.watres.2014.05.019.

Zhang, J., Lü, F., Shao, L., et al., 2014. The use of biochar-amended composting to improve the humification and degradation of sewage sludge. Bioresource Technology 168, 252–258. https://doi.org/10.1016/j.biortech.2014.02.080.

Zhang, J., Chen, M., Sui, Q., et al., 2016. Impacts of addition of natural zeolite or a nitrification inhibitor on antibiotic resistance genes during sludge composting. Water Research 91, 339–349. https://doi.org/10.1016/j.watres.2016.01.010.

Zhang, L., Gu, J., Wang, X., et al., 2017. Behavior of antibiotic resistance genes during co-composting of swine manure with Chinese medicinal herbal residues. Bioresource Technology 244, 252–260. https://doi.org/10.1016/j.biortech.2017.07.035.

Zhang, R., Gu, J., Wang, X., et al., 2018. Contributions of the microbial community and environmental variables to antibiotic resistance genes during co-composting with swine manure and cotton stalks. Journal of Hazardous Materials 358, 82–91. https://doi.org/10.1016/j.jhazmat.2018.06.052.

Zhang, H., Li, J., Zhang, Y., et al., 2020. Quality of vermicompost and microbial community diversity affected by the contrasting temperature during vermicomposting of dewatered sludge. International Journal of Environmental Research and Public Health 17 (5), 1748. https://doi.org/10.3390/ijerph17051748.

PART
IV

Organic waste

11

Critical influencing factors for decreasing the antibiotic resistance genes during anaerobic digestion of organic wastes

Ananthanarayanan Yuvaraj[1,a], Muniyandi Biruntha[3,a], Natchimuthu Karmegam[2], J. Christina Oviya[4] and Balasubramani Ravindran[5]

[1]*VERMITECHNOLOGY AND ECOTOXICOLOGY LABORATORY, DEPARTMENT OF ZOOLOGY, SCHOOL OF LIFE SCIENCES, PERIYAR UNIVERSITY, SALEM, TAMIL NADU, INDIA;* [2]*PG AND RESEARCH DEPARTMENT OF BOTANY, GOVERNMENT ARTS COLLEGE (AUTONOMOUS), SALEM, TAMIL NADU, INDIA;* [3]*VERMICULTURE TECHNOLOGY LABORATORY, DEPARTMENT OF ANIMAL HEALTH AND MANAGEMENT, ALAGAPPA UNIVERSITY, KARAIKUDI, TAMIL NADU, INDIA;* [4]*DEPARTMENT OF BIOTECHNOLOGY, ST. JOSEPH'S COLLEGE OF ENGINEERING, CHENNAI, TAMIL NADU, INDIA;* [5]*DEPARTMENT OF ENVIRONMENTAL ENERGY AND ENGINEERING, KYONGGI UNIVERSITY YEONGTONG-GU, SUWON, GYEONGGI-DO, REPUBLIC OF KOREA*

1. Introduction

In the past several decades, there has been a rapid generation of various organic waste materials, leading to environmental pollution and disposal problems. Due to the fast development of various modernization activities, different organic solid waste materials are released unethically into the environment. Notable organic waste materials such as agricultural waste, municipal solid waste, household food waste, and different animal wastes are produced worldwide. For example, 2.01 billion tons of municipal waste are produced, and it has been predicted that 3.4 billion tons of municipal waste will be generated each year by 2050 (Chen et al., 2020; Wainaina et al., 2020). The annual production of municipal solid waste in India is approximately 62 million tons; however, only 20% of the waste materials are treated (CPCB, 2017–18). Over the course of the past

[a]Equal contributions.

Occurrence and Behavior of Emerging Contaminants in Organic Wastes and Their Control Strategies
https://doi.org/10.1016/B978-0-443-13585-9.00011-2

few decades, these waste materials have been rapidly incinerated or dumped on land (Mussatto et al., 2012). Nevertheless, the incineration method is one of the highly expensive disposal techniques that generates serious air pollution. Rapid dumping of various organic waste materials in agricultural lands is broken down by various soil microbial communities, resulting in the release of leachate-containing pollutants that migrate into water bodies during rainy seasons and create water pollution (Eco-Cycle, USA, 2011). On the other hand, the microbial breakdown of organic solid waste materials releases a great quantity of greenhouse gases like methane and others (Sánchez et al., 2015). Unscientific disposal practices of organic waste materials lead to environmental contamination, leading to the deterioration of the environment and human health (Giuntini et al., 2006). However, rich amounts of essential proteins, carbohydrates, and mineral nutrients in different organic solid wastes trigger the growth of microbial populations.

Recently, various antibiotic resistance genes (ARGs) have been detected in different organic solid waste materials, viz., livestock manure, municipal solid waste, activated sludge, and food waste materials (Cui et al., 2020). In fact, about 284 ARGs with a relative abundance of 0.08–3.07 gene copies/16S rRNA were recorded in livestock manure, including cattle and chicken manure. Chicken manure contains a rich amount of essential macronutrients and various ARGs. Most of the farmers apply chicken manure for crop production, whereby ARGs easily migrate into water bodies and other terrestrial environments, resulting in soil/water contamination. In addition, livestock manure and activated sludge contain a great quantity of ARGs. For example, Su et al. (2015) reported 56 ARGs with 10^{11} gene copies/g (dry substrate) through a polymerase chain reaction. About 25 ARGs with a typical presence of 5.1×10^{11} gene copies/g in sewage sludge are well documented by Liao et al. (2018). Food waste materials contain 4×10^{-4} gene copies/16S rRNA (ARGs), while agricultural soil consists of 10^{-4} gene copies/16S rRNA (Lee et al., 2017). Improper disposal practices of ARGs containing waste materials can enter the food chain, resulting in a deleterious impact on living beings, particularly humans. At present, various waste treatment methods, including anaerobic digestion, composting, and biodrying, have been employed for the effective degradation of organic solid waste materials. Among them, anaerobic digestion (AD) is one of the suitable methods to reduce organic waste materials and ARGs. AD involves various microbial communities such as methane bacteria, acetic acid bacteria, and hydrogen-generating bacteria. Various factors, such as temperature, pH, treatment period, and types of substrate, strongly influence the removal of ARGs during anaerobic digestion. Most environmental workers have reported on the AD of various organic waste materials that impact the reduction of ARGs. However, various environmental factors that influence the decrease of ARGs from organic waste materials are not well addressed. The main objectives of this work are: (i) To investigate the production of organic waste materials and their mitigation via AD; and (ii) to explore the various factors influencing the reduction of ARGs from waste materials.

2. Generation of organic waste materials

Organic garbage is one of the categories of biodegradable waste, and its production may originate from a wide variety of sources (Fig. 11.1).

These waste materials come from a wide variety of animals and plants, especially by means of anthropogenic activities. Higher amounts of organic waste materials are produced by developed countries than under developing countries. For example, European countries generated roughly 25.38 million tons of waste in 2016 (Wainaina et al., 2020). A wide variety of microbial communities and soil invertebrates are capable of efficiently breaking down organic waste materials and lowering the amount of pollutants and ARGs in the environment, which is believed to reduce the impact of these wastes and ARGs.

2.1 Agriculture-related organic wastes

Throughout the crop production, a large amount of organic waste materials (e.g., waste biomass) were produced worldwide. A huge quantity of agricultural-based waste biomass (500 metric tons/year) is reported to be generated in India (Bhuvaneshwari et al., 2019). Particularly, lignocellulose-based materials such as fruit peel, husk, straw, stem, and bagasse are produced during agricultural and agro-based industrial activities. These wastes contain significant amounts of lignin (25%–30%), hemicellulose (25%–30%), and cellulose (35%–50%), as documented by Behera and Ray (2016) and described in Table 11.1.

The phenylpropane components are responsible for the development of the complex lignin contents, where hemicellulose is one of the polymerlike structures made up of five different sugar constituents, whereas glucose is the accountable sugar in cellulose

FIGURE 11.1 Organic waste originates from a wide range of sources.

Table 11.1 Various properties (%) of agricultural waste materials (overall biomass).

Agricultural waste	Hemicellulose	Cellulose	Lignin	Others	References
Barely straw	25.7	37.8	18.9	17.6	Murray et al. (2010)
Corn stover	26	30.6	29	14.4	Eylen et al. (2011)
Rapeseed straw	11	27.9	14.5	46.6	Wi et al. (2011)
Rice straw	25.4	15	10.6	49	Kumar et al. (2013)
Miscanthus	21.2	45.3	25	1.5	Kang et al. (2013)
Wheat straw	29.2	49.2	16.3	4.3	Amini et al. (2014)

materials (Mussatto et al., 2012). Based on the available evidences agricultural-based organic waste materials are composed of various components that have the ability to produce different types of valuable enzymes. However, the burning of agricultural waste materials is currently practiced by the majority of farmers. These kinds of activities are significant contributors to serious air pollution and are deleterious to the environment, as they are also responsible for causing large quantities of inorganic salts to bio-accumulate in the soil system, which ultimately leads to contamination of the soil (Kapoor et al., 2020). Therefore, it is a pressing necessity for the treatment methods of degrading organic agricultural waste materials.

2.2 Municipal organic waste

At present, the rapid development of the human population worldwide and urbanization lead to an increase in the release of municipal organic waste every year. Municipal solid waste comprises waste originating from gardens, various parks, kitchens, and other sources (Al Seadi et al., 2013). The release and composition of municipal solid waste materials are based on season, local food habits, social conditions, and geographical region (Hansen et al., 2007a,b; VALORGAS, 2010). The annual generation of municipal solid wastes around the world is estimated to be 1300 million tons, and it is anticipated that by the year 2025, the production will have increased to 2200 million tons annually (Al Seadi et al., 2013). In fact, the treatment of municipal solid waste materials in underdeveloped countries presents very substantial challenges, especially in periurban and urban regions (Diener et al., 2011). The municipal waste includes a number of parameters, such as the total contents of nitrogen, potassium, and phosphorus, as well as other parameters like solids and humidity. These criteria are necessarily required for the accurate prediction of the existence of organic materials and their nutritional status in order to effectively practice decomposition. The composition of nutritive components in municipal waste plays a significant role in assessing the probability of recovering essential nutrients (Jansen et al., 2004; Buffiere et al., 2006). During the course of the past few years, a significant amount of municipal waste materials have been placed in landfills, which has led to considerable environmental problems, including soil and water contamination. By this means, organic wastes are eliminated in an untreated state or by incineration at an appropriate temperature. These procedures have only been used to reduce waste volume, but vital nutrients may also be eliminated (Thygesen et al., 2021). Due to pollution concerns, several countries have banned the disposal of waste in landfills. For instance, the European Union can be pointed out in promoting recycling and reusing of waste resources (EC Council, 2018).

2.3 Industrial related organic wastes

In recent decades, rapid industrialization has led to the generation of a vast quantity of organic waste materials. The main industrial waste materials are activated sludge, industrial food waste, slaughterhouse waste, plant waste from industry, and others (Zhang et al., 2020). Worldwide, there is a massive amount of waste production related to green and food waste, which contributes to 44% of total waste materials, as documented by Kaza et al. (2018). In fact, the treatment of industrial wastes is one of the most important challenges in both developed and developing countries, as these wastes contain a large quantity of various hazardous components. Most of these wastes were generated by the urban population compared to the nonurban areas. As a consequence, high rates of urbanization may appear in many developed countries, leading to greater rates of waste production per capita. Over 34% of the world's waste is produced by developed countries, while only 5% is released by underdeveloped countries (Kaza et al., 2018). It is anticipated that the amount of waste generated in developing countries will increase by a three-fold amount by the year 2050. In general, the consumption of raw materials during the initial stage of production, the utilization of waste, and the process of recycling are all playing a role in the development of industrial waste. These waste materials consist of metals, plastic substances, broken glass, leather, rubber, and others (Cointreau et al., 2006). Currently, waste management techniques are facing challenges and improper waste treatment procedures that can lead to the contamination of air, soil, and water (Boruah et al., 2020).

3. Antibiotic resistance genes in organic waste materials

In recent years, there has been a rapid increase in the resistance capacity of various pathogenic microorganisms to different beneficial antibiotics, which has heightened the importance of antimicrobial treatment practices. By the year 2050, it is anticipated that antibiotic-resistant diseases caused by microbes will be responsible for the deaths of 10 million individuals annually (O'Neill, 2014). Mainly, microbes that were resistant to antibiotics moved from different animals to people through the food chain. On the other hand, crop development and various human medicines are required to limit the use of various antibiotics and to take multiple severe efforts to prevent antimicrobial resistance in humans and different animals (consumed as food), as documented by several workers (Berendonk et al., 2015; Ter Kuile et al., 2016; Hernando-Amado et al., 2019).

Nonetheless, several antibiotics are not completely absorbed and metabolized by different animals, resulting in a large proportion of antibiotics being eliminated through waste (i.e., urine and feces of the animals). Apart from this, multiple pathways can considerably remove antibiotics, as depicted in Fig. 11.2.

At this point, environmental workers are paying a significant amount of consideration to ARGs due to the fact that these genes produce a variety of problems that affect both humans and the environment. According to Lee et al. (2017) findings, antibiotic-resistant genes can be found in substantial quantities in food waste products. Qian et al. (2018) reported that various ARGs (e.g., *erm*Q, *tet*C, aac(6)-Ib-cr, *tet*X, and *tet*G) were detected in cattle manure. Hu et al. (2019) provided evidence that swine manure contained a variety of ARGs. According to the research done by Munir et al. (2011), sewage sludge contains a substantial amount of ARGs such as *tet*W and *sul*1. Importantly, the horizontal spread of antibiotic-resistant genes across bacteria in the environment is facilitated greatly by plasmids (Vrancianu et al., 2019; Liu et al., 2020). The spread of plasmids and ARGs among discharged and native bacterial populations in activated sludge was strongly suggested by Soda et al. (2008).

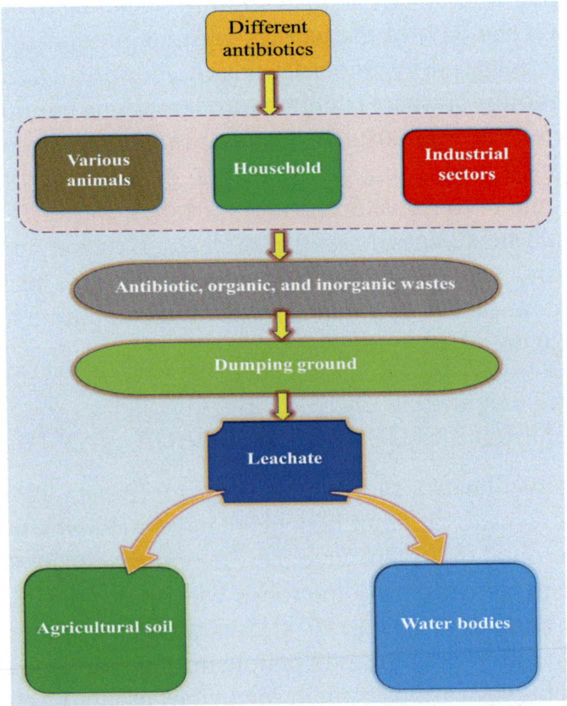

FIGURE 11.2 Antibiotics are obtained from a variety of sources. *Adopted from Anand, U., Reddy, B., Singh, V.K., Singh, A.K., Kesari, K.K., Tripathi, P., Kumar, P., Tripathi, V., Simal-Gandara, J., 2021. Potential environmental and human health risks caused by antibiotic-resistant bacteria (ARB), antibiotic resistance genes (ARGs) and emerging contaminants (ECs) from municipal SolidWaste (MSW) landfill. Antibiotics 10, 374.*

4. Anaerobic digestion of organic waste materials

Due to the rapid production of organic waste products, which were eventually dumped on agricultural soil, both the water and soil systems became contaminated. Among the waste management techniques, AD is one of the most effective methods for reducing organic waste and harmful contaminants (Ariunbaatar et al., 2014). The AD process contains various microorganisms to develop the different valve-added materials. Further, it is divided into four main stages: (i) The hydrolysis process; (ii) acidogenesis; (iii) acetogenesis; and (iv) methanogenesis (Fig. 11.3).

(i) Hydrolysis process

In the course of the hydrolysis process, the various microbial communities significantly contribute to the hydrolysis of organic polymers like lipids, carbohydrates, and proteins into glycerol, sugars, and amino acids (Deng et al., 2014). The various microbes, including *Micrococcus, Bacteriodes, Clostridium, Staphylococcus, Proteus vulgaris*, and *Bacillus*, can effectively release extracellular enzymes such as cellobiase, xylanase, cellulose, lipase, protease, and amylase that degrade macromolecules, as documented by Tiehm et al. (2001). For example, lipase degrades lipid-based materials, beta-glucosidase, xylanase, and cellulose decompose polysaccharides, and protease-related enzymes degrade proteins (Azman et al., 2015; Mshandete et al., 2005). The Stickland process (or

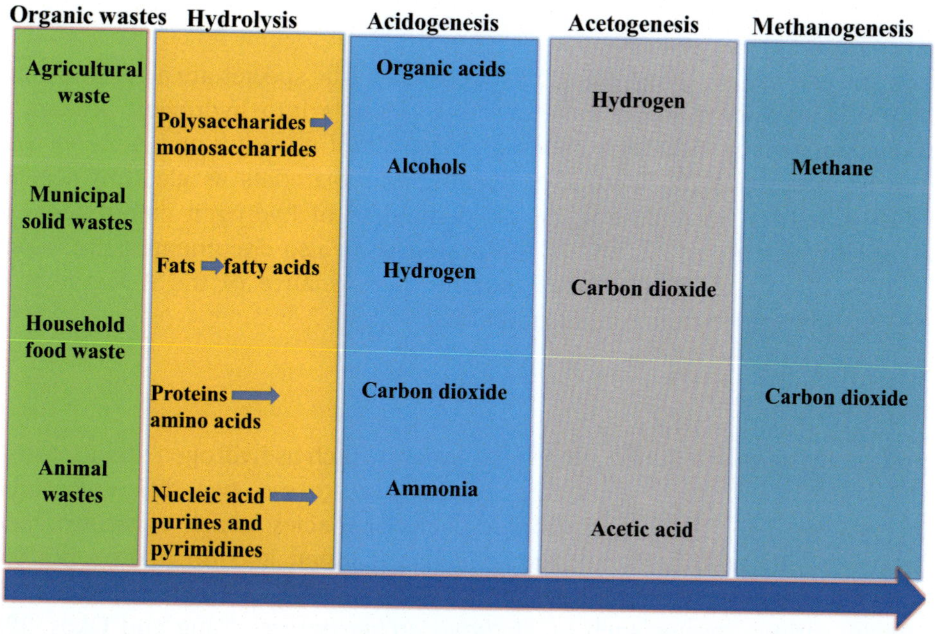

FIGURE 11.3 The overall course of AD. *Adapted from Ravindran, B., Karmegam, N., Yuvaraj, A., Thangaraj, R., Chang, S.W., Zhang, Z., Awasthi, M.K., 2021. Cleaner production of agriculturally valuable benignant materials from industry generated bio-wastes: a review. Bioresource Technology 320 (2021) 124281.*

oxidation process) is interestingly implicated with protease enzyme activity. Include converting amino acids into organic acids through reduction and oxidation, with one amino acid acting as an electron acceptor and another as an electron donor during the Stickland process (Thanarasu et al., 2022). The study by Klimiuk et al. (2010) confirmed that volatile carboxylic acids can be produced via the oxidation process. In general, hydrolysis may be affected by strong chemical interactions between hemicellulose and cellulose in lignocellulosic materials. Furthermore, the presence of lignin prevents waste products from enzymatic activity.

(ii) Acidogenesis

Some microorganisms (e.g., *Eubacterium limosum, Desulforomonas, Sarcina, Escherichia coli, Staphylococcus, Lactobacillus, Desulfobacter, Desulforomonas, Veillonella,* and *Veillonella*) participate actively in the acidogenesis process, which is necessary for the breakdown of the end products of the hydrolysis process (Peng et al., 2020; Juliana et al., 2020). Various materials, such as alcohol, valerate, lactate, hydrogen, and carbon dioxide, are rapidly formed during acidogenesis (Kim et al., 2017). But the deposition of high amounts of volatile fatty acids leads to a reduction in the process (Thanarasu et al., 2022). According to findings released by Komemoto et al. (2009), there is compelling evidence that enormous quantities of lipids in substrate materials efficiently produce ammonia and ammonium, which inhabit the breakdown process.

(iii) Acetogenesis

During the process of acetogenesis, acetogenic bacterial species are actively involved. These bacteria transform acidic materials and alcohols into hydrogen, acetate, and carbon dioxide. Notable acetogens, such as *Clostridium aceticum* and *Acetobacterium woodii* (Srivastava et al., 2020), use acidogenic end materials as electron acceptors (Rawoof et al., 2021). In addition, the rapid increase of hydrogen components that severally modify the metabolic rate of the acetogens is also documented by Tsapekos et al. (2017). Kim et al. (2006) found that the performance of the AD system was enhanced by optimal environmental conditions.

(iv) Methanogenesis

Important byproducts of the acetogenesis process, such as hydrogen and acetate, are utilized by methanogenic bacterial species. In this process, two different kinds of methanogens are involved: (i) Acetotrophic bacterial species, which transform acetate into carbon dioxide and methane in an effective manner; and (ii) hydrogenotrophic bacteria, which utilize the hydrogen to produce methane (Rawoof et al., 2021). Interestingly, various species such as *Methanosarcina barkeri* (Yang and Okos, 1987), *Methanosarcina* sp. MSTA-1 (Clarens and Moletta, 1990), *Methanosarcina* spp.

(Mladenovska and Ahring, 2000), *Methanosarcina thermophile* (Zinder et al., 1985), *Methanosarcina* CHTI55 (Touzel et al., 1985), *Methanospirillum hungatei* (Robinson and Tiedje, 1984), and *Methanomicrobium paynteri* (Dubach and Bachofen, 1985) can produce a significant quantity at the end of the process. Acetate was responsible for the vast majority of methane production, with carbon dioxide and hydrogen accounting for the remaining amounts (Angelidaki et al., 2011).

4.1 Reduction of ARGs by anaerobic digestion and different influencing factors

Various microorganisms are actively involved in the AD process and have a major influence on ARGs. The AD of organic waste materials with ARGs has been found to be drastically reduced at the end of the process. For example, mesophilic AD of cattle manure for 60 days led to a dramatic decrease in ARGs (e.g., *aph*A2, *erm*B, and $bla_{TEM\text{-}1}$), as reported by Resende et al. (2014). Diehl et al. (2010) observed that AD of activated sludge can significantly reduce ARGs such as *tet*Q, *tet*L, *tet*X, *tet*A, and *tet*W. Similarly, Zhang et al. (2016) determined that the combination of food waste materials and activated sludge in an AD process led to a rapid reduction of ARGs. The temperature of the anaerobic system is critical for the efficient elimination of ARGs. For instance, Zhang et al. (2018) demonstrated that ARGs were more effectively eliminated during the AD process of swine manure at 55°C than at 37°C.

In fact, a number of factors play a significant role in the process of AD, which leads to a rapid reduction of ARGs. The notable factors can be categorized into four groups: (i) Composition of substrate materials (i.e., microbes, metals, and level of solidity), as recommended by Miller et al. (2016), Yin et al. (2017), and Sun et al. (2019). In fact, the microbial population of organic waste materials can have a significant impact on the reduction of ARGs. Miller et al. (2016) showed that AD of sludge caused a quick fall in ARGs and arrived at the conclusion that the microbial population had a significant impact on this decline. In addition, Zhang et al. (2019) showed that AD of diverse biowaste materials, which results in a variety of chemical characteristics, has a significant impact on combating the ARGs that are present in the waste materials. (ii) Preprocessing (e.g., thermal process and alteration of pH of the substrate) of waste materials is required to reduce various ARGs found in organic waste materials (Huang et al., 2017; Tong et al., 2018). The system temperature plays a crucial role in the degradation of ARGs prior to the AD process (Cui et al., 2020). Further, (iii) group-related factors like different amendment materials such as biochar (Sun et al., 2018), natural zeolite (Zhang et al., 2018), and red mud (Wang et al., 2016). The various amendment materials can trigger the degradation of ARGs during anaerobic digestion. According to Sun et al. (2018), adding biochar to the AD of cattle manure reduced the amount of resistance genes. (iv) AD system factors such as process period (Wu et al., 2016) and temperature (Huang et al., 2019; Sun et al., 2016). Anaerobic system conditional parameters can play an essential

FIGURE 11.4 Several strategies for enhancing the decomposition of ARGs within the AD system.

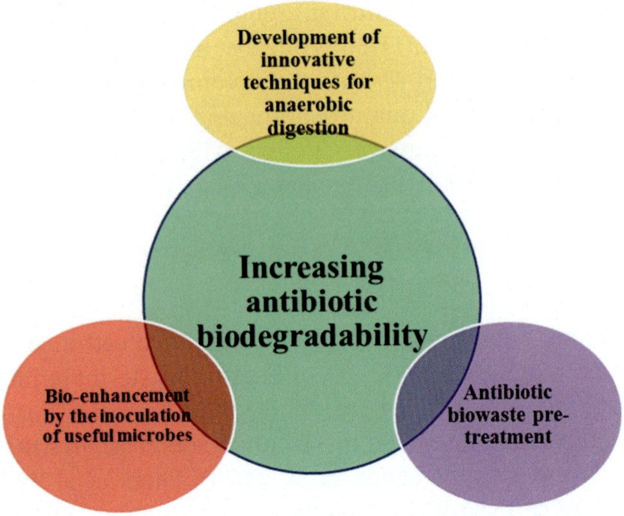

role in the AD of diverse organic waste products while also reducing ARGs. The study by Liao et al. (2018) concluded that a significant amount of ARGs were eliminated at 90°C while compared to the 49% and 89% composting units. In addition to this, the authors strongly advocate a variety of beneficial strategies, as described in Fig. 11.4.

An ideal condition of the anaerobic system and the presence of distinct microbial communities could trigger the antibiotic ARGs.

5. Conclusions

The tremendous growth of the human population and the industrial sector in the last several decades has resulted in the emission of enormous quantities of organic waste. There is a high concentration of ARGs in these wastes. Improper disposal procedures might produce major environmental threats. For instance, the majority of organic waste items that are dumped on agricultural lands and ARGs have the potential to enter the soil system, which would result in contamination of the soil. The accumulation of ARGs in crops can then impact humans via the food chain. ARGs can be effectively combated by the AD of organic waste. During the AD process, several different processes, including hydrolysis, acidogenesis, acetogenesis, and methanogenesis, are carried out in an effective manner. Further, multiple factors such as substrate material composition, preprocess, amendment materials, and process period effectively contribute to the reduction of ARGs during AD. In addition, more research demands to be done on the AD of organic wastes, and there are several factors that must be studied to combat ARGs.

References

Al Seadi, T., Owen, N., Hellström, H., Kang, H., 2013. Source Separation of MSW: An Overview of the Source Separation and Separate Collection of the Digestible Fraction of Household Waste, and Other Similar Wastes from Municipalities, Aimed to Be Used as Feedstock for Anaerobic Digestion in Biogas Plants. IEA Bioenergy Task 37 Energy from Biogas Report.

Amiri, H., Karimi, K., Zilouei, H., 2014. Organosolv pre treatment rice straw for efficient acetone, butanol, and ethanol production. Bioresource Technology 152, 450–456.

Angelidaki, I., Karakashev, D., Batstone, D.J., Plugge, C.M., Stams, A.J.M., 2011. Biomethanation and its potential. In: Rosenzweig, A.C., Ragsdale, S.W. (Eds.), Methods in Enzymology: Methods in Methane Metabolism, first ed., vol 494. Elsevier Academic Press, pp. 327–351.

Ariunbaatar, J., Panico, A., Esposito, G., Pirozzi, F., Lens, P.N.L., 2014. Pretreatment methods to enhance anaerobic digestion of organic solid waste. Applied Energy 123, 143–156.

Azman, S., Khadem, A.F., Vaner, J.B., Zeeman, G., Plugge, C.M., 2015. Presence and role of anaerobic hydrolytic microbes in conversion of lignocellulosic biomass for biogas production. Critical Reviews in Environmental Science and Technology 3389, 2523–2564.

Behera, S.S., Ray, R.C., 2016. Solid state fermentation for production of microbial cellulases: recent advances and improvement strategies. International Journal of Biological Macromolecules 86, 656–669.

Berendonk, T.U., Manaia, C.M., Merlin, C., Fatta-Kassinos, D., Cytryn, E., Walsh, F., Bürgmann, H., Sørum, H., Norström, M., Pons, M.-N., Kreuzinger, N., Huovinen, P., Stefani, S., Schwartz, T., Kisand, V., Baquero, F., Martinez, J.L., 2015. Tackling antibiotic resistance: the environmental framework. Nature Reviews Microbiology 13, 310.

Bhuvaneshwari, S., Hettiarachchi, H., Meegoda, J.N., 2019. Crop residue burning in India: policy challenges and potential solutions. International Journal of Environmental Research and Public Health 16, 832.

Boruah, T., Morang, A., Deka, H., 2020. Current understanding on industrial organic waste management by employing *Eisenia fetida*. International Journal of Environment and Pollution 67 (No. 1).

Buffiere, P., Loisel, D., Bernet, N., Delgenes, J.P., 2006. Towards new indicators for the prediction of solid waste anaerobic digestion properties. Water Science and Technology 53 (8), 233–241.

Central Pollution Control Board (CPCB), 2017–18. Annual Report. Ministry of Environment, Forest, and Climate Change. Government of India.

Chen, T., Zhang, S., Yuan, Z., 2020. Adoption of solid organic waste composting products: a critical review. Journal of Cleaner Production 272, 122712.

Clarens, M., Moletta, R., 1990. Kinetic studies of acetate fermentation by Methanosarcina sp. MSTA-1. Applied Microbiology and Biotechnology 33, 239–244.

Cointreau, S., 2006. Occupational and environmental health issues of solid waste management: special emphasis on middle-and lower-income countries. Urban Papers 1 (2), 1–48.

Cui, G., Lü, F., Zhang, H., Shao, L., He, P., 2020. Critical Insight into the Fate of Antibiotic Resistance Genes During Biological Treatment of Typical Biowastes. Bioresource Technology.

Deng, L., Yang, H., Liu, G., Zheng, D., Chen, Z., Liu, Y., Pu, X., Song, L., Wang, Z., Lei, Y., 2014. Kinetics of temperature effects and its significance to the heating strategy for anaerobic digestion of swine wastewater. Applied Energy 134, 349–355.

Diehl, D.L., LaPara, T.M., 2010. Effect of temperature on the fate of genes encoding tetracycline resistance and the integrase of class 1 integrons within anaerobic and aerobic digesters treating municipal wastewater solids. Environmental Science & Technology 44 (23), 9128–9133.

Diener, S., Zurbrugg, C., Roa Gutiérrez, F., Nguyen, H.D., Morel, A., Koottatep, T., Tockner, K., 2011. Black soldier fly larvae for organic waste treatment—prospects and constraints. Paper Presented at the WasteSafe 2011, 2nd International Conference on Solid Waste Management in Developing Countries, Khulna, Bangladesh, 13–15 February.

Dubach, A.C., Bachofen, R., 1985. Methanogens: a short taxonomic review. Experentia 41, 441–446.

EC. Council, 2018. Establishing Best Available Techniques (BAT) Conclusions for Waste Treatment, Under Directive 2010/75/EU of the European Parliament and of the Council.

Eco-Cycle, 2011. Waste of Energy-Why Incineration Is Bad for Our Economy. Environment and Community; Eco-Cycle, Boulder County, CO, USA, pp. 1–20.

Eylen, D.V., Dongen, F.V., Kabel, M., Bont, J.A., 2011. Corn fiber, cobs and stover: enzyme- aided saccharification and co-fermentation after dilute acid pretreatment. Bioresource Technology 102, 5995–6004.

Giuntini, E., Bazzicalupo, M., Castaldini, M., Fabiani, A., Miclaus, N., Piccolo, R., Ranalli, G., Santomassimo, F., Zanobini, S., Mengoni, A., 2006. Genetic diversity of dinitrogen-fixing bacterial communities in soil amended with olive husks. Annals of Microbiology 56, 83–88.

Hansen, T.L., Jansen, J.C., Davidsson, A., Christensen, T.H., 2007a. Effects of pretreatment technologies on quantity and quality of source-sorted municipal organic waste for biogas recovery. Waste Management 27 (3), 398–405.

Hansen, T.L., Jansen, J.C., Spliid, H., Davidsson, A., Christensen, T.H., 2007b. Composition of source sorted municipal organic waste collected in Danish cities. Waste Management 27 (4), 510–518.

Hernando-Amado, S., Coque, T.M., Baquero, F., Martínez, J.L., 2019. Defining and combating antibiotic resistance from one health and global health perspectives. Nature Microbiology 4, 1432–1442.

Hu, T., Wang, X., Zhen, L., Gu, J., Zhang, K., Wang, Q., Ma, J., Peng, H., 2019. Effects of inoculation with lignocellulose-degrading microorganisms on antibiotic resistance genes and the bacterial community during co-composting of swine manure with spent mushroom substrate. Environment and Pollution 252, 110–118.

Huang, H., Zheng, X., Chen, Y., Liu, H., Wan, R., Su, Y., 2017. Alkaline fermentation of waste sludge causes a significant reduction of antibiotic resistance genes in anaerobic reactors. Science of the Total Environment 580, 380–387.

Huang, X., Zheng, J., Tian, S., Liu, C., Liu, L., Wei, L., Fan, H., Zhang, T., Wang, T., Wang, L., Zhu, G., Xu, K., 2019. Higher temperatures do not always achieve better antibiotic resistance gene removal in anaerobic digestion of swine manure. Applied and Environmental Microbiology 85 (7) e02878-18.

Jansen, J.C., Spliid, H., Hansen, T.L., Svärd, A., Christensen, T.H., 2004. Assessment of sampling and chemical analysis of source-separated organic household waste. Waste Management 24 (6), 541–549.

Juliana, F.S., Tássia, C.C., Izelmar, T., Flávio, D.M., Marcio, A.M., 2020. Dark fermentative biohydrogen production from lignocellulosic biomass: technological challenges and future prospects. Renewable and Sustainable Energy Reviews 117, 109484.

Kapoor, R., Ghosh, P., Kumar, M., Sengupta, S., Gupta, A., Kumar, S.S., Vijay, V., Kumar, V., Vijay, V.V., Pant, D., 2020. Valorization of agricultural waste for biogas based circular economy in India: a research outlook. Bioresource Technology 304 (2020), 123036.

Kang, K.E., Han, M., Moon, S.K., Kang, H.W., Kim, Y., Cha, Y.L., et al., 2013. Optimization of alkali-extrusion pretreatment with twin-screw for bioethanol production from miscanthus. Fuel 109, 520–526.

Kaza, S., Yao, L., Bhada-Tata, P., Van Woerden, F., 2018. What a Waste 2.0: A Global Snapshot of Solid Waste Management to 2050. World Bank Publications.

Kim, J.K., Oh, B.R., Chun, Y.N., Kim, S.W., 2006. Effects of temperature and hydraulic retention time on anaerobic digestion of food waste. Journal of Bioscience and Bioengineering 102, 328–332.

Kim, M.S., Kim, D.H., Yun, Y.M., 2017. Effect of operation temperature on anaerobic digestion of food waste: performance and microbial analysis. Fuel 209, 598–605.

Klimiuk, E., Pokój, T., Budzyński, W., Dubis, B., 2010. Theoretical and observed biogas production from plant biomass of different fibre contents. Bioresource Technology 101, 9527–9535.

Komemoto, K., Lim, Y.G., Nagao, N., Onoue, Y., Niwa, C., Toda, T., 2009. Effect of temperature on VFA's and biogas production in anaerobic solubilization of food waste. Waste Management 29, 2950–2955.

Kumar, S., Gupta, R., Kumar, G., Sahoo, D., Kuhad, R.C., 2013. Bioethanol production from Gracilaria verrucosa, a red alga, in a biorefinery approach. Bioresource Technology 135, 150–156.

Lee, J., Shin, S.G., Jang, H.M., Kim, Y.B., Lee, J., Kim, Y.M., 2017. Characterization of antibiotic resistance genes in representative organic solid wastes: food wasterecycling wastewater, manure, and sewage sludge. Science of the Total Environment 579, 1692–1698.

Liao, H., Lu, X., Rensing, C., Friman, V.P., Geisen, S., Chen, Z., Yu, Z., Wei, Z., Zhou, S., Zhu, Y., 2018. Hyperthermophilic composting accelerates the removal of antibiotic resistance genes and mobile genetic elements in sewage sludge. Environmental Science & Technology 52 (1), 266–276.

Liu, Y., Tong, Z., Shi, J., Jia, Y., Yang, K., Wang, Z., 2020. Correlation between exogenous compounds and the horizontal transfer of plasmid-borne antibiotic resistance genes. Microorganisms 8, 1211.

Miller, J.H., Novak, J.T., Knocke, W.R., Pruden, A., 2016. Survival of antibiotic resistant bacteria and horizontal gene transfer control antibiotic resistance gene content in anaerobic digesters. Frontiers in Microbiology 7, 263.

Mladenovska, Z., Ahring, B.K., 2000. Growth kinetics of thermophilic Methanosarcina spp. isolated from full-scale biogas plants treating animal manures. FEMS Microbiology Ecology 31 (3), 225–230.

Mshandete, A., Bjornsson, L., Kivaisi, A.K., Rubindamayugi, S.T., Mattiasson, B., 2005. Enhancement of anaerobic batch digestion of sisal pulp waste by mesophilic aerobic pre-treatment. Water Research 39, 1569–1575.

Munir, M., Wong, K., Xagoraraki, I., 2011. Release of antibiotic resistant bacteria and genes in the effluent and biosolids of five wastewater utilities in Michigan. Water Research 45 (2), 681–693.

Murray, D., Parsons, S.A., Jarvis, P., Jefferson, B., 2010. The impact of barley straw conditioning on the inhibition of scenedesmus using chemostats. Water Research 44, 1373–1380.

Mussatto, S.I., Ballesteros, L.F., Martins, S., Teixeira, J.A., 2012. Use of agro-industrial wastes in solid-state fermentation processes. In: Industrial Waste. InTech, Rijeka, Croatia, pp. 121–141.

O'Neill, J., 2014. Antimicrobial Resistance: Tackling a Crisis for the Health and Wealth of Nations. The Review on Antimicrobial Resistance. https://amr-review.org/sites/default/files/AMR%20Review%20Paper%20-%20Tackling%20a%20crisis%20for%20the%20health%20and%20wealth%20of%20nations_1.pdf.

Peng, L., Fu, D., Chu, H., Wang, Z., Qi, H., 2020. Biofuel production from microalgae: a review. Environmental Chemistry Letters 18, 285–329.

Qian, X., Gu, J., Sun, W., Wang, X., Su, J., Stedfeld, R., 2018. Diversity, abundance, and persistence of antibiotic resistance genes in various types of animal manure following industrial composting. Journal of Hazardous Materials 344, 716–722.

Rawoof, S.A.A., Senthil Kumar, P., Vo, D.V.N., Subramanian, S., 2021. Sequential production of hydrogen and methane by anaerobic digestion of organic wastes: a review. Environmental Chemistry Letters 19 (2021), 1043–1063.

Resende, J.A., Diniz, C.G., Silva, V.L., Otenio, M.H., Bonnafous, A., Arcuri, P.B., Godon, J.J., 2014. Dynamics of antibiotic resistance genes and presence of putative pathogens during ambient temperature anaerobic digestion. Journal of Applied Microbiology 117 (6), 1689–1699.

Robinson, J.A., Tiedje, J.M., 1984. Competition between sulfate-reducing and methanogenic bacteria for H2 under resting and growing conditions. Archives of Microbiology 137, 26–32.

Sánchez, A., Artola, A., Font, X., Gea, T., Barrena, R., Gabriel, D., Sánchez-Monedero, M.Á., Roig, A., Cayuela, M.L., Mondini, C., 2015. Greenhouse gas emissions from organic waste composting. Environmental Chemistry Letters 13, 223–238.

Soda, S., Otsuki, H., Inoue, D., Tsutsui, H., Sei, K., Ike, M., 2008. Transfer of antibiotic multiresistant plasmid RP4 from Escherichia coli to activated sludge bacteria. Journal of Bioscience and Bioengineering 106, 292–296.

Srivastava, R.K., Shetti, N.P., Reddy, K.R., Aminabhavi, T.M., 2020. Biofuels, biodiesel and biohydrogen production using bioprocesses. Environmental Chemistry Letters, A Review 18, 1049–1072.

Su, J., Wei, B., Ouyang, W., Huang, F., Zhao, Y., Xu, H., Zhu, Y., 2015. Antibiotic resistome and its association with bacterial communities during sewage sludge composting. Environmental Science & Technology 49 (12), 7356–7363.

Sun, W., Gu, J., Wang, X., Qian, X., Peng, H., 2019. Solid-state anaerobic digestion facilitates the removal of antibiotic resistance genes and mobile genetic elements from cattle manure. Bioresource Technology 274, 287–295.

Sun, W., Gu, J., Wang, X., Qian, X., Tuo, X., 2018. Impacts of biochar on the environmental risk of antibiotic resistance genes and mobile genetic elements during anaerobic digestion of cattle farm wastewater. Bioresource Technology 256, 342–349.

Sun, W., Qian, X., Gu, J., Wang, X.-J., Duan, M.-L., 2016. Mechanism and effect of temperature on variations in antibiotics resistance genes during anaerobic digestion of dairy manure. Scientific Reports 6, 30237.

Ter Kuile, B.H., Kraupner, N., Brul, S., 2016. The risk of low concentrations of antibiotics in agriculture for resistance in human health care. FEMS Microbiology Letters 363.

Thanarasu, A., Periyasamy, K., Subramanian, S., 2022. An integrated anaerobic digestion and microbial electrolysis system for the enhancement of methane production from organic waste: fundamentals, innovative design and scale-up deliberation. Chemosphere 287 (2022), 131886.

Thygesen, A., Tsapekos, P., Alvarado-Morales, M., Angelidaki, I., 2021. Valorization of municipal organic waste into purified lactic acid. Bioresource Technology 342 (2021), 125933.

Tiehm, A., Nickel, K., Zellhorn, M., Neis, U., 2001. Ultrasonic waste activated sludge disintegration for improving anaerobic stabilization. Water Research 35, 2003–2009.

Tong, J., Lu, X., Zhang, J., Angelidaki, I., Wei, Y., 2018. Factors influencing the fate of antibiotic resistance genes during thermochemical pretreatment and anaerobic digestion of pharmaceutical waste sludge. Environment and Pollution 243, 1403–1413.

Touzel, J.P., Petroff, D., Albagnac, G., 1985. Isolation and characterization of a new thermophilic Methanosarcina, the strain CHTI55. Systematic & Applied Microbiology 6, 66–71.

Tsapekos, P., Kougias, P.G., Vasileiou, S.A., Treu, L., Campanaro, S., Lyberatos, G., Angelidaki, I., 2017. Bioaugmentation with hydrolytic microbes to improve the anaerobic biodegradability of lignocellulosic agricultural residues. Bioresource Technology 234, 350–359.

VALORGAS, 2010. Compositional Analysis of Food Waste from Study Sites in Geographically Distinct Regions of Europe. MTT Agrifood Research Finland (Maa Ja Elintarviketalouden Tutkimuskeskus). VALORGAS Project, Finland. http://www.valorgas.soton.ac.uk/deliverables.htm.

Vrancianu, C.O., Popa, L.I., Bleotu, C., Chifiriuc, M.C., 2019. Targeting plasmids to limit acquisition and transmission of antimicrobial resistance. Frontiers in Microbiology 11, 761.

Wainaina, S., Awasthi, M.K., Sarsaiya, S., Chen, H., Singh, E., Kumar, A., Ravindran, B., Awasthi, S.K., Liu, T., Duan, Y., Kumar, S., Zhang, Z., Taherzadeh, M.J., 2020. Resource recovery and circular economy from organic solid waste using aerobic and anaerobic digestion technologies. Bioresource Technology 301, 122778.

Wang, R., Zhang, J., Sui, Q., Wan, H., Tong, J., Chen, M., Wei, Y., Wei, D., 2016. Effect of red mud addition on tetracycline and copper resistance genes and microbial community during the full scale swine manure composting. Bioresource Technology 216, 1049–1057.

Wi, S.G., Chung, B.Y., Lee, Y.G., Yang, D.J., Bae, H.J., 2011. Enhanced enzymatic hydrolysis of rapeseed straw by popping pretreatment for bioethanol production. Bioresource Technology 102, 5788–5793.

Wu, Y., Cui, E., Zuo, Y., Cheng, W., Rensing, C., Chen, H., 2016. Influence of twophase anaerobic digestion on fate of selected antibiotic resistance genes and class I integrons in municipal wastewater sludge. Bioresource Technology 211, 414–421.

Yang, S.T., Okos, M.R., 1987. Kinetic study and mathematical modelling of methanogenesis of acetate using pure cultures of methanogens. Biotechnology and Bioengineering 30 (5), 661–667.

Yin, Y., Gu, J., Wang, X., Song, W., Zhang, K., Sun, W., Zhang, X., Zhang, Y., Li, H., 2017. Effects of copper addition on copper resistance, antibiotic resistance genes, and *intI*1 during swine manure composting. Frontiers in Microbiology 8, 344.

Zhang, J., Chen, M., Sui, Q., Wang, R., Tong, J., Wei, Y., 2016. Fate of antibiotic resistance genes and its drivers during anaerobic co-digestion of food waste and sewage sludge based on microwave pretreatment. Bioresource Technology 217, 28–36.

Zhang, J., Lu, T., Shen, P., Sui, Q., Zhong, H., Liu, J., Tong, J., Wei, Y., 2019. The role of substrate types and substrate microbial community on the fate of antibiotic resistance genes during anaerobic digestion. Chemosphere 229, 461–470.

Zhang, J., Sui, Q., Zhong, H., Meng, X., Wang, Z., Wang, Y., Wei, Y., 2018. Impacts of zero valent iron, natural zeolite and Dnase on the fate of antibiotic resistance genes during thermophilic and mesophilic anaerobic digestion of swine manure. Bioresource Technology 258, 135–141.

Zhang, L., Loh, K.C., Kuroki, A., Dai, Y., Tong, Y.W., 2020. Microbial biodiesel production from industrial organic wastes by oleaginous microorganisms: current status and prospects. Journal of Hazardous Materials 123543. https://doi.org/10.1016/j.jhazmat.2020.123543.

Zinder, S.H., Sowers, K.R., Ferry, J.G., 1985. Methanosarcina thermophila sp. nov., a thermophilic, acetotrophic, methane- producing bacterium. International Journal of Systematic and Evolutionary Microbiology 35, 522–523.

12

Occurrence, behavior, and fate of microplastics in agricultural and livestock wastes and their impact on farmers fields

Sirat Sandil

NATIONAL LABORATORY FOR WATER SCIENCE AND WATER SECURITY, INSTITUTE OF AQUATIC ECOLOGY, CENTRE FOR ECOLOGICAL RESEARCH, BUDAPEST, HUNGARY

1. Introduction

Plastic pollution has become a major environmental problem in recent years. Plastics are low-cost, lightweight, flexible, and highly durable synthetic polymers with applications in copious sectors (Surendran et al., 2023). They are used comprehensively in the packaging industry, production of essential domestic items and personal care products, textile manufacturing, agriculture-based industries like fertilizer, pesticide, mulching sheets, and pipes, and construction of roads (Allouzi et al., 2021). Due to the huge demand for plastics, its production rose from 1.5 MT in the 1950s to 322 MT in 2015 and is expected to further increase to 670 MT by 2040 (Weithmann et al., 2018; Surendran et al., 2023). Despite its numerous benefits, a substantial amount of plastic waste is released, which pollutes the environment. This is due to the fact that 50% of plastics are designed with the intention of single-usage, there are a few end-of-life treatment options available, and there is a dearth of reuse or recycling options. Plastics are difficult to degrade, chemically stable, and can persist in the environment for hundreds of years (Li et al., 2023).

Plastics accounted for 1% of municipal solid waste in the 1960s, which has now increased to more than 10%. About 6300 million tons of plastic were manufactured in 2015, of which around 79% ended up in landfills. Large quantities of improperly disposed plastics have become a major environmental hazard, and if the scenario remains the same in the future, the impact on the environment would be hazardous (Maaβ et al., 2017; Bigalke et al., 2022; Surendran et al., 2023). Plastics have been detected in almost all environmental

Occurrence and Behavior of Emerging Contaminants in Organic Wastes and Their Control Strategies
https://doi.org/10.1016/B978-0-443-13585-9.00001-X

media, including marine, freshwater, terrestrial, and atmospheric environments (Qi et al., 2018; Liu et al., 2018; Huang et al., 2020; Almeshal et al., 2022). Although plastics are not easily degradable, larger ones can break down into smaller ones in the soil through physical crushing, photodegradation, biological activity, etc. Plastics typically fragment into macro- (>2 cm), meso- (5 mm–2 cm), and microplastics (MPs) (<5 mm). Plastic particles with sizes ranging from a few μm to up to 5 mm are defined as MPs (van den Berg et al., 2020; Li et al., 2021; Sahasa et al., 2023). MPs are a major environmental concern because of their small size, which makes the particles more accessible to soil organisms; they can be easily ingested by biota, act as vectors for contaminants and pathogens in the environment, and release additives used during the process of plastic manufacturing (Li et al., 2019; Zhang et al., 2022). To date, the majority of research has focused on plastics and derived MPs in the marine environment, and information regarding the terrestrial environment is scant. This is despite the fact that only 1% of global plastic waste is released directly into the oceans, and the majority of waste is released into the soils and freshwater environments (Zhang and Liu, 2018; Sharma et al., 2023).

Soil aids in a multitude of services, including biogeochemical cycles and carbon sequestration, and acts as a biodiversity host (Abel et al., 2018). It is considered to be the direct and main sink of MPs because it accumulates more MPs than aquatic environments. The amount of plastic released annually into the soil has been calculated to be approximately 4–23 times higher than that released into the marine environment (Gong et al., 2021; Bigalke et al., 2022). In the past decade, a number of studies have reported on the abundance of MPs in agricultural soil: 890–1109 MPs/g (Huerta et al., 2023), 888–2242 MPs/kg (van Schothorst et al., 2021), 7100–42,960 MPs/kg (Zhang and Liu, 2018), and 158,100–292,400 MPs/ha (Piehl et al., 2018). The abundance of MPs in the soil is mainly caused by the usage of plastics in agriculture (Weber et al., 2022).

Agricultural ecosystems are vulnerable to MP contamination due to the manifold sources of plastics employed in agricultural practices. MPs enter the terrestrial environment primarily through the use of plastic film mulch, organic manures, sludge from wastewater treatment plants (WWTPs), irrigation with untreated and treated wastewater, waste dumping, surface runoff, and atmospheric deposition (Yang et al., 2021; Kim et al., 2021; Kumar and Sheela, 2021; Zhang et al., 2020). Agricultural plastic mulching is a significant source of MPs in the terrestrial environment owing to its extensive usage and improper disposal. The plastic film debris is acted upon by various physical, chemical, and biological forces and fragmented into an abundant number of smaller MP particles (Zhang and Liu, 2018; Ding et al., 2020). Another emerging contributor of MP particles in agricultural soils is organic fertilizers, including livestock (cow, goat, sheep, and pig), poultry manure, and domestic waste. They are environment-friendly and are routinely applied for soil improvement as they can regulate soil quality, add humus, nutrients, and trace elements to the soil, and act as crucial sources of plant nutrients (Yang et al., 2021; Zhang et al., 2022). However, organic fertilizers are contaminated with plastics, and MPs remain in the compost even after manual sorting and sieving of plastics (Zhang et al., 2022). The use of sewage sludge as fertilizer is another important source of MPs, as it adds significant amounts of MPs to the agricultural soil (Crossman et al., 2020; Weber et al., 2022). Sludge is the end product of WWTPs, and more than 90%–99% of the MPs removed by WWTPs are retained in the sludge (Prendergast-miller et al., 2019; Corradini et al., 2019). Sludge application can add up to 3.8×10^9 MPs particles to the agricultural field annually (Crossman et al., 2020), and the repeated application of sludge can cause increased accumulation of MPs in the soil (Corradini et al., 2019).

MPs in the environment pose a threat to soil health, microorganisms, plants, and animals (Huerta Lwanga et al., 2017; Lian et al., 2021; Gudeta et al., 2023) (Fig. 12.1). In soil, MPs threaten the soil ecosystem and affect soil productivity, texture, and quality (Zhang and Liu, 2018; Li et al., 2019). MPs alter the soil's physicochemical properties, lower soil pH, inhibit soil enzyme activity, impact nutrient cycling and sorption processes, and compete with microorganisms for ecological niches (Yu et al., 2020a). MPs are easily ingested by soil organisms and result in increased mortality rates, decreased growth, and diminished rates of reproduction (Prendergast-miller et al., 2019; van den Berg et al., 2020; Kim et al., 2021). MPs uptaken by earthworms and nematodes can hinder their development, delay growth, reduce feeding, and cause genotoxicity, oxidative stress, and even death (Rout et al., 2022; Huerta et al., 2023). In plants, they affect the roots and plant growth, prevent seed germination, reduce plant height, decrease crop yield, and inhibit nutrient uptake (Gong et al., 2021; Rout et al., 2022; Pinto-Poblete et al., 2023). In animals, ingestion of MPs affects the digestive system, causes intestinal obstruction, stunts development, disrupts reproduction, causes weight loss, affects the nervous system, alters gut microbes, and increases mortality (Beriot et al., 2021; Yuan et al., 2022). MPs can also pose exposure risks to human beings through

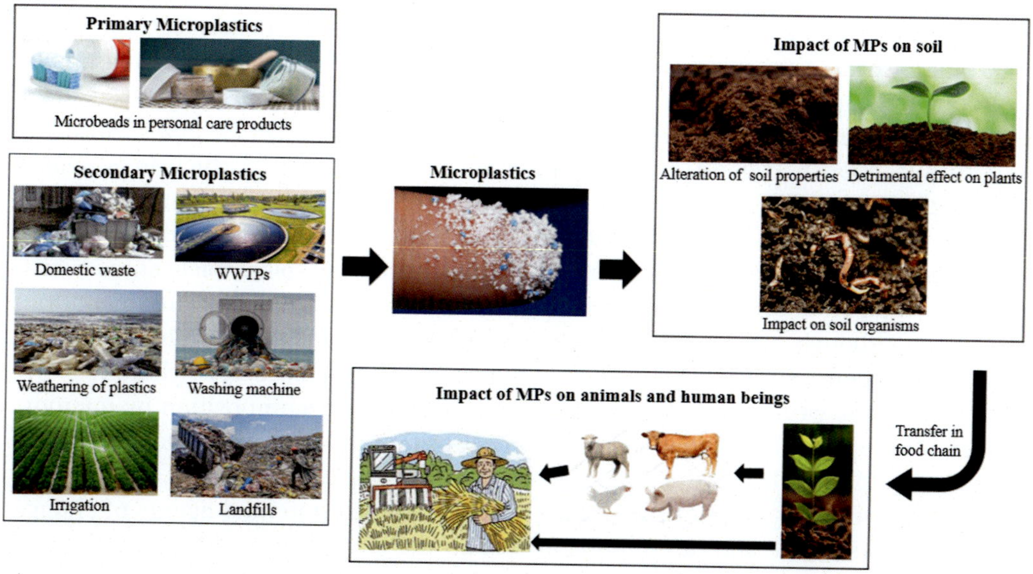

FIGURE 12.1 Sources and impacts of microplastics in the environment.

accumulation and translocation in the food chain (Li et al., 2019; Huerta Lwanga et al., 2017). In human beings, MPs may cause inflammation and stress responses, modify the gut microbiome, alter digestive functions, and cause cancer (Beriot et al., 2021).

Additionally, in the agricultural fields, MPs release chemical pollutants, including phthalate plasticizers, antioxidants, polybrominated diphenyl ether flame retardants, and bisphenols (Kumar and Sheela, 2021). Due to their small size, high bioavailability, and large surface area-to-volume ratio, they act as potential vectors for contaminants like heavy metals (HMs), organic pollutants, pharmaceuticals, and pathogens. MPs can thus affect the transportation and transformation of other contaminants in the terrestrial environment and enable their transfer to soil organisms, causing severe toxic effects (Wang et al., 2020; Yu et al., 2020b; Kumar and Sheela, 2021; Du et al., 2023; Satarupa Dey et al., 2023). Despite the studies conducted to date on MPs, information regarding the occurrence of MPs in the terrestrial environment, particularly in agricultural fields is limited. This chapter sheds light on the sources, abundance, behavior, transport, and fate of MPs in the farmers' fields. It also attempts to provide an overview of how MPs in the soil affect soil characteristics, soil microorganisms, and agricultural crops.

2. Microplastic production and uses

Plastics are ubiquitous in the terrestrial environment due to day-to-day anthropogenic activities (Maaβ et al., 2017; Huang et al., 2020). They are highly sought-after in numerous industries due to their durability, versatility, low manufacturing cost, chemical and corrosion resistance, and lightweight nature (Chen et al., 2020; Choi et al., 2021). Plastics have applications in multiple sectors, such as industries, packaging, and agriculture. Global plastic production is currently about 460 MT annually and is predicted to increase by 4% every year. Among all countries, China has the highest plastic production, accounting for 23.9% of the world's total plastic production. In line with the increased demand and thus the production of plastics, the amount of discarded plastic waste has also increased and is expected to further increase in the near future (Li et al., 2020; Kim et al., 2021; Dey et al., 2023). In 2015, about 6.3 billion tons of plastic waste were generated, of which an estimated 79% ended up in landfills or the natural environment (Dong et al., 2020). A study in Korea estimated that with the current MP quantity in soil and agricultural practices about 300 trillion pieces of MP particles would be present in the agricultural soil of solely Korea, which was greater than the amount of plastic pieces (>0.33 mm) in the marine environment (Kim et al., 2021). Although plastics are biochemically inert, they can be gradually fragmented into MPs by various physical, chemical, and biological processes, such as photolysis, pyrolysis, hydrolysis, weathering, erosion, biodegradation, and decomposition by microorganisms. Anthropogenic activities such as crop cultivation can also cause the mechanical degradation of plastics (Choi et al., 2021; Kumar and Sheela, 2021).

Among the different sizes of plastic, MPs are considered potential agents of change due to their persistence and detrimental effects on the environment. Microplastics can enter the environment as primary or secondary MPs. Primary MPs are manufactured in the size range of 1 μm–5 mm for use as raw materials in industrial abrasives, plastic preproduction pellets, abrasives in personal care products (toothpaste, soaps, facial cleaners, scrubs), exfoliating agents, etc., while secondary MPs are formed by the degradation of larger plastic debris (Chen et al., 2020; Choi et al., 2021; Crossman et al., 2020; Sholokhova et al., 2022). MPs in the environment are composed of different carbon-based polymers such as polyethylene (PE), polyethylene terephthalate (PET), polypropylene (PP), polyvinyl chloride (PVC), polytetrafluoroethylene, polystyrene (PS), and polyamide (PA) (Dey et al., 2023). PE is used in plastic mulch, greenhouses, drainage pipes, and tunnel foils; PA is used in textile and packaging production; and PS is used in manufacturing, packaging industries, and cosmetics (Bigalke et al., 2022). Another prominent usage of plastics is in the textile industry, where about two-thirds of all textiles are synthetic and made of oil-based polymers. These include polyester (PES), PA, and acrylic (Chen et al., 2020).

MPs enter the soil environment from point sources or diffuse pollution, such as plastic waste disposal, landfill sites, industrial plastic production, atmospheric deposition, precipitation, wastewater irrigation, and agricultural activities (Li et al., 2021). On the basis of current manufacturing rates and disposal strategies about 12,000 MT of plastic are expected to be deposited in landfills by 2050. The crop production and livestock sectors in agriculture are the largest consumers of plastic products and are expected to clock up a yearly 9.5 million tons by 2030 (Sahasa et al., 2023). In agricultural soils, MPs are introduced through a multitude of anthropogenic routes, including the production of livestock and agricultural waste, the application of sewage sludge, the use of thin plastic mulch, road dust, and atmospheric deposition (Dey et al., 2023; Du et al., 2023). The MP concentration in soil is increasing with time due to increased MP deposition and a lower rate of migration (Dey et al., 2023).

3. Primary sources of MP in agricultural fields

Soils are crucial components of the ecosystem that experience immense pollution. Agricultural soils are said to harbor more MPs than the marine environment (Kim and An, 2019). About 7% of the plastics produced globally are utilized in agriculture (Kim et al., 2021). MPs can enter agricultural soils from multiple sources, including composting, the addition of manure, the application of sewage sludge, plastic mulching, irrigation, soil amendment with products containing MPs, and the primary use of plastics in the field and agricultural equipment such as plastic strings, sunshade nets, and greenhouse material (van den Berg et al., 2020; Zhou et al., 2020; Beriot et al., 2021; Sahasa et al., 2023).

a. *Application of livestock waste*: The application of livestock and poultry manure in agricultural fields represents an entry point of MPs in the agricultural soil. Livestock and poultry manure are extensively used as organic fertilizers in both raw and treated forms in many countries due to the growing demand for organic products. In 2018, about 71.5 million ha of the global agricultural area was under organic farming (Wu et al., 2021; Zhang et al., 2022; Sheriff et al., 2023). Organic waste undergoes biological treatments such as composting, anaerobic treatment, or fermentation to reduce the amount of biodegradable waste discarded in landfills as well as the need for chemical fertilizers. These treated organic wastes are nutrient-rich, and their application on land restores the nutrients, trace elements, and humus in the soil (Sholokhova et al., 2022). Long-term application of organic fertilizers has also been found to improve and increase the abundance of soil microorganisms such as bacteria, fungi, etc. (Yang et al., 2021). Animal manure has been reported to contain numerous pollutants (Sheriff et al., 2023), including MPs (Huerta Lwanga et al., 2017; Beriot et al., 2021). In pig, poultry, and cow manure, Wu et al. (2021) reported MP abundances of 902, 667, and 74 items/kg, respectively. Plastics are widely employed in all aspects of breeding, including manure scrapers, water taps, and bowls; pipelines for transporting feed; and plastic bottles for holding medicines (Wu et al., 2021). The source and amount of MPs in organic manure depend on the raw materials used in production, the type of biological treatment applied, and the surrounding environment where composting is carried out. In compost made of green and household wastes, household biowaste, and commercial biowaste, Weithmann et al. (2018) reported MP abundances of 20–24, 70–146, and 895 particles/kg, respectively. In compost made of green waste and food waste, Sholokhova et al. (2022) reported MP abundances of 5733–6433 and 3783–4066 particles/kg, respectively. In agricultural fields receiving organic manure, a hMP burden is also added. Yang et al. (2021) reported the average annual abundance of MPs in pig manure as 1250 particles/kg and estimated the average accumulation rate of MPs in agricultural soil with long-term application of pig manure to be 3.50 million particles/ha/annum (Yang et al., 2021).

b. *Application of sewage sludge*: The application of sewage sludge produced by WWTPs as a fertilizer in agricultural fields is a significant pathway for MPs to enter the agricultural soils (Corradini et al., 2019; Zhang et al., 2020). The WWTPs receive large volumes of MPs (synthetic microfibers from the laundering of synthetic textile materials, microbeads from personal care products, and plastic fragments and films arising from the degradation of larger plastics) from household drainage, industrial effluent, and stormwater (Corradini et al., 2019; Weber et al., 2022; Naderi Beni et al., 2023). Several studies have determined that the WWTPs are able to efficiently remove up to 99% of MPs from the wastewater, thereby accruing them in the sewage sludge (Crossman et al., 2020; Zhang et al., 2020). Given the colossal amounts of wastewater processed daily, a correspondingly enormous amount of sewage sludge is produced. However, some of the MP particles are too small to be removed by the WWTPs and end up in the wastewater effluent, which is used for agricultural irrigation, causing the accretion of MPs in vegetable and agricultural fields (Corradini et al., 2019; Chen et al., 2020). Sewage sludge is used as a fertilizer and soil conditioner in agricultural fields all over the world (Corradini et al., 2019) because it contains a high amount of organic and inorganic nutrients (van den Berg et al., 2020; Naderi Beni et al., 2023). But other than the valuable nutrients, they also contain HMs, organic pollutants, pathogens, and MPs (Naderi Beni et al., 2023). In North America and Europe about 44,000–430,000 tons of MPs are annually introduced on agricultural land (Zhou et al., 2020). The application of sewage sludge in fields is also advantageous economically since it decreases the amount of land required for its disposal. However, the sewage sludge contains enormous amounts of MP particles. In sewage sludge obtained from WWTPs in Spain, van den Berg et al. (2020) reported the presence of a light-density plastic load of 18,000 MPs/kg and a heavy-density plastic load of 32,070 MPs/kg. This amount of MPs would result in the entry of 3.78×10^8 light-density MPs/ha and 6.74×10^8 heavy-density MPs/ha of agricultural fields during each application (van den Berg et al., 2020). In sewage sludge used as fertilizer in China, the concentration of fibers in the sludge was up to 38,080 MPs particles/kg (Zhang and Liu, 2018). MPs in sewage arise mainly from washing clothes in washing machines and personal care products and are retained by the WWTPs. While the nutrients contained in the sludge are used up soon, the vast amounts of MP particles in the sludge remain in the soil for extended periods (van den Berg et al., 2020).

c. *Plastic mulching and shedding from greenhouses*: Plastic mulching is the technique of covering soils with plastic sheets, mainly PE, in order to improve crop production. It aids in increasing soil temperature, improves water content and water use efficiency, minimizes the use of nitrogen fertilizers, suppresses weed growth, stimulates microbial activity, and improves the physical condition of the soil. Plastic mulching has been a widely used agricultural practice for decades and has increased at a rate of 5%–10% annually in recent years. China utilizes the highest amount of plastic film mulch, covering over 19.8 million ha of agricultural land. Often, it is not possible to remove the plastic mulch from the fields after harvest because the films are very thin (0.01–0.03 mm). Abiotic factors like mechanical degradation (wind, water, plowing), thermal and photodegradation, oxidation, and hydrolysis, as well as biological factors, cause the plastic mulch remaining in the field to fragment into smaller pieces, resulting in large quantities of MPs that disburse in the topsoil. As a result, these fields wind up becoming hotspots for plastic debris, which is then consumed by animals that graze on agricultural remnants (Zhu et al., 2019; Qi et al., 2018; Li et al., 2020; Beriot et al., 2021; Kim et al., 2021; Kumar and Sheela, 2021). Beriot et al. (2021) reported the presence of approximately 2×10^3 MP particles/kg in agricultural soil and $\sim 10^3$ MP particles/kg in the feces of sheep that grazed on that soil. The study indicated that the animals ingested the MP particles present in the soil and then themselves became a source of MP contamination in other farms and grasslands (Huerta Lwanga et al., 2017; Beriot et al., 2021).

d. *Other sources*: These include MPs from agricultural equipment and practices (water pipes, plastic strings, tunnel foils, shade nets, fertilizer bags, irrigation, and polymer-coated fertilizers) as well as natural events like atmospheric deposition, soil erosion, and runoff (Chen et al., 2020; Lian et al., 2021; Lwanga et al., 2022). Plastic nets are used for protecting crops; plastic pipes enable irrigation; and plastic string provides support to plants (Zhang and Liu, 2018; Beriot et al., 2021). These plastic products are also degraded by the abiotic and biotic factors mentioned earlier. Plastic strings are commonly used in vegetable production and could be an important source of MP fibers. Zhang and Liu (2018) noted that although plastic strings contributed less than 0.3% of MPs, the amount of fiber they released was enormous. They also reported that irrigation with wastewater was another major source of MPs in agricultural soil. Zhou et al. (2020) also found irrigation water to be a major pathway for MPs, particularly fibers, into the soil. These fibers are mainly discharged from washing machines during the laundering of synthetic fabrics and persist in wastewater effluent. Atmospheric deposition of MPs appears to be a significant source of fibers in continental environments, and rainfall seems to be an important factor influencing the fallout flux (Zhang and Liu, 2018). In soils without any amendment, the MP abundance was 0.9 particles/g, which could have originated from wet/dry deposition or wind transport (Naderi Beni et al., 2023).

4. Microplastics distribution in agricultural fields around the world

Agricultural ecosystems are exposed to a cocktail of MPs arising from the sources aforementioned. Depending upon the different sources of plastics, varying amounts of MPs have been detected in agricultural soil around the world (Table 12.1), such as in China (Ding et al., 2020), the USA (Naderi Beni et al., 2023), Germany (Frei et al., 2019), India (Kumar and Sheela, 2021), Chile (Corradini et al., 2019), Spain (van den Berg et al., 2020), and the Netherlands (Huerta et al., 2023).

Mulching sheets have been found to contribute significantly to MP abundance in the soil in multiple studies. In China, Li et al. (2020) witnessed a higher abundance of MPs (40.35 mg/kg) in the 0.9–2.0 mm particle size soil sample that had been continuously mulched for about 30 years as compared to the soil that had been continuously mulched for 5 years (7.1 mg/kg). Further, the abundance of mesoplastics was also noted to increase with time, and these mesoplastics will degrade with time and further increase the MP pollution. As the time period of mulching or greenhouse usage increases, the amount of MPs in agricultural soil also increases. The distribution of plastic residues in the surface soil after mulching for 2, 4, 6, and 10 years was noted to be 15.11%, 21.50%, 25.42%, and 37.97%, respectively (Kumar and Sheela, 2021). The MP abundance was also found to increase in the deeper soil layers after a long period of mulching (Kumar and Sheela, 2021). A similar increase in MPs in soil was reported by Huang et al. (2020), wherein the MP concentration was 80.3, 308, and 1075.6 pieces/kg in fields mulched for 5, 15, and 24 years, respectively. The variation in MP abundance contributed by

Table 12.1 Microplastic input from prominent sources across the world.

Land use type and location	Source of MPs/ baseline assessment	MP abundance in the input source	Soil depth	MP abundance in agricultural soil	Size range	Shape and polymer	References
Vegetable farmland, China	Mulching sheets	–	0–3 and 3–6 cm	78 items/kg and 62.5 items/kg	<1 mm	Fragment, film, pellet; PP, PE	Liu et al. (2018)
Agricultural field, China	Mulching sheets	–	0–20 cm	40.35 mg/kg	0.9–2.0 mm	Fiber, fragment; PE	Li et al. (2020)
Agricultural field, India	Mulching sheets	–	0–−10, 11–20, and 21–30 cm	0.092–4.96 g/kg, 0.075–3.45 g/kg, and 0.01–2.81 g/kg	–	Film and fibers	Kumar and Sheela (2021)
Cropping land, China	Mulching sheets	–	0–40 cm	80.3–1075.6 particles/kg	–	Fragment; PE	Huang et al. (2020)
Agricultural soil, China	Mulching sheets and irrigation	3.9 pieces/L in irrigation water	0–10 cm	263–571 pieces/kg	1–3 mm	Fragment fiber, film; PE, PP, PES, Rayon, Acrylic, PA	Zhou et al. (2020)
Agricultural land, Republic of Korea	Greenhouse (GH) and Mulching sheets	–	0–5 cm	1880 items/kg (inside GH), 1302 items/kg (outside GH), 160 items/kg (Rice field), 81 items/kg (mulching fields)	0.1–0.2 mm (inside GH), 0.2–0.3 mm (outside GH), 0.2–0.3 mm (rice field), 1–2 mm (mulching field)	Fiber, sheet, fragment; PE, PET, PP (GH), PE, PP (rice fields), and PE, PP, PS, PET (mulching field)	Kim et al. (2021)
Farmland and grassland, China	Plastic mulch and greenhouse	–	0–3 and 3–6 cm	53.2 and 43.9 items/kg	<0.5 (66.3%) and <0.5 (78.3%)	Film, fragment, fiber, foam, spheres; PE, PP	Feng et al. (2021)
Vegetable fields, Netherlands	Plastic mulch and compost	–	0–30 cm	1109 and 890 MP/g	50–150 μm	PE, PS, PP	Huerta et al. (2023)
Vegetable fields, Spain and Netherlands	Plastic mulch and compost	2800 MPs/kg (municipal organic waste compost) and 1253 MPs/kg (garden and greenhouse compost)	0–10 and 10–30 cm	888 and 2242 MPs/kg	–	–	van Schothorst et al. (2021)

Continued

Table 12.1 Microplastic input from prominent sources across the world.—cont'd

Land use type and location	Source of MPs/ baseline assessment	MP abundance in the input source	Soil depth	MP abundance in agricultural soil	Size range	Shape and polymer	References
Agricultural soil, USA	Cattle manure and municipal sewage	1.5 particles/g and 9.1 particles/g	0–5 and 5–15 cm	1.1 particles/g and 2.6 particles/g	–	Fiber, fragment, film, foam; PE, PET, PP	Naderi Beni et al. (2023)
Agricultural soil, China	Pig manure	1250 particles/kg	–	43.8 particles/kg	–	–	Yang et al. (2021)
Agricultural soil, Spain	Sheep manure	997 particles/kg	0–10 cm	2116 particles/kg	–	–	Beriot et al. (2021)
Agricultural fields, China	Chicken manure; Sludge; Domestic waste	14,720, 8600, and 11,640 items/kg	–	2733, 2289, and 2462 items/kg	–	–	Zhang et al. (2022)
Agricultural fields, Spain	Sewage sludge	50,070 MPs/kg	0–10 and 10–30 cm	5190 MPs/kg	–	Fragment, fiber	van den Berg et al. (2020)
Agricultural field, Canada	Sewage sludge	8678–14,407 MP,/kg	0–5, 5–10, and 10 –15 cm	541 MP/kg	–	Fiber, Fragment; PE, PES, PP, Acrylic	Crossman et al. (2020)
Agricultural soil, China	Sludge-based fertilizer	250–5160 items/kg	0–5, 5–15, and 15 –25 cm	87.6–545.9 items/kg	0.5–5 mm	Flakes, fibers, films; PE, PP, PET	Zhang et al. (2020)
Agricultural field, Chile	Sewage sludge	22–53 mg/kg	0–25 cm	1.79–10.3 mg/kg	<5 mm	Fibers, films	Corradini et al. (2019)
Agricultural field, Germany	Sewage sludge	–	0–30, 30 –60, and 60–90 cm	0-56 particles/kg (after 34 years of sludge application)	0.3–4.9 mm	Fragments, pellets, filaments, films; Chlorinated PE, rubber, high-density PE	Weber et al. (2022)
Cropped area, China	Soil amendments and wastewater irrigation	–	0–5 and 5–10 cm	7100–42960 particles,/kg	0.05–0.25 mm (82%)	Fiber, fragments, films	Zhang and Liu (2018)
Vegetable farmland, China	Baseline assessment	–	0–5 cm	320–12560 items/kg	<0.2 mm (70%)	Fibers, microbeads; PA, PP, PS, PE, PVC	Chen et al. (2020)

Agricultural farmland, Germany	Baseline assessment	–	0–5 cm	0.34 items/kg	1–5 mm	Film, fragment; PE, PS	Piehl et al. (2018)
Agricultural land, Germany	Baseline assessment	–	0–10 cm	0-218 MPs/kg	1–5 mm	Fragment, film, fiber; PE, PP, Nylon, PA	Harms et al. (2021)
Agricultural field, China	Baseline assessment	–	0–10 cm	1430–3410 items/kg	0–0.49 mm (81%)	Fibers, film, fragment; PS, PE, PP, HDPE, PVC, PET	Ding et al. (2020)
Vegetable fields, China	Baseline assessment	–	0–3 and 3–6 cm	78 and 62.5 items/kg	<1 mm (48.8% and 59.8%)	Fiber, fragment, film; PP, PE	Liu et al. (2018)
Vegetable fields, China	Baseline assessment	–	0–5, 5–10, and 10–25 cm	275–4165, 179–4169, and 415–4322 items/kg	<0.5 (65%)	Fragment, film, fiber, pellet, foam; PP, EPC, PE, PS, PES, PU, rayon	Yu et al. (2021)

mulching sheets or greenhouses over different durations depends on the application time, corresponding to the exposure duration for weathering by physical, chemical, biological, and anthropogenic influences. The PE is degraded by UV exposure, abrasion with soil particles, and wind erosion (Li et al., 2020; Kim et al., 2021).

The extensive usage of organic manures in agriculture is considered an important source of MPs in soil. Zhang et al. (2022) explored the MP accumulation in an agricultural soil treated with organic fertilizers (chicken manure, sludge, and domestic waste composts) for a period of 13 years. The MPs concentrations in chicken manure, sludge, and domestic waste composts were 14,720, 8600, and 11,640 items/kg, respectively, and the MPs concentrations in soils after 13 years were 2733, 2289, and 2462 items/kg, respectively. The authors noted that the MP concentrations in soils with compost were significantly higher than in the control soils. In addition, the MPs in the long-term compost-amended soils ranged from 3.63×10^9 to 4.99×10^9 items/ha, which were about 3–4 times higher than those input by composts, indicating that the MPs derived from composts had gradually degraded into smaller fragments with time (Zhang et al., 2022). In a study by Yang et al. (2021), pig manure was used as fertilizer, and the MP abundance in control soils and soils amended with pig manure for 22 years was found to be 16.4 and 43.8 particles/kg, respectively. The authors estimated that the long-term application of pig manure would add 3.50 million MP particles/ha (Yang et al., 2021). Livestock feces have also been found to be an indirect source of MPs in agricultural fields. In Spain, plastic mulch was widely used on intensive vegetable farming land, and sheep were allowed to graze on the vegetable remnants. The sheep ingested the macroplastic debris present in the crops and silage, and their feces became a significant source of MPs (997 MP particles/kg) in the agricultural fields (2116 MP particles/kg) (Beriot et al., 2021).

In Spain, van den Berg et al. (2020) studied 16 agricultural fields for the presence of MPs and noted that sewage sludge application resulted in the accumulation of MPs in agricultural soils. The majority of the fields were fertilized with sewage sludge with 0–8 sewage applications of 20–22 tons/ha per application. In the fields not treated with sewage sludge, the average MP load was 2030 MPs/kg, while in the fields that received sewage sludge, the average MP load was 5190 MPs/kg. The majority of MPs were noted to be of higher density (>1 g/cm3). Successive sludge application significantly increased the plastic load, and the soils with a history of sewage application contained 256% higher MPs content than those without (van den Berg et al., 2020). Even in soils that had formerly received sewage sludge (34 years earlier), the MP load was significantly higher compared to surrounding soils without a history of sewage application (Weber et al., 2022).

Some studies have investigated the MP contribution from multiple sources. In an agricultural field, Naderi Beni et al. (2023) applied cow manure and sewage sludge separately and reported that the MPs in the topsoil of the control, manure-amended, and sewage-amended areas were 0.9 particle/g, 1.1 particle/g, and 2.6 particle/g, respectively. The lower concentration of MP in cow manure suggests that animal manure should be preferred over sewage sludge for fertilizing agricultural fields. In similar research, Huerta et al. (2023), while maintaining all other parameters in adjacent fields at comparable levels, calculated the quantity of MPs produced by plastic mulch and compost over a 5-year period. Plastic mulch was found to contribute a higher quantity of MPs (1109 MP/g) as compared to compost (890 MP/g). Contrastingly, Kim et al. (2021) reported the lowest contribution of MPs in soil by mulching sheets (81 items/kg) as compared to greenhouse soil (1302–1880 items/kg) or even rice fields (160 items/kg). The difference in MPs contribution was due to the variation in time duration of PE sheet application (<1 year for mulching and >3 years for greenhouse) (Kim et al., 2021).

A high abundance of MPs has also been observed in soils without any prominent sources of plastic. Ding et al. (2020) studied soil obtained from nine agricultural sites in China without any known MP sources and reported that all the sites were contaminated with MPs (1430–3410 items/kg), possibly arising from agricultural activities. In vegetable farmland in Wuhan, the mean MP concentration was 2020 items/kg (320–12,560 items/kg). The possible sources of MPs were building materials, roads and vehicles, wastewater irrigation, as well as PE nets in the field (Chen et al., 2020). Another study in vegetable soils reported MPs in a much higher range of 7100–42,960 particles/kg, arising from long-standing degradation of plastic film, sewage sludge application, and wastewater irrigation (Zhang and Liu, 2018). While MP abundance in agricultural soil with known sources of MPs can be documented easily, it is difficult to do so in the case of fields where known sources are absent. The high concentration of MPs in such cases is a result of anthropogenic activities and high population densities (Chen et al., 2020). The MPs could arrive in these agricultural fields by migration from surrounding plastic-contaminated farmlands and residential and urban areas. They could also be deposited by surface runoff and atmospheric transport (Chen et al., 2020; Li et al., 2020; Kumar and Sheela, 2021; Zhang et al., 2020).

Other than the sources of MPs in agricultural soil, factors like soil texture, planting age, and irrigation methods also regulate their abundance in the soil (Yu et al., 2021). The variations in the abundance of MP particles in different studies can be explained by the differences in MP source, treatment type, number of applications, seasonal variability, and variations in social and geographic locations (van den Berg et al., 2020; Kumar and Sheela, 2021).

5. Characteristics and behavior of MPs in the soil environment

Soil acts as a sink for MPs and hosts MPs, particularly in the smaller size range (Yu et al., 2021). The MPs observed in the agricultural and vegetable fields were generally in the following size range: <0.5 mm (Feng et al., 2021), 150–250 μm (van den Berg et al., 2020), 200–250 μm (Li et al., 2019), 0.8–0.3 mm (Li et al., 2020), and 0–0.49 mm (Ding et al., 2020). In vegetable soils, Zhang and Liu (2018) reported that 95% of the MP particles were in the 0.05–1 mm size range, and the most frequent MP particle size was in the 0.05–0.25 mm range. Similarly, in vegetable farms in China, Chen et al. (2020) and Yu et al. (2021) noted the dominance of MPs smaller than 0.2 and 0.5 mm, respectively. The higher abundance of MPs in agricultural soils arises from the decomposition of large MPs by weathering (Ding et al., 2020). The size of MPs decreased gradually with an increase in soil residence time. In fields with extended periods of mulching (5, 15, or 30 years), the particle size of MPs decreased from 791 to 317 μm. This was probably due to the degradation rate of the film being slow over a short period, which leads to the formation of larger MP particles. An increase in the weathering period and long-term cultivation enhances the damage and the rate of degradation of MPs, thus reducing MP particle size (Li et al., 2020; Horton et al., 2017). MPs also exhibit an inverse relationship between concentration and particle size, and the proportion of MPs in a particular size range decreases as the size enlarges (Ding et al., 2020). A gradual decrease in the size of MPs will enhance their accessibility to soil organisms and their potential as possible vectors for contaminants and pathogens in soil environments (Li et al., 2020). The fragmentation of MPs in agricultural soils with time poses an increasing risk to agricultural ecosystems. Another possible contributing factor to their prevalence may be the manufacture of small MPs in the micrometer size range for industrial usage (Chen et al., 2020).

The dominant shape of MPs in agricultural soil depends on the source of MPs and the extent of weathering. In soils that received sewage sludge, fibers were dominant, reiterating that synthetic fibers could be used as an indicator of sewage sludge application (van den Berg et al., 2020; Kim et al., 2021). This is because mainly fibers (released from the laundry of synthetic fabrics) and a smaller amount of fragments predominate (80% –90%) in the sludge (Corradini et al., 2019; Naderi Beni et al., 2023). In agricultural soils that received successive applications of sewage sludge, Corradini et al. (2019) noted that 97% of the MPs were fibers. A similar observation was made by Zhang and Liu (2018) in vegetable soils, where 92% of MPs were fibers. A smaller percentage of fragments and films were also present probably because of multiple sources of MPs. In soils without any known addition of plastics or in greenhouse soils, fragments were found to be the most dominant type of MPs. In such soils, the amount of fiber and films was minimal (Li et al., 2019; van den Berg et al., 2020). Conversely, in another baseline study in vegetable soils, microbeads (48%) and fibers (37%) were dominant, followed by fragments (15%) (Chen et al., 2020). The morphology of MPs indicates that they have been weathered by physical, mechanical, chemical, and biological forces. Factors like mechanical erosion due to crop cultivation, chemical oxidation by UV rays, and biological degradation along with the properties of plastic material are responsible for the change in the surface morphology of MPs (Li et al., 2020). As the residence time of MPs in soil increases (from 0 to 30 years), the surface of MP particles begins showing visible cracks with round holes, and the surface roughness increases. The diameter of the round holes also increases from 8 to 20 μm (Li et al., 2020).

Several studies have also reported on the prevalent polymer type and color of MPs in agricultural fields. The dominant polymers have been identified as PE, PET, PP, PVC, PA, PS, and PC. The occurrence of plastic polymers in the soil environment differs depending on the land use type and agricultural practices (Wu et al., 2021; Surendran et al., 2023). The MPs in agricultural soils occur in various colors, such as translucent, white, black, red, green, blue, and yellow. Among these, translucent and white have been reported to be the dominant colors (Li et al., 2019; Yu et al., 2021; Piehl et al., 2018).

6. Migration of MPs

MPs in the environment migrate from one matrix to another depending upon their inherent physical and chemical properties and environmental conditions. In terrestrial environments, MPs in soil migrate vertically and horizontally due to water (rainfall, surface runoff, erosion, infiltration, and groundwater), wind, and soil organisms. In agricultural soils, additional factors like agricultural practices (irrigation, plowing, and drainage) and plants regulate the movement of MPs (Piehl et al., 2018; Kim et al., 2021; Lwanga et al., 2022; Weber et al., 2022; Huerta et al., 2023).

In an agricultural field in India, Kumar and Sheela (2021) noted the maximum plastic debris in the surface layer (0–10 cm) and a decrease in the distribution of plastic residues with increasing depth. The distribution of plastic residues at 0–10, 11–20, and 21–30 cm was 37.97%, 35.07%, and 36.99%, respectively. A similar observation was made by Feng et al. (2021) on farmland in China and Harms et al. (2021) on agricultural land in Germany. Conversely, Yu et al. (2021) and Zhang et al. (2020) noted a higher amount of MPs in the 10–25 cm layer of soil as compared to the 0–10 cm layer. It indicates that the size of MPs regulates their transport in the soil; compared to bigger MP particles, the smaller ones (<350 μm) are highly mobile and can quickly reach the deeper soil layers/ groundwater/drainage water. The MPs with the smallest sizes were observed to be highly mobile and capable of deeper penetration in the soil (Yu et al., 2021; Kim et al., 2021; Bigalke et al., 2022). Additionally, the shape of MPs plays an important role in their transport. In an uncontaminated soil sample, Zhang and Liu (2018) noted that the concentration of plastic films and fragments was higher in the 0–5 cm layer than that in the 5–10 cm layers, although the amount of fibers did not vary significantly. It implied that the mobility of films and fragments was lower than that of fibers. However, in some cases, no variations in the MP concentration between different soil layers were observed. Agricultural practices like tilling or crop cultivation up to a certain depth tend to blend the MPs in the different soil layers (van den Berg et al., 2020; Huerta Lwanga et al., 2017).

MPs in the form of beads, films, and foams in the sludge tend to have a rougher surface as compared to fibers and fragments, which facilitates the growth of biofilms on their surfaces. This surface roughness also controls MP movement in the soil, as fibers and fragments with lower roughness experience greater transport during runoff events (Naderi Beni et al., 2023). Rainfall events have also been observed to mobilize the MPs in the agricultural soil and to initiate their transfer to surface waters. Naderi Beni et al. (2023) noted 10–14, 16–31, and 8–20 particles/L in control, sewage-amended, and manure-amended plots, respectively, after five rainfall events. In the sewage-amended plot, about 7000 MPs were present in the run-off, which was 0.4% of the total MPs applied. MP transport by runoff is believed to be controlled by environmental factors such as soil texture, vegetation, slope, and rainfall frequency, which regulate the movement of other contaminants as well (Piehl et al., 2018; Naderi Beni et al., 2023).

MPs can also move to deeper soil layers through biogenic activities. Soil organisms like earthworms, collembolans, and springtails form biopores via which long-distance movement of MPs occurs (Rillig et al., 2017; Maaβ et al., 2017; Yu et al., 2019; Kim and An, 2019). Rillig et al. (2017) noted that earthworms transported significant amounts of MPs to greater depths by incorporating MPs in the soil through burrows, casts, egestion, and adherence to their exterior. The smaller MPs were noted to travel the greatest distances. A similar observation was made by Yu et al. (2019), wherein LDPE MPs were introduced by earthworms to deeper soil layers, with the amount of smaller MPs (<250 μm) increasing with soil depth. Another mode of MP transport from one field to another and fallow lands was reported by Beriot et al. (2021), where sheep that consumed plastic debris from agricultural fields transported the smaller fragments to other lands through feces. Based on a herd of 1000 animals and an MP content of ~500 particles/kg of feces, the MP transported was estimated to be $\sim 0.36 \times 10^6$ particles/ha/year. Crops with deep and coarse root systems are also known to aid in the vertical downward/upward movement of MPs in the soil. Additionally, conventional tillage practices like plowing and hoeing can also aid in the formation and transport of MPs to soil subsurface layers (Kumar and Sheela, 2021; Naderi Beni et al., 2023).

7. Effect on soil properties and soil microorganisms

Soils represent a large MP reservoir due to a number of MP sources. The transit of MPs to the aquatic environment is delayed by soils, which act as temporary or permanent MP sinks. Intensive agricultural practices can lead to decreased soil health, declining organic matter, and deteriorating soil structure. MPs buildup in soil poses a hazard to soil properties, soil organisms, and the entire soil ecosystem (Zhang and Liu, 2018; van den Berg et al., 2020; Boots et al., 2019) (Fig. 12.2). Soil exposed to MPs faces a scarcity of soil organisms, a reduction in soil biodiversity, and inhibition of soil ecosystem functions like nutrient cycling, water infiltration, and percolation (Abel et al., 2018; Huerta et al., 2023).

Exposure to MPs affects several soil properties, including soil pH, cation exchange capacity, soil aggregation, bulk density, water holding capacity, infiltration, nutrient levels, and the carbon cycle (Abel et al., 2018; Boots et al., 2019; Yu et al., 2020a; Wang et al., 2020, 2021; Huerta et al., 2023; Li et al., 2023). Exposure to different shapes and polymers of MPs significantly reduced the soil pH, and the pH of the soil exposed to HDPE was significantly lower (0.62 units less than the control). It is probable that the HDPE with a large, possibly reactive surface area modified the cation exchange in the

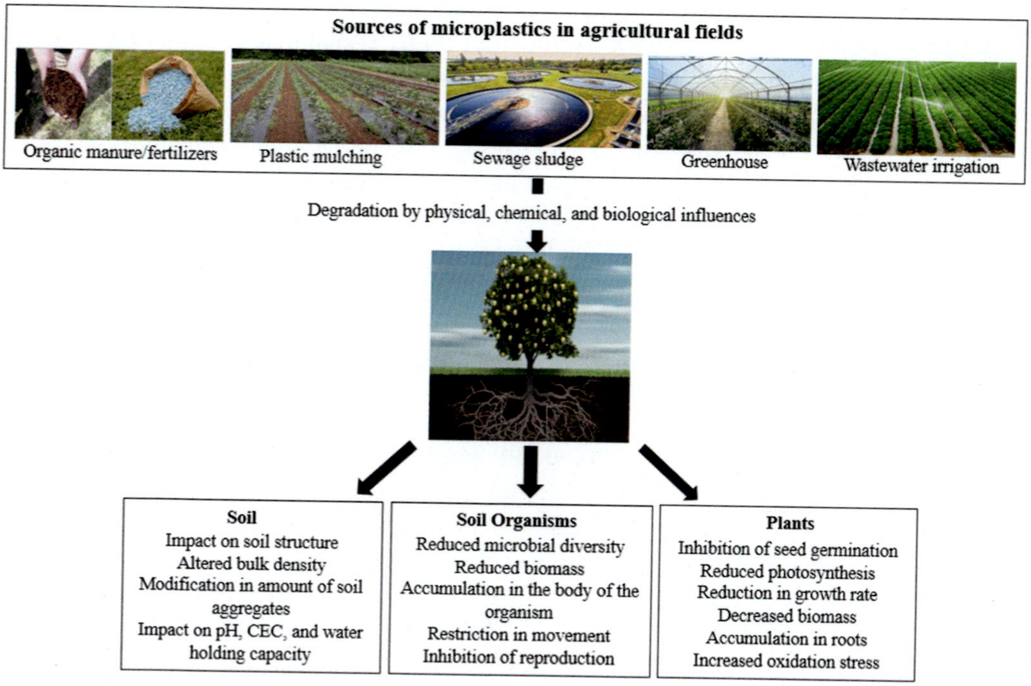

FIGURE 12.2 Effect of microplastics on soil, soil organisms, and plants.

soil and allowed free exchange of protons in the soil water (Boots et al., 2019). The presence of MPs of different shapes and compositions (PA beads, PE fragments, PES, and polyacrylic fibers) in the soil was found to decrease the soil bulk density, possibly due to the lower density of plastics compared to that of natural soil minerals (Abel et al., 2018; Li et al., 2023). Soil is a complex matrix containing micro/nanosized clay minerals and organic matter, which bind the soil into stable aggregates. The MPs interact with clay minerals and organic matter and become enfolded in the soil aggregates (Li et al., 2019, 2020; Kumar and Sheela, 2021). Abel et al. (2018) observed that polyacrylic and PES fibers formed the backbone of larger soil clumps. In a study by Zhang and Liu (2018), it was estimated that 72% of plastic particles in soil were associated with soil aggregates. The MPs are incorporated into the soil aggregates, probably due to their rough surfaces formed as a result of extreme weathering. The rough surfaces of MPs allow easy attachment and increased adsorption of clay particles, hence increasing soil aggregation. This modifies the degradation rate of MPs by either shielding them from any external physical degradation factors or enhancing biological degradation courtesy of the microbes present in the aggregates (Boots et al., 2019; Li et al., 2020). A significant alteration in the size distribution of the soil aggregates when exposed to different types of MPs was observed by Boots et al. (2019). The control soil had 24%, 35%, and 28% higher mean weight diameters compared to the soil exposed to fibers, HDPE, and PLA treatments, respectively. The water-stable soil aggregates were also significantly different when exposed to MPs; the control soil had 60% and 53% more large macroaggregates (>2000 μm) compared to soil with HDPE and PLA, respectively, while the microaggregates (63–250 μm) were higher in soil exposed to all types of MPs (Boots et al., 2019). A similar decrease in water-stable aggregates due to an increase in PES fibers was also noted by Abel et al. (2018). The MPs are thought to change the physical properties of the soil by increasing or decreasing the cohesiveness between the aggregate-forming particles (Boots et al., 2019). MPs were observed to enhance the water-holding capacity of the soil, with PES fibers causing the highest increase. The increase in water holding capacity can affect soil moisture, evapotranspiration, and thereby ecosystem services (Abel et al., 2018). Among MPs, PES fibers were found to produce the most extreme changes in soil properties because their linear shape and flexibility allowed them to blend more effectively in the soil matrix (Abel et al., 2018). The type and amount of MP also influence the carbon allocation in soil, carbon emissions, and the C-flow through the soil-plant system (Zang et al., 2020). PP was found to highly enhance the total organic carbon of the soil, likely due to its high carbon content (Li et al., 2023).

Soil microorganisms catalyze a number of soil functions and are affected by any physical or chemical alteration in the soil environment. Soil pore space configuration, aggregates, and hydrological properties impact the microbial metabolic rates and decomposition of organic matter (Abel et al., 2018). Furthermore, MPs cause alterations in the soil's physicochemical properties and nutritional substrates, thereby affecting microbial habitats and inhibiting enzyme activities (Kumar and Sheela, 2021; Huerta et al., 2023). The addition of virgin and aged PP and PS in soil was found to diminish soil enzyme activity as well as alter the microbial community (Choi et al., 2021). Yu et al. (2020a) reported that MPs inhibited soil enzyme activities by altering the level of soil nutrients (nitrogen, phosphorous, potassium, organic carbon, organic nitrogen) through adsorption or competed with microorganisms for physicochemical niches and diminished both microbial activity and soil enzyme activity. However, MPs have also been reported to increase the soil microbial biomass (+43.6%) and modify the structure and metabolic status of the microbial community (Zang et al., 2020), as well as increase the soil enzyme activity (Li et al., 2023). Soil-dwelling organisms are crucial for sustaining healthy soil. They aid in the decomposition of soil organic matter and cycle major nutrients, which are vital for plant growth and development. The rise in MP pollution causes extreme stress to the soil microorganisms because MPs due to their small sizes can be easily ingested by soil microorganisms, affecting their livelihood and survival. MPs are harmful to soil organisms as they reduce biomass, restrict movement, diminish reproduction, and increase mortality (Boots et al., 2019; Huerta et al., 2023; Kim and An, 2019; Schöpfer et al., 2020). Several studies have noted the detrimental effect of MPs on soil organisms (Table 12.2).

Earthworms are the most significant ecosystem engineers and indicators of environmental quality. They decompose organic materials, recycle nutrients, and create formations like burrows and middens that influence soil structure, water availability, and nutrient availability. They contribute to the formation of soil aggregates through their

Table 12.2 Effect of MPs on soil fauna.

Microorganism	Type of MPs	MPs concentration	Exposure duration (days)	Effects	References
Earthworm (*Lumbricus terrestris*)	Polyester microfibers	0%, 0.1%, and 1% w/w	35	Decrease in cast production and increase in transcriptional response related to stress	Prendergast-miller et al. (2019)
Earthworm (*Aporrectodea rosea*)	HDPE, Biodegradable PLA, Acrylic, and nylon synthetic fibers	0.1%w/w, 0.1%w/w, 0.001% w/w	30	Reduced biomass, particularly on exposure to HDPE; no mortality	Boots et al. (2019)
Springtails (*Lobella sokamensis*)	PS and PE	8 mg/kg	–	Disturbed movement and immobilization of organisms	Kim and An (2019)
Collembolan species (*Folsomia candida* and *Proisotoma minuta*)	Urea-formaldehyde MPs	5 mg of 100–200 μm and 2.5 mg of <100 μm	7	*F. candida* transferred larger particles faster and to a greater distance than *P. minuta*	Maaβ et al. (2017)
Caenorhabditis elegans	LDPE and biodegradable polylactide and poly (butylene adipate-co-terephthalate)	1, 10, and 100 mg MP/L	6	Reduction in the number of offsprings with a greater decrease at higher concentrations; Negative effect on body length of nematode	Schöpfer et al. (2020)

feeding, burrowing, and casting activities. These characteristics also enable the earthworms to ingest, transport, and bioconcentrate MPs in the soil. Earthworms can influence the characteristics of belowground plant parts and diminish the impairments produced by MPs, thereby affecting the above-ground plant parts (Boots et al., 2019; Huerta Lwanga et al., 2017; Qi et al., 2018; Rillig et al., 2017; Prendergast-miller et al., 2019). Boots et al. (2019) determined the effect of different types of MPs (acrylic and nylon synthetic fibers), high-density polyethylene (HDPE), and polylactic acid (PLA) in soil on an endogeic earthworm, *Aporrectodea rosea*, and noted that all types of MPs, particularly HDPE, lowered the earthworms' biomass. Microfibers (PES) were also found to lower cast production by 1.5 times (Prendergast-miller et al., 2019). The decrease in biomass may be a result of defense mechanisms such as blockage and abrasion of the digestive tract, which would restrict the bioavailability and absorption of nutrients, thereby causing stunted development and even death. The adverse effects could also be due to intestinal damage, oxidative stress, and an abnormal immune system (Kim and An, 2019). Additionally, long-term exposure to MPs has been reported to eliminate earthworms from the soil (Huerta et al., 2023).

8. Effect on plants

Higher plants are a crucial part of the terrestrial ecosystem (Li et al., 2023). The effect of MPs on a number of plants, like tomatoes, beans, wheat, rice, maize, etc., has been documented in the last few years (Table 12.3). In intensive agriculture, increased crop

Table 12.3 Effect of MPs on different plants.

Plant	Type of MP	Concentration of MPs	Exposure duration (days)	Physical and chemical effects on the plant	References
Lolium perenne (perennial ryegrass)	HDPE, biodegradable PLA, acrylic and nylon synthetic fibers	0.1%, 0.1%, and 0.001% w/w	30	Reduction in shoot height and biomass; reduced seed germination; increase in root biomass	Boots et al. (2019)
Lettuce (*Lactuca sativa* L.)	PE	0.25, 0.50, and 1.00 mg/mL,	14 and 28	Reduction in growth, chlorophyll content, and photosynthetic parameters; higher damage to roots; an increase in oxidative stress; enhanced the negative effect of di-n-butyl phthalate on plants	Gao et al. (2019)
Broad beans (*Vicia faba*	PS	10, 50, 100 mg/L	48 h	Decrease in root biomass and catalase activity; increased superoxide dismutase and peroxidase activities; oxidative damage; blockage of cell wall pores causing inhibition of nutrient transport	Jiang et al. (2019)
Cicer arietinum L.	PS (carboxyl-PS, aminated-PS, nonfunctionalized-PS) of sizes 11 and 12 μm	10, 50, and 100 mg/L	10	Dose-dependent reduction in root and shoot length of seedlings; hindered plant growth; less growth in	Dey et al. (2023)

Table 12.3 Effect of MPs on different plants.—cont'd

Plant	Type of MP	Concentration of MPs	Exposure duration (days)	Physical and chemical effects on the plant	References
				the presence of larger-sized MP; increase in chl content and fluorescence; increase in oxidative stress	
Citrus aurantium	Low-density PE, PP, PS	1% w/w	2, 4, 6, 8 months	Greater decrease in root biomass; LDPE displayed the most negative effect; LDPE+PP+PS increased the phytochemical content	Enyoh et al. (2020)
Italian lettuce, radish, wheat, corn	PS	1 and 10 mg/L	7	The toxicity of MP was species-dependent; more detrimental effects on lettuce and corn; inhibition of seed germination and root growth	Gong et al. (2021)
Black gram (*Vigna mungo* L.) and tomato (*Solanum lycopersicum* L.)	PE	0.25%, 0.50%, 0.75%, and 1.0%	24 h	Significant reduction in germination in black gram but minimal in tomato; reduced root length in black gram; minimal influence on all parameters of tomato	Sahasa et al. (2023)
Soybean (*Glycine max*)	PE, PVC	7% in dry soil	75	Decreased photosynthetic rate; reduced PSII activity; oxidative stress; PVC caused the most severe oxidative stress	Li et al. (2023)
Garden cress (*Lepidium sativum*)	PP, PE, PVC	0.02% (w/w)	6	Oxidative burst in the plant; PVC was more toxic than other MPs	Pignattelli et al. (2020)
Strawberry (*Fragaria* × *ananassa*)	HDPE	0.2 g/kg	48 h	Decrease in plant height, stem diameter, root surface area, inflorescence, number of fruits, fruit weight, and aerial biomass	Pinto-Poblete et al. (2023)
Barley	PS and PMMA	2 g/mL	14	Limited development of rootlets; higher concentrations of H_2O_2	Li et al. (2021)

Continued

Table 12.3 Effect of MPs on different plants.—cont'd

Plant	Type of MP	Concentration of MPs	Exposure duration (days)	Physical and chemical effects on the plant	References
				and O_2 in roots; increased activities of glutathione reductase and phosphofructokinase; decreased activities of cell wall peroxidase, sucrose synthase, and glucose-6-phosphate dehydrogenase in roots.	
Wheat (*Triticum aestivum*)	Low-density PE (LDPE) and starch-based biodegradable plastic	1% (w/w)	14–139	Effect on above- and below-ground parts of wheat during vegetative and reproductive growth by both macro- and microplastics; stronger negative effect of biodegradable plastic mulch	Qi et al. (2018)
Rice (*Oryza sativa*) seedlings	PS and PTFE	0.04, 0.1, and 0.2 g/L	10	Decrease in rice biomass; Inhibited root activity and RuBisCO activity; inhibition of photosynthesis, reduced chlorophyll fluorescence, photosynthetic rate, and chl-a content; reduction of arsenic absorption and uptake by plant	Dong et al. (2020)

production and fertilizer/amendment application give rise to multiple entrance pathways for MPs (Li et al., 2021). MPs can affect plants either directly or indirectly by impacting soil properties and soil microorganisms (Swarnali Dey et al., 2023). In plants, MPs affect germination, reduce growth and biomass, inhibit photosynthetic machinery and synthesis of photosynthetic pigments, alter nutrient uptake, generate oxidative stress, and injure cell structure (Qi et al., 2018; Jiang et al., 2019; Pignattelli et al., 2020; Li et al., 2023; Sahasa et al., 2023). However, the effects of MPs in plants are dependent on the plant species and cultivar and the exposure duration (Gong et al., 2021; Lian et al., 2021; Sahasa et al., 2023).

The concentration of MPs in soil and their characteristics (size, shape, polymer type, surface functionalization) regulate the physiological response and its magnitude in plants (Swarnali Dey et al., 2023). A decrease in the overall growth of *Cicer arietinum* L. and the biomass of *Vicia faba* was observed with an increase in the concentration of PS MPs. Further, a dose-dependent increase in oxidative stress was also observed in both. The larger-sized MPs produced enhanced oxidative stress as compared to the smaller ones, mostly because larger-sized MPs adhered to the root surface, caused more mechanical membrane damage, and enhanced the production of reactive oxygen species (ROS) (Jiang et al., 2019; Dey et al., 2023). An increase in oxidative stress was also noted in barley exposed to PS MPs (Li et al., 2021). Different shapes and polymers of MPs produce different effects in plants (Boots et al., 2019; Enyoh et al., 2020). In *Lolium perenne*, the growth responses were altered when exposed to synthetic fibers, HDPE, and PLA. On exposure to fibers, HDPE, and PLA, the germination rate decreased by 7%, 2%, and 6%, respectively, while the shoot length decreased by 16%, 14%, and 32%, respectively. Fewer seeds germinated, particularly in the presence of fibers, and the average shoot length was reduced on exposure to PLA. The decrease in germination and shoot length may be a result of the blockage of pores in the seed capsule. Furthermore, the biomass of roots exposed to HDPE was 45% higher than that exposed to PLA, indicating stress due to MPs in plants. Plants develop an elaborate root system to enable the absorption of more nutrients and water to adapt to stressful environments. The stress was further corroborated by an elevation in the chlorophyll a/chlorophyll b ratio, which is an indicator of photosynthetic activity. It is probable that the availability of macro- and micronutrients to the plant was altered as a result of the soil's biophysical changes, which had a domino effect on the plant's ability to photosynthesize (Boots et al., 2019). Plants can absorb MPs through the crack-entry mode, leading to detrimental effects on plant growth. MPs can enter the roots by penetrating the stele and translocating into the different plant parts via the transpiration stream (Li et al., 2020, 2021). In barley, Li et al. (2021) reported the absorption of MPs, which was indicated by a higher intensity of red fluorescence resulting from the uptake of fluorescently labeled polyethylmethacrylate.

9. Association of MPs with other contaminants in the soil

MPs are agricultural pollutants due to their small size, polarity, and large surface area-to-volume ratio, which enhances their capacity to absorb, release, and transport other contaminants. They are also highly bioavailable and act as potential vectors of contaminants in the food chain. Over time, the MP particles age, become rough, develop cracks, and break down into smaller pieces, further enhancing the surface area of the MPs. This facilitates their interaction with contaminants, leading to increased adsorption of HMs and organic pollutants on the surface of the MPs (Li et al., 2020; Wang et al., 2020; Du et al., 2023). Wang et al. (2020) reported that MPs arising from plastic mulch are a good carrier of pesticides in the agricultural field. Ramos et al. (2015) reported a higher presence of pesticides on plastic mulch film (584–2284 μg pesticide/g of plastic) as compared to soil

(13–32 μg pesticide/g of soil), along with the bidirectional movement of pesticide between the soil and plastic mulch. MPs decreased the bioavailability of copper, chromium, and nickel in soil and increased the organic-bound metal fraction by either adsorbing and coprecipitating the HMs or by altering the physical and chemical properties of soil (Yu et al., 2020b). The cooccurrence of MPs and HMs/organic pollutants impacts plant growth, chlorophyll content, photosynthetic activity, and the production of ROS. MPs have also been reported to assist in enhancing bioaccumulation as well as exacerbating the negative effect of HMs/organic pollutants in plants (Gao et al., 2019; Wang et al., 2021; Du et al., 2023; Pinto-Poblete et al., 2023). The addition of MPs was found to restrict arsenic (As) absorption and uptake in rice seedlings, probably by direct adsorption of As or by competition between As and MPs for adsorption sites on the root surface. MPs, together with As decreased the plant biomass, inhibited root activity, inhibited photosynthesis, and produced ROS, thereby destroying the cell membrane (Dong et al., 2020). Conversely, MPs were reported to enhance the bioavailability of cadmium (Cd) in soil and its concentration in lettuce, thus exacerbating the toxicity (Wang et al., 2021). MPs and HMs in the soil also diminish the activity of the soil enzymes and organisms (Du et al., 2023; Pinto-Poblete et al., 2023). MPs in the presence of Cd were observed to significantly influence the species composition and abundance of arbuscular mycorrhizal fungi (Wang et al., 2020). Colored plastics also release plastic additives and dyes, such as titanium dioxide, iron oxide, etc., into the soil and water, which are then transferred to the organisms (Chen et al., 2020). Plasticizers like phthalic acid esters in the plastic film mulch have been associated with fruit and vegetable contamination (Surendran et al., 2023).

10. Conclusion

The study on the occurrence, behavior, and fate of MPs in agricultural fields sheds light on the growing concern about MP pollution in the terrestrial environment. Existing research demonstrates that MPs are present in agricultural and livestock wastes and can potentially impact soil quality, plant growth, and human health. Due to the fragmentation and breakdown of plastics, it is anticipated that MPs in agricultural soils will continue to rise over time, causing further harm to soil ecosystems (Li et al., 2020). It is clear that MP pollution is a complex and multifaceted problem that requires a coordinated effort from all stakeholders, including farmers, policymakers, and researchers. Mitigation strategies such as reducing the use of plastic materials in agriculture, improving waste management practices, and promoting sustainable farming practices can help reduce the release of MPs into the environment. Further research is needed to fully understand the impact of MPs on soil quality, plant growth, and human health. It is crucial to establish standardized testing methods to accurately assess the occurrence and behavior of MPs in agricultural and livestock wastes. Developing effective and sustainable solutions to this issue will require collaboration between various sectors and a commitment to protecting the environment for future generations.

Acknowledgment

This research was supported by the National Laboratory for Water Science and Water Security (RRF-2.3.1-21-2022-00008).

References

Abel, A., Souza, S De, Lau, C.W., Till, J., Kloas, W., Lehmann, A., et al., 2018. Characterization of Natural and Affected Environments Impacts of Microplastics on the Soil Biophysical Environment. https://doi.org/10.1021/acs.est.8b02212.

Allouzi, M.M.A., Tang, D.Y.Y., Chew, K.W., Rinklebe, J., Bolan, N., Allouzi, S.M.A., et al., 2021. Micro (nano) plastic pollution: the ecological influence on soil-plant system and human health. Science of the Total Environment 788, 147815. https://doi.org/10.1016/j.scitotenv.2021.147815.

Almeshal, W., Tak, A., Sandil, S., Dobosy, P., Gyula, Z., 2022. Comparison of Freshwater Mussels Unio tumidus and Unio crassus as Biomonitors of Microplastic Contamination of Tisza River (Hungary).

Beriot, N., Peek, J., Zornoza, R., Geissen, V., Huerta, E., 2021. Science of the total environment low density-microplastics detected in sheep faeces and soil: a case study from the intensive vegetable farming in Southeast Spain. Science of the Total Environment 755, 142653. https://doi.org/10.1016/j.scitotenv.2020.142653.

Bigalke, M., Fieber, M., Foetisch, A., Reynes, J., Tollan, P., 2022. Microplastics in agricultural drainage water: a link between terrestrial and aquatic microplastic pollution. Science of the Total Environment 806, 150709. https://doi.org/10.1016/j.scitotenv.2021.150709.

Boots, B., Russell, C.W., Green, D.S., 2019. Effects of Microplastics in Soil Ecosystems: Above and Below Ground. https://doi.org/10.1021/acs.est.9b03304.

Chen, Y., Leng, Y., Liu, X., Wang, J., 2020. Microplastic pollution in vegetable farmlands of suburb Wuhan, central China. Environmental Pollution 257, 113449. https://doi.org/10.1016/j.envpol.2019.113449.

Choi, H.J., Ju, W.J., An, J., 2021. Impact of the virgin and aged polystyrene and polypropylene microfibers on the soil enzyme activity and the microbial community structure. Water, Air, and Soil Pollution 232, 1–9. https://doi.org/10.1007/s11270-021-05252-7.

Corradini, F., Meza, P., Eguiluz, R., Casado, F., Huerta-Lwanga, E., Geissen, V., 2019. Evidence of microplastic accumulation in agricultural soils from sewage sludge disposal. Science of the Total Environment 671, 411–420. https://doi.org/10.1016/j.scitotenv.2019.03.368.

Crossman, J., Hurley, R.R., Futter, M., Nizzetto, L., 2020. Transfer and transport of microplastics from biosolids to agricultural soils and the wider environment. Science of the Total Environment 724, 138334. https://doi.org/10.1016/j.scitotenv.2020.138334.

Dey, S., Anand, U., Kumar, V., Kumar, S., Ghorai, M., Ghosh, A., et al., 2023. Microbial strategies for degradation of microplastics generated from COVID-19 healthcare waste. Environmental Research 216, 114438. https://doi.org/10.1016/j.envres.2022.114438.

Dey, S., Guha, T., Barman, F., Natarajan, L., Kundu, R., Mukherjee, A., et al., 2023. Surface functionalization and size of polystyrene microplastics concomitantly regulate growth, photosynthesis and anti-oxidant status of Cicer arietinum L. Plant Physiology and Biochemistry 194, 41–51. https://doi.org/10.1016/j.plaphy.2022.11.004.

Ding, L., Zhang, S., Wang, X., Yang, X., Zhang, C., Qi, Y., et al., 2020. The occurrence and distribution characteristics of microplastics in the agricultural soils of Shaanxi Province, in north-western China. Science of the Total Environment 720, 137525. https://doi.org/10.1016/j.scitotenv.2020.137525.

Dong, Y., Gao, M., Song, Z., Qiu, W., 2020. Microplastic particles increase arsenic toxicity to rice seedlings. Environmental Pollution 259, 113892. https://doi.org/10.1016/j.envpol.2019.113892.

Du, R., Wu, Y., Lin, H., Sun, J., Li, W., Pan, Z., et al., 2023. Microplastics may act as a vector for potentially hazardous metals in rural soils in Xiamen, China. Journal of Soils and Sediments. https://doi.org/10.1007/s11368-023-03489-9.

Enyoh, C.E., Verla, A.W., Verla, E.N., Enyoh, E.C., 2020. Effect of macro- and micro-plastics in soil on quantitative phytochemicals in different part of Juvenile Lime tree (Citrus aurantium). International Journal of Environmental Research 14, 705–726. https://doi.org/10.1007/s41742-020-00292-z.

Feng, S., Lu, H., Liu, Y., 2021. The occurrence of microplastics in farmland and grassland soils in the Qinghai-Tibet plateau: different land use and mulching time in facility agriculture. Environmental Pollution 279, 116939. https://doi.org/10.1016/j.envpol.2021.116939.

Frei, S., Piehl, S., Gilfedder, B.S., Löder, M.G.J., Krutzke, J., Wilhelm, L., et al., 2019. Occurence of microplastics in the hyporheic zone of rivers. Scientific Reports 9, 1–11. https://doi.org/10.1038/s41598-019-51741-5.

Gao, M., Liu, Y., Song, Z., 2019. Effects of polyethylene microplastic on the phytotoxicity of di-n-butyl phthalate in lettuce (Lactuca sativa L. var. ramosa Hort). Chemosphere 237, 124482. https://doi.org/10.1016/j.chemosphere.2019.124482.

Gong, W., Zhang, W., Jiang, M., Li, S., Liang, G., Bu, Q., et al., 2021. Species-dependent response of food crops to polystyrene nanoplastics and microplastics. Science of the Total Environment 796. https://doi.org/10.1016/j.scitotenv.2021.148750.

Gudeta, K., Kumar, V., Bhagat, A., Julka, J.M., Bhat, S.A., Ameen, F., et al., 2023. Ecological adaptation of earthworms for coping with plant polyphenols, heavy metals, and microplastics in the soil: a review. Heliyon 9, e14572. https://doi.org/10.1016/j.heliyon.2023.e14572.

Harms, I.K., Diekötter, T., Troegel, S., Lenz, M., 2021. Amount, distribution and composition of large microplastics in typical agricultural soils in Northern Germany. Science of the Total Environment 758, 143615. https://doi.org/10.1016/j.scitotenv.2020.143615.

Horton, A.A., Walton, A., Spurgeon, D.J., Lahive, E., Svendsen, C., 2017. Microplastics in freshwater and terrestrial environments: evaluating the current understanding to identify the knowledge gaps and future research priorities. Science of the Total Environment 586, 127–141. https://doi.org/10.1016/j.scitotenv.2017.01.190.

Huang, Y., Liu, Q., Jia, W., Yan, C., Wang, J., 2020. Agricultural plastic mulching as a source of microplastics in the terrestrial environment. Environmental Pollution 260, 114096. https://doi.org/10.1016/j.envpol.2020.114096.

Huerta, E., Roshum, I Van, Munhoz, D.R., Meng, K., Rezaei, M., Goossens, D., et al., 2023. Microplastic appraisal of soil, water, ditch sediment and airborne dust: the case of agricultural systems. Environmental Pollution 316, 120513. https://doi.org/10.1016/j.envpol.2022.120513.

Huerta Lwanga, E., Mendoza Vega, J., Ku Quej, V., Chi J de los, A., Sanchez del Cid, L., Chi, C., et al., 2017. Field evidence for transfer of plastic debris along a terrestrial food chain. Scientific Reports 7, 1–7. https://doi.org/10.1038/s41598-017-14588-2.

Jiang, X., Chen, H., Liao, Y., Ye, Z., Li, M., Klobučar, G., 2019. Ecotoxicity and genotoxicity of polystyrene microplastics on higher plant Vicia faba. Environmental Pollution 250, 831–838. https://doi.org/10.1016/j.envpol.2019.04.055.

Kim, S.K., Kim, J.S., Lee, H., Lee, H.J., 2021. Abundance and characteristics of microplastics in soils with different agricultural practices: importance of sources with internal origin and environmental fate. Journal of Hazardous Materials 403, 123997. https://doi.org/10.1016/j.jhazmat.2020.123997.

Kim, S.W., An, Y., 2019. Soil microplastics inhibit the movement of springtail species. Environment International 126, 699–706. https://doi.org/10.1016/j.envint.2019.02.067.

Kumar, M.V., Sheela, A.M., 2021. Effect of plastic film mulching on the distribution of plastic residues in agricultural fields. Chemosphere 273. https://doi.org/10.1016/j.chemosphere.2020.128590.

Li, H., Song, F., Song, X., Zhu, K., Lin, Q., Zhang, J., et al., 2023. Single and composite damage mechanisms of soil polyethylene/polyvinyl chloride microplastics to the photosynthetic performance of soybean (Glycine max [L.] merr.). Frontiers of Plant Science 13, 1–13. https://doi.org/10.3389/fpls.2022.1100291.

Li, J., Yu, Y., Zhang, Z., Cui, M., 2023. The positive effects of polypropylene and polyvinyl chloride microplastics on agricultural soil quality. Journal of Soils and Sediments 23, 1304–1314. https://doi.org/10.1007/s11368-022-03387-6.

Li, L., Luo, Y., Li, R., Zhou, Q., Peijnenburg, W.J.G.M., Yin, N., et al., 2020. Effective uptake of submicrometre plastics by crop plants via a crack-entry mode. Nature Sustainability 3, 929–937. https://doi.org/10.1038/s41893-020-0567-9.

Li, Q., Wu, J., Zhao, X., Gu, X., Ji, R., 2019. Separation and identification of microplastics from soil and sewage. Environmental Pollution 254, 113076. https://doi.org/10.1016/j.envpol.2019.113076.

Li, S., Wang, T., Guo, J., Dong, Y., Wang, Z., Gong, L., et al., 2021. Polystyrene microplastics disturb the redox homeostasis, carbohydrate metabolism and phytohormone regulatory network in barley. Journal of Hazardous Materials 415. https://doi.org/10.1016/j.jhazmat.2021.125614.

Li, W., Wufuer, R., Duo, J., Wang, S., Luo, Y., Zhang, D., et al., 2020. Microplastics in agricultural soils: extraction and characterization after different periods of polythene film mulching in an arid region. Science of the Total Environment 749, 141420. https://doi.org/10.1016/j.scitotenv.2020.141420.

Lian, J., Liu, W., Meng, L., Wu, J., Zeb, A., Cheng, L., et al., 2021. Effects of microplastics derived from polymer-coated fertilizer on maize growth, rhizosphere, and soil properties. Journal of Cleaner Production 318, 128571. https://doi.org/10.1016/j.jclepro.2021.128571.

Liu, M., Lu, S., Song, Y., Lei, L., Hu, J., Lv, W., et al., 2018. Microplastic and mesoplastic pollution in farmland soils in suburbs of Shanghai, China. Environmental Pollution. https://doi.org/10.1016/j.envpol.2018.07.051.

Lwanga, E.H., Beriot, N., Corradini, F., Silva, V., Yang, X., Baartman, J., et al., 2022. Review of microplastic sources, transport pathways and correlations with other soil stressors: a journey from agricultural sites into the environment. Chemical and Biological Technologies in Agriculture 9, 1–20. https://doi.org/10.1186/s40538-021-00278-9.

Maaß, S., Daphi, D., Lehmann, A., Rillig, M.C., 2017. Transport of microplastics by two collembolan species. Environmental Pollution 3–6. https://doi.org/10.1016/j.envpol.2017.03.009.

Naderi Beni, N., Karimifard, S., Gilley, J., Messer, T., Schmidt, A., Bartelt-Hunt, S., 2023. Higher concentrations of microplastics in runoff from biosolid-amended croplands than manure-amended croplands. Communications Earth & Environment 4. https://doi.org/10.1038/s43247-023-00691-y.

Piehl, S., Leibner, A., Löder, M.G.J., Dris, R., Bogner, C., Laforsch, C., 2018. Identification and quantification of macro- and microplastics on an agricultural farmland. Scientific Reports 8, 1–9. https://doi.org/10.1038/s41598-018-36172-y.

Pignattelli, S., Broccoli, A., Renzi, M., 2020. Physiological responses of garden cress (L. sativum) to different types of microplastics. Science of the Total Environment 727, 138609. https://doi.org/10.1016/j.scitotenv.2020.138609.

Pinto-Poblete, A., Retamal-Salgado, J., Zapata, N., Sierra-Almeida, A., Schoebitz, M., 2023. Impact of polyethylene microplastics and copper nanoparticles: responses of soil microbiological properties and strawberry growth. Applied Soil Ecology 184. https://doi.org/10.1016/j.apsoil.2022.104773.

Prendergast-miller, M.T., Katsiamides, A., Abbass, M., Sturzenbaum, S.R., Thorpe, K.L., Hodson, M.E., 2019. Polyester-derived micro fi bre impacts on the soil-dwelling earthworm. Environmental Pollution 251, 453–459. https://doi.org/10.1016/j.envpol.2019.05.037.

Qi, Y., Yang, X., Mejia, A., Huerta, E., Beriot, N., Gertsen, H., et al., 2018. Science of the Total Environment Macro- and micro- plastics in soil-plant system: effects of plastic mulch fi lm residues on wheat

(*Triticum aestivum*) growth. Science of the Total Environment 645, 1048–1056. https://doi.org/10.1016/j.scitotenv.2018.07.229.

Ramos, L., Berenstein, G., Hughes, E.A., Zalts, A., Montserrat, J.M., 2015. Polyethylene film incorporation into the horticultural soil of small periurban production units in Argentina. Science of the Total Environment 523, 74–81. https://doi.org/10.1016/j.scitotenv.2015.03.142.

Rillig, M.C., Ziersch, L., Hempel, S., 2017. Microplastic transport in soil by earthworms. Scientific Reports 1–6. https://doi.org/10.1038/s41598-017-01594-7.

Rout, P.R., Mohanty, A., Aastha, Sharma, A., Miglani, M., Liu, D., et al., 2022. Micro- and nanoplastics removal mechanisms in wastewater treatment plants: a review. Journal of Hazardous Materials Advances 6, 100070. https://doi.org/10.1016/j.hazadv.2022.100070.

Sahasa, R.G.K., Dhevagi, P., Poornima, R., Ramya, A., Moorthy, P.S., Alagirisamy, B., et al., 2023. Effect of polyethylene microplastics on seed germination of Blackgram (Vigna mungo L.) and Tomato (Solanum lycopersicum L.). Environmental Advances 11, 100349. https://doi.org/10.1016/j.envadv.2023.100349.

Schöpfer, L., Menzel, R., Schnepf, U., Ruess, L., Marhan, S., Brümmer, F., et al., 2020. Microplastics effects on reproduction and body length of the soil-dwelling nematode. Caenorhabditis Elegans 8, 1–9. https://doi.org/10.3389/fenvs.2020.00041.

Sharma, U., Sharma, S., Rana, V.S., Rana, N., Kumar, V., Sharma, S., et al., 2023. Assessment of microplastics pollution on soil health and eco-toxicological risk in horticulture. Soil Systems 7, 7. https://doi.org/10.3390/soilsystems7010007.

Sheriff, I., Yusoff, M.S., Manan, T.S.B.A., Koroma, M., 2023. Microplastics in manure: sources, analytical methods, toxicodynamic, and toxicokinetic endpoints in livestock and poultry. Environmental Advances 12, 100372. https://doi.org/10.1016/j.envadv.2023.100372.

Sholokhova, A., Ceponkus, J., Sablinskas, V., Denafas, G., 2022. Abundance and characteristics of microplastics in treated organic wastes of Kaunas and Alytus regional waste management centres, Lithuania. Environmental Science & Pollution Research 29, 20665–20674. https://doi.org/10.1007/s11356-021-17378-6.

Surendran, U., Jayakumar, M., Raja, P., Gopinath, G., Chellam, P.V., 2023. Microplastics in terrestrial ecosystem: sources and migration in soil environment. Chemosphere 318, 137946. https://doi.org/10.1016/J.CHEMOSPHERE.2023.137946.

Wang, F., Wang, X., Song, N., 2021. Polyethylene microplastics increase cadmium uptake in lettuce (Lactuca sativa L.) by altering the soil microenvironment. Science of the Total Environment 784, 147133. https://doi.org/10.1016/j.scitotenv.2021.147133.

Wang, F., Zhang, X., Zhang, S., Zhang, S., Sun, Y., 2020. Interactions of microplastics and cadmium on plant growth and arbuscular mycorrhizal fungal communities in an agricultural soil. Chemosphere 254, 126791. https://doi.org/10.1016/j.chemosphere.2020.126791.

Wang, T., Yu, C., Chu, Q., Wang, F., Lan, T., Wang, J., 2020. Adsorption behavior and mechanism of five pesticides on microplastics from agricultural polyethylene films. Chemosphere 244, 125491. https://doi.org/10.1016/j.chemosphere.2019.125491.

Weber, C.J., Santowski, A., Chifflard, P., 2022. Investigating the dispersal of macro- and microplastics on agricultural fields 30 years after sewage sludge application. Scientific Reports 12, 1–13. https://doi.org/10.1038/s41598-022-10294-w.

Weithmann, N., Möller, J.N., Löder, M.G.J., Piehl, S., Laforsch, C., Freitag, R., 2018. Organic fertilizer as a vehicle for the entry of microplastic into the environment. Science Advances 4, 1–8. https://doi.org/10.1126/sciadv.aap8060.

Wu, R.T., Cai, Y.F., Chen, Y.X., Yang, Y.W., Xing, S.C., Liao, X Di, 2021. Occurrence of microplastic in livestock and poultry manure in South China. Environmental Pollution 277, 116790. https://doi.org/10.1016/j.envpol.2021.116790.

Yang, J., Li, R., Zhou, Q., Li, L., Li, Y., Tu, C., et al., 2021. Abundance and morphology of microplastics in an agricultural soil following long-term repeated application of pig manure. Environmental Pollution 272, 116028. https://doi.org/10.1016/J.ENVPOL.2020.116028.

Yu, H., Fan, P., Hou, J., Dang, Q., Cui, D., Xi, B., et al., 2020a. Inhibitory effect of microplastics on soil extracellular enzymatic activities by changing soil properties and direct adsorption: an investigation at the aggregate-fraction level. Environmental Pollution 267, 115544. https://doi.org/10.1016/j.envpol.2020.115544.

Yu, H., Hou, J., Dang, Q., Cui, D., Xi, B., Tan, W., 2020b. Decrease in bioavailability of soil heavy metals caused by the presence of microplastics varies across aggregate levels. Journal of Hazardous Materials 395, 122690. https://doi.org/10.1016/j.jhazmat.2020.122690.

Yu, L., Zhang, J Di, Liu, Y., Chen, L.Y., Tao, S., Liu, W.X., 2021. Distribution characteristics of microplastics in agricultural soils from the largest vegetable production base in China. Science of the Total Environment 756, 1–9. https://doi.org/10.1016/j.scitotenv.2020.143860.

Yu, M., Van Der Ploeg, B.M., Lwanga, E.E.H., Yang, X., Zhang, B.S., Ma, D.X., et al., 2019. Leaching of Microplastics by Preferential Flow in Earthworm (Lumbricus Terrestris) Burrows, pp. 31–40.

Yuan, Y., Qin, Y., Wang, M., Xu, W., Chen, Y., Zheng, L., et al., 2022. Microplastics from agricultural plastic mulch films: a mini-review of their impacts on the animal reproductive system. Ecotoxicology and Environmental Safety 244, 114030. https://doi.org/10.1016/j.ecoenv.2022.114030.

van den Berg, P., Huerta-Lwanga, E., Corradini, F., Geissen, V., 2020. Sewage sludge application as a vehicle for microplastics in eastern Spanish agricultural soils. Environmental Pollution 261, 114198. https://doi.org/10.1016/j.envpol.2020.114198.

van Schothorst, B., Beriot, N., Huerta Lwanga, E., Geissen, V., 2021. Sources of light density microplastic related to two agricultural practices: the use of compost and plastic mulch. Environments – MDPI 8, 1–12. https://doi.org/10.3390/ENVIRONMENTS8040036.

Zang, H., Zhou, J., Marshall, M.R., Chadwick, D.R., Wen, Y., Jones, D.L., 2020. Microplastics in the agroecosystem: are they an emerging threat to the plant-soil system? Soil Biology and Biochemistry 148, 107926. https://doi.org/10.1016/j.soilbio.2020.107926.

Zhang, G.S., Liu, Y.F., 2018. The distribution of microplastics in soil aggregate fractions in southwestern China. Science of the Total Environment 642, 12–20. https://doi.org/10.1016/j.scitotenv.2018.06.004.

Zhang, J., Wang, X., Xue, W., Xu, L., Ding, W., Zhao, M., et al., 2022. Microplastics pollution in soil increases dramatically with long-term application of organic composts in a wheat–maize rotation. Journal of Cleaner Production 356, 131889. https://doi.org/10.1016/J.JCLEPRO.2022.131889.

Zhang, L., Xie, Y., Liu, J., Zhong, S., Qian, Y., Gao, P., 2020. An overlooked entry pathway of microplastics into agricultural soils from application of sludge-based fertilizers. Environmental Science and Technology 54, 4248–4255. https://doi.org/10.1021/acs.est.9b07905.

Zhou, B., Wang, J., Zhang, H., Shi, H., Fei, Y., Huang, S., et al., 2020. Microplastics in agricultural soils on the coastal plain of Hangzhou Bay, east China: multiple sources other than plastic mulching film. Journal of Hazardous Materials 388, 121814. https://doi.org/10.1016/j.jhazmat.2019.121814.

Zhu, F., Zhu, C., Wang, C., Gu, C., 2019. Occurrence and ecological impacts of microplastics in soil systems: a review. Bulletin of Environmental Contamination and Toxicology 102, 741–749. https://doi.org/10.1007/s00128-019-02623-z.

13

Organic waste management and health

Gea Oliveri Conti[1], Eloise Pulvirenti[1,2], Antonio Cristaldi[1] and Margherita Ferrante[1,3]

[1]DEPARTMENT OF MEDICAL, SURGICAL SCIENCES, AND ADVANCED TECHNOLOGIES "G.F. INGRASSIA", UNIVERSITY OF CATANIA, CATANIA, ITALY; [2]DEPARTMENT OF BIOLOGICAL, GEOLOGICAL AND ENVIRONMENTAL SCIENCES, UNIVERSITY OF CATANIA, CATANIA, ITALY; [3]CRIAB—CENTRO DI RICERCA INTERDIPARTIMENTALE PER L'IMPLEMENTAZIONE DEI PROCESSI DI MONITORAGGIO FISICO, CHIMICO E BIOLOGICO NEI SISTEMI DI BIORISANAMENTO E DI ACQUACOLTURA E DI BIORISANAMENTO, OF DEPARTMENT "G.F. INGRASSIA", UNIVERSITY OF CATANIA, CATANIA, ITALY

1. Waste definition in EU and extra-EU countries

Waste can be defined differently in EU and nonEU countries.

Waste management, including recycling, is important for promoting sustainability and addressing environmental challenges. The EU Waste Framework Directive is an important pillar in guiding governments and waste management and recycling enterprises toward sustainable waste management practices. The EU directive establishes important objectives and requirements to reduce waste generation, promote recycling and recovery, and minimize the impact that this waste could cause on the environment and consequently on human health (Table 13.1).

The EU Waste Framework Directive defines waste as "*any substance, material, or object which the holder discards or intends to discard or is required to discard*".

Disposing of waste means managing waste in the following steps: collection, sorting, transport, and treatment of waste and its deposit above or below ground, as well as the

Table 13.1 Waste management procedures and related diseases.

Waste management	Diseases
Dump	Respiratory and cardiovascular diseases
Recycling	Neurological disorders
Mechanical-biological treatments	Infectious diseases
Old plant for waste-to-energy	*Cancer and reproductive diseases*

Occurrence and Behavior of Emerging Contaminants in Organic Wastes and Their Control Strategies
https://doi.org/10.1016/B978-0-443-13585-9.00013-6

transformation operations necessary for its reuse, recovery, or recycling (The European Parliament and the Council of the European Union, 2008/98/EC).

Waste in the United Nations Environment Program is defined according to the Basel Convention as "*wastes are substances or objects that are disposed of or are intended to be disposed of or are required to be disposed of by the provisions of national law*" (Basel Convention, 1989).

Waste is described by the UNSD Glossary of Environment Statistics as "*materials that are not prime products (that is, products produced for the market) for which the generator has no further use in terms of his/her own purposes of production, transformation, or consumption, and of which he/she wants to dispose. Waste may be generated during the extraction of raw materials, the processing of raw materials into intermediate and final products, the consumption of final products, and other human activities. Residuals recycled or reused at the place of generation are excluded*" (UNSD, 1997).

2. Waste management and health-related risks

Waste can follow various routes of management; it can be recycled, placed in landfills, mechanical-biologically treated (e.g., composting), or thermally treated.

All these methods make it possible to manage waste aimed at ensuring the reduction of its effects on human health and the environment (Fig. 13.1).

Waste management involves various phases: collection, transport, and treatment that allow the recovery or disposal of waste materials, and then finally the reuse and recycling of waste materials.

The EU directives make it possible to control the entire waste cycle and have environmental sustainability as a priority, the principle being to minimize the negative impacts on the environment and on humans.

In the Indian state of Gujarat in March 2009, 240 people contracted hepatitis B after undergoing medical treatment, only after it was discovered that the syringes had been acquired through the black market of unregulated medical waste (Solberg, 2009).

In Kabul, Afghanistan, in October 2008, by-products of a mass polio vaccination campaign of 1.6 million people were dumped in local municipal waste, causing infectious injuries to people scavenging for reusable items in landfills (Reuters News, 2008).

Public health is severely affected by the incorrect management of waste in many cities and urbanized regions of the EU but also in Africa, Asia, and the Middle East (Harhay et al., 2009).

Due to their increased sensitivity to the harmful effects of chemicals, physical and microbiological exposure to pollutants from waste can cause severe acute and chronic health problems in children.

For example, illegal chemical dumping and its potential impact on children's health have been investigated in Pasir Gudang, Malaysia. The dumping that occurred in Kim Kim River provoked 975 nearby students symptoms and signs of respiratory disease

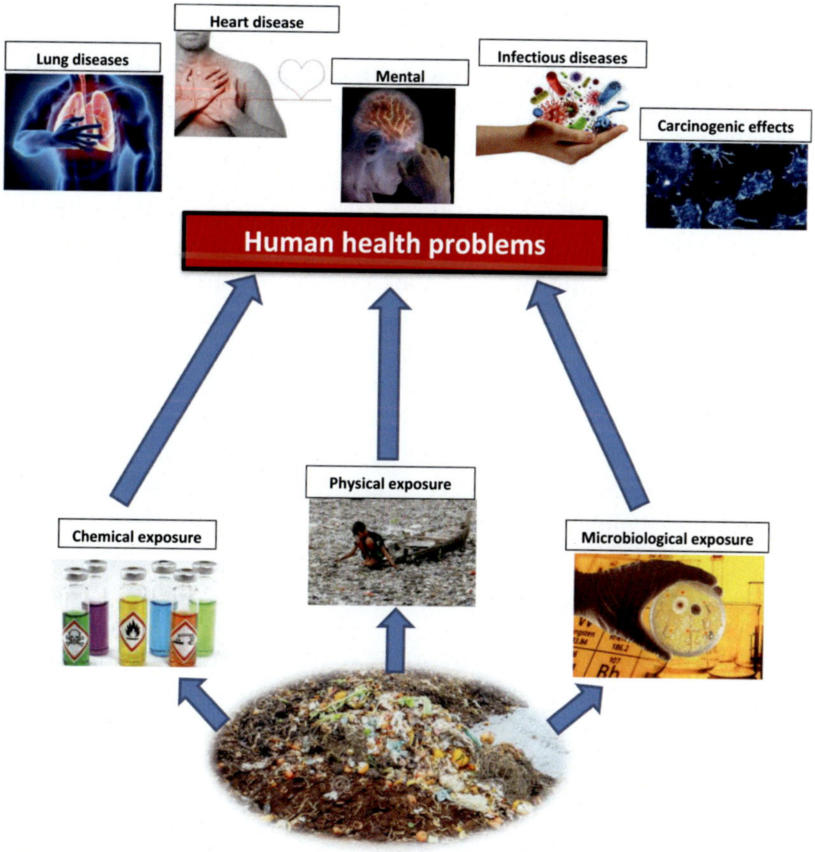

FIGURE 13.1 Waste and its effect on human health.

caused by chemical poisoning by benzene, acrolein, acrylonitrile, hydrochloric acid, methane (CH_4), toluene, xylene, ethylbenzene, and d-limonene.

In low- and middle-income countries, landfills remain the dominant means of waste disposal.

Many landfills cause serious hazards without proper management, and formal waste and recycling systems often fail. Around 15 million people worldwide are involved in the informal recycling of waste.

The informal waste sector has a variety of health problems. Working in landfills and dumps is inherently dangerous, and there are frequent accidents such as fires and debris landslides, as happened in Shenzhen.

Organic Waste workers, but also the general population residing near waste treatment plants can inhale gaseous emissions, bioaerosols, and microorganisms. These microorganisms include viruses, fungal spores, bacteria, and protozoa. Human health can also be affected by landfill gas emissions such as CH_4 and hydrogen sulfide, in addition to vehicular emissions related to the management of waste collection and its delivery.

These air emissions can cause respiratory, skin, and eye problems, infections in the lungs such as bronchitis and gastric ulcers, as well as parasitic diseases.

Landfills are known as favorable habitats for bacteria and vector-borne diseases such as dengue, leishmaniasis, diarrhea, typhoid, anthrax, cholera, malaria, and many skin diseases.

Efforts have been made to unite the formal and informal waste sectors. Studies in Brazil, Egypt, and India have shown that integration can help public health by reducing illness in people exposed to waste management and in residents in areas with the presence of dumping. The application of the "InteRa" model to integrate the informal sector into waste management systems in low- and middle-income countries showed that this would have a positive impact on everyone (Yang et al., 2018).

Chromium, nickel, mercury, cadmium, copper, manganese, arsenic, aluminum, zinc, and iron are some of the most common pollutants found in organic wastes. Premature birth, low birth weight, congenital disorders, thyroid disorders, mental and neurobehavioral disorders, cancer, genotoxicity, and irreversible DNA damage are all side effects of these metals (Balali-Mood et al., 2021; Tchounwou et al., 2012).

2.1 Organic waste incineration

More and more people are using organic waste incineration to generate electricity and reduce the volume of waste. More and more people are using organic waste incineration to generate electricity and reduce the volume of waste. The workers of public health were concerned about the health effects of incinerators.

A significant number of adverse health effects have been discovered, including significant correlations with some cancers, congenital anomalies, infant deaths, and miscarriage, but not for other diseases from old technologies of waste incineration.

Many old incinerators, in fact, were linked to cancer, reproductive problems, and other noncommunicable diseases (Tait et al., 2020). According to past numerous studies, residents may be exposed to carcinogenic pollutants that come from waste incinerator emissions.

However, these studies are not very good because many of them ignore important details about the type of incinerator design, specific standards for identifying residents (e.g., distance of residence by waste treatment plant, historical residence in the area studied, questionnaire use, self-reported diagnosis without certification of the same diagnosis by the doctor or hospital, and details describing the analysis of biases and confounders, including lifestyles). In fact, some updated meta-nalysis and systematic reviews, carried out with a scientific and qualitative approach, showed that the risks for human health are near zero for the latest generation incinerator, which in the EU have very low emission limits known as the Best Available Techniques (BAT).

Detrimental health effects have been linked to outdated incinerator technology and an infrequent maintenance schedule, often present in extra-EU countries (China, Mexico, etc.).

Malignancy findings, in fact, were generally inconsistent. Not surprisingly, many people use proximity to the incinerator as an independent variable, despite the limitations of this method. Additionally, most of the documents omitted pertinent details about the incinerator design, and several statistically significant results were irrelevant when approaching unity (Gatti et al., 2017; Galise et al., 2012; Domingo et al., 2002).

The incineration of organic wastes is a "necessary" piece of a most complicated waste cycle; we cannot describe "perfect waste management if we do not consider the necessary contribution of this approach in the management of the not reusable residues of wastes".

2.2 Organic waste disposal in landfills

Many municipalities in the world use landfills for disposing of organic materials.

Common types of organic waste include mulch, weeds, leaves, tree branches, flowers, biodegradable food waste, grass trimmings, etc.

Pathogenic bacteria (*Bacilli, Bacillus,* and *Burkholderia-Paraburkholderia*) and allergenic fungi (*Aspergillus, Cladosporium,* and *Curvularia*) prevail in landfills.

Biological risk refers to the presence of microorganisms (bacteria and viruses) in waste, which can infect humans. Bacteria and fungi are released into the air, and smell disturbance is a constant of this specific waste treatment plant. It is not enough for waste to be contaminated by pathogenic microorganisms to cause infection; a number of other circumstances are required, such as the infectious dose, the virulence of the germ, the route of penetration, and other factors. During waste disposal processes, antibiotic-resistant pathogenic microbes can thrive on organic waste as nutrients and can be dispersed in aerosols and transported to nearby areas. Most bioaerosols can be inhaled into the bronchi, which range in size from 2.1–3.3 μm to 3.3–4.7 μm.

Campylobacter, Salmonella, Listeria monocytogenes, verocytotoxic *Escherichia coli,* and *Yersinia* are some of the major pathogens. In addition, *Clostridium botulinum,* norovirus, toxoplasma, vibrios, hepatitis A virus, and some coagulase-positive staphylococci are dangerous microorganisms for residents and general people exposed to airborne landfill dust or polluted soils, especially for children (Kozajda et al., 2019).

Odor problems in landfills have been mainly caused by high amounts of microbial volatile organic compounds (VOCs) producing bioaerosols (Piccardo et al., 2022). Landfills significantly contribute to the release of biogas (CH_4) and carbon dioxide. The by-products of organic solid wastes deposited in landfills adversely affect the communities's health, like nitrogen oxides and particulate matter that cause heart and lung disease (Signorelli et al., 2019). Several studies have shown that harmful chemicals such as Pb, Cd, As, and Hg prevail in landfills. These can cause endemic malnutrition and increase susceptibility to mutagenic substances, increasing the prevalence of cancer in developing countries (Anetore, 2016).

They also trigger or worsen chronic conditions, such as asthma, heart attacks, bronchitis, and other respiratory problems (Corrêa et al., 2011). The level of exposure to the bioaerosol correlates with reduced lung function as a short-term or long-term effect. Gastrointestinal problems, skin and eye irritation, as well as organic dust toxic syndrome symptoms, have been reported. Notably, surrounding areas have higher antibiotic resistance genes (ARGs) than inland dumps, where tetC, acrB, acrF, mdtF, and bacA are the dominant genes. Most ARGs are related to the bacterial community, while environmental factors mainly influence the presence of fungi (Liang et al., 2023).

Average exposure to fungi, bacteria, and endotoxins across all countries and waste types was approximately 104 colony-forming units (cfu)/m^3, 104 cfu/m^3, and 10 EU/m^3. However, some studies have found that average exposure levels were up to 10 or more times higher or lower than those seen in other studies.

Exposure to fungi and endotoxins is often associated with an inflammatory and allergic response in humans and, in the worst case, cancer (Jeon et al., 2015).

However, to arrive at more reliable risk assessments, a better understanding of the impact of particular microbial species on health outcomes is needed. Exposure to biological agents is expected to increase due to both the growing recycling effort and the effects of global warming. Accordingly, it is crucial to think ahead and prepare strategies to prevent both long- and short-term adverse health effects (Madsen et al., 2021).

2.3 Mechanical and biological treatment of organic waste

Several health risks of the organic fraction of waste mechanics and biological treatment are known. Odors, noise, parasites, bioaerosol (organic dust containing bacterial or fungal spores, emission of VOCs), and potential food web contamination from use on land are among the most remarkable health disadvantages of this option. Adverse health effects of emitted VOCs were detected, e.g., benzene and 1,3-butadiene are potentially inductive agents of leukemia; formaldehyde has a nasal carcinogen potential; certain PAHs are potentially promotors of cancer; and polychlorinated dibenzo-p-dioxins and polychlorinated dibenzo-p-furans (PCDD/Fs) are reported as carcinogenic. Among the adverse effects of VOCs, annoyance by odors must also be included, as it can be caused by diverse indirect health effects such as nausea and vomiting, reactions of hypersensitivity, and even alterations in respiratory activity. VOCs are relevant also because they are capable of inducing renal, hematological, neurological, and hepatic alterations, as well as mucosal irritations. It is well known that benzene can cause aplastic anemia and polycythemia; dichloromethane produces carboxyhemoglobin; dichloromethane, toluene, styrene, trichloroethylene, and tetrachloroethylene are neurotoxic; and naphthalene and styrene produce irritations of the mucosal membranes (Domingo and Nadal, 2009). The presence of micro and macrorganisms, and/or substances generated in atmospheres can cause adverse effects on both workers and residents near the plants. In humans, they can be infectious, allergenic, toxic, and carcinogenic. The workers can be easily exposed to infectious viruses, microorganisms, bacteria that generate allergenic endotoxins, fungi, and parasitic protozoa. The workers of MSW composting facilities are not only potentially subjected to this exposure but also to that concerning organic dust, which causes several general health problems such as pulmonary inflammation, occupational asthma, and chronic bronchitis. Moreover, the following adverse effects can also occur specifically from exposure to organic dust: gastrointestinal diseases, fevers, ocular infections, and irritations, but also in the ear and skin.

3. Physical risks in the management of organic wastes

The presence of MPs, and nanoplastic particles, and metal particles constitutes a physical hazard associated with waste management.

Plastic is a material that is used in every productive sector and human activity. It is available in a variety of sizes (macro-, micro-, and nanoplastics) and shapes (fragments, spheres, and fibers) based on its end use or technological cost.

Microplastic (MP) is a combination of various materials of various shapes and sizes ranging from 1 μm to 5 mm.

The European Community recently started a dialog with member countries regarding the introduction of MPs as an emerging contaminant in water and food. The pharmaceutical sector and all other economic sectors that are directly or indirectly involved with this contaminant will be next to be interested in this problem.

Concerns regarding the adverse public health impact of exposure to MPs may stem in part from the chemical constituents of the polymer chains; some of these components can be present in free form and, after ingestion, can interact with biological molecules.

Several impacts were described by MPs exposure, such as inflammatory, mutagen, and toxicity effects both in animals and humans (Pulvirenti et al., 2022; Ferrante et al., 2021; Oliveri Conti et al., 2020; Najahi et al., 2022). The concern could stem from the fact that some components of plastic polymers are known to be toxic. Furthermore, residues of other (potentially) harmful chemicals used as catalysts in the plastic polymerization process can be found in the final product and its components. MPs can absorb and concentrate contaminants from the environment (Cristaldi et al., 2020).

In the waste management industry, heavy metals are known to pose a significant health hazard. Many industrial products (such as batteries, electrical equipment, stainless steel, or plastic products) that contain heavy metals are still waste and are hardly recycled. Also, heavy metals can be released during waste processing. Determining the true sources of the heavy metals found in a variety of waste types and waste products is still difficult. Over 30 different metals have been found in the ashes of unsorted municipal waste; most of these metals, such as arsenic, cadmium, chromium, lead, and mercury, are harmful to human health. Many of these metals also have combined effects (Laitinen and Rantio, 2014). Organic wastes can also contain metallic microparticles (Yu et al., 2012). For example, a large application of titanium dioxide as nano- and microparticles of E171 on common products, including transformed food, creates a concern for human health; in fact, it has been considered no longer safe for use in foods by the European Food Safety Authority due to concerns over genotoxicity

(Kirkland et al., 2022). However, the current Kirkland study (2022) demonstrated that it was positive for chromosome damage mainly at levels where reactive oxygen or other cellular toxicity was prevalent, but it was negative for point mutations in vivo (Kirkland et al., 2022). Currently, several flaws in knowledge about the effects of microparticles on human health are debated; for example, the last study by Wu demonstrated the presence of pigment (metallic) microparticles and MPs found in human thrombi (Wu et al., 2023).

4. Myths and realities of human health and waste management

Today some waste treatment systems are demonized. Despite the recurrent myths and false legends, the best systems for managing waste are the latest generation waste-to-energy (WtE) plants coupled with the residual disposal in landfills and recycling of recyclable materials, permitting the closure of the waste cycle.

For example, the opinions are often conflicting since not all the necessary data are always taken into consideration, or the data taken into consideration are not sufficient to establish whether the WtE plant is in fact an optimal method for the management of waste.

The debate about the health impact of waste burning for energy generation is becoming more heated due to the EU's plans to become climate-neutral by mid-century. Germany is the largest consumer in the sector, providing a significant share of the EU's total energy supply.

Like other power plants that use coal, oil, or natural gas, WtE plants produce energy using household waste as fuel. Heated water and steam from burning waste drive a turbine to produce electricity.

The Department for Environment, Food and Rural Affairs states that 'when considering the relative environmental benefits of landfills and energy from waste, the most important factor is their potential contribution to climate change. Different amounts of greenhouse gases would be released if the same waste were burned or buried.

WtE is considered the lesser evil of the two; both landfilling and the burning of mixed, untreated waste will release carbon into the atmosphere in different amounts. While energy from waste produces only carbon dioxide, landfills produce about the same amount of CH_4. WtE is thought to have a less negative overall impact on the environment because CH_4 is approximately 25 times more harmful to the atmosphere (DEFRA, 2014).

Waste incineration also allows us to bury less waste in the ground.

For many years, waste incineration was used extensively in the United States, Europe, and East Asia due to the perceived advantages of this technological management over landfills. Additionally, several WtE plants are currently in the planning stages in Australia.

However, communities often wrongly reject incinerators out of concern for air pollution. Many people still think that the carbon emissions released through high technological burning are too high.

5. Current emergencies

Among the emerging issues of organic waste, particular attention is given to MPs contained in waste.

Primary MPs are very small plastics designed to be used in the production of consumer products such as cosmetics, packaging, and fertilizers. Secondary MPs derive from larger plastic materials (such as tires, fishing nets, silicones, synthetic fabrics, and plastic packaging) that deteriorate, fragment, or melt due to light, thermo-oxidative processes, or mechanical movements.

The presence of MPs in food products is a much-discussed but also very constructive problem. The results led to the conclusion that foods accumulate plastic through the complex food chain, either through direct absorption from the environment or through direct contamination during product processing.

To fully understand the problem and its health effects, researchers closely observed the presence of MPs in the gastric contents of fish and other marine species, such as mussels and crustaceans. Obviously, the international literature has recently collected enough data to support the claim that MPs (fragments smaller than 0.5 mm) are the main contaminants in the human food chain.

However, searching for plastic fragments in gastric contents is useless because it does not provide new information about the problem. Instead, it is necessary to understand whether these MPs and nanoplastics (smaller than 1 μm to 1 μm in size) are able to move from the gastric sac and intestinal tube of aquatic organisms into nearby tissues and thus reach the muscle.

Some studies in Italy have discovered that the 10 μm MPs exposed to Zebrafish larvae during growth (a fish internationally recognized as an experimental model of ecotoxicological studies in vivo) have crossed not only the tissues of the larvae but also the muscles and sensory organs such as the eyes and brain. This has rendered many of these small fish blind (De Marco et al., 2022).

Furthermore, the 3 μm MPs to which the common mussel *M. galloprovincialis* had been exposed were shown to have reached many essential organs, including the ovaries. This altered the redox balance toward a prooxidant and, therefore, an inflammatory condition.

Therefore, after demonstrating the ability of these particles to transfer and concentrate in edible tissues, a research group from Catania, for example, conducted an international study to evaluate the concentration of MPs in the common edible marine fish species of the Mediterranean to evaluate the nutritional risk they may entail for humans. The results therefore show a general contamination of the edible tissues of the sampled marine species that had a significant content of MPs with a size of less than 3 μm.

The contribution of the terrestrial food chain, in particular as regards the consumption of fruit and vegetables, is even more significant (Ferrante et al., 2022; Cappello et al., 2021).

Indeed, an Italian study discovered MPs in vegetables (carrots, potatoes, lettuce, and broccoli) and fruit (apples and pears) from both organic and industrial supply chains. The smallest particles were 1.51 μm in diameter in carrot root hairs and 2.52 μm in lettuce. Fruits, such as apples and pears, are among the most "polluted" vegetables. This is due to the higher concentration of MPs in the fruit, which is made up of a complex vascular network that, unlike plants, stores and concentrates greater quantities of plastic. For example, adults had an EDI of 4.62E+05 and children had an EDI of 1.41E+06.

In the case of vegetables, contamination occurs directly from the removal of root systems from the contaminated soil. This often happens with the addition of sewage sludge used as soil improvers, which are also the main source of MPs in the soil (Oliveri Conti et al., 2020).

Therefore, the unmanaged or incorrect management of organic wastes can contain MPs that can be returned to humans.

The first FAO international report ("MPs in food commodities: A food safety review on human exposure through dietary sources", 2022) collected data from the international literature on this topic to stimulate new legislative acts by competent bodies throughout Europe.

From the point of view of the relationship between environment and health, these studies seem to open up and support scenarios not considered before. The MPs present in the human body can be a risk factor for chronic degenerative and inflammatory diseases in various human organs.

The current Italian-Tunisian study underlined the mutagenic and epigenetic capabilities of polyethylene terephthalate (PET) MPs of environmental origin, which measure about 1.6 μm in size and are capable of modifying cell differentiation into stem cells. Although initially intended to become chondrocytes, these cells transform into adipocytes after being exposed to PET.

The striking results of these studies warn the consumer about any health risks; therefore, they are the only way to protect the population, especially pregnant women and children, from the already known mutagenic, inflammatory, and oncological effects (Pulvirenti et al., 2022; Ferrante et al., 2021; Garrido Gamarro and Costanzo, 2022; Najahi et al., 2022).

6. Conclusions

Waste causes pollution and harmful substances, and the by-products of materials can pollute the air, water, and land. Waste brings a social cost, steals space, and requires human and economic resources not only for its treatment but also to remedy the damage it causes to the environment and health.

Waste is therefore an environmental pollutant that must be eliminated in an appropriate manner and with due precautions in order not to cause a progressive and exponential deterioration of our ecosystem. All strategies of reuse, recycling, circular economy, and mechanical-biological treatments such as composting or thermal treatments using incinerators facilitate the concept of waste reduction and pollution reduction, which are essential factors to keep the environment healthier and the health of residents.

Waste management is closely related to health; incorrect waste management causes and will cause, if not managed correctly, an increase in environmental and human damage in terms of disease and therefore mortality.

The current resolutions and intentions seem positive, but they will have to be put into practice right away in order not to risk an unstoppable decline of our ecosystem and an increase in human health problems caused by infections and degenerative diseases.

References

Anetor, G.O., 2016. Waste dumps in local communities in developing countries and hidden danger to health. Perspectives in Public Health 136 (4), 245–251. https://doi.org/10.1177/1757913915626192.

Balali-Mood, M., Naseri, K., Tahergorabi, Z., Khazdair, M.R., Sadeghi, M., 2021. Toxic mechanisms of five heavy metals: mercury, lead, chromium, cadmium, and arsenic. Frontiers in Pharmacology 12, 643972. https://doi.org/10.3389/fphar.2021.643972.

Basel Convention, 1989. Basel Convention on the Control of Transboundary Movements of Hazardous Wastes and Their Disposal.

Cappello, T., De Marco, G., Oliveri Conti, G., Giannetto, A., Ferrante, M., Mauceri, A., Maisano, M., 2021. Time-dependent metabolic disorders induced by short-term exposure to polystyrene microplastics in the Mediterranean mussel *Mytilus galloprovincialis*. Ecotoxicology and Environmental Safety 209, 111780. https://doi.org/10.1016/j.ecoenv.2020.111780.

Corrêa, C.R., Abrahão, C.E., Carpintero, M.doC., Anaruma Filho, F., 2011. Landfills as risk factors for respiratory disease in children. Jornal de Pediatria 87 (4), 319–324. https://doi.org/10.2223/JPED.2098.

Cristaldi, A., Fiore, M., Zuccarello, P., Oliveri Conti, G., Grasso, A., Nicolosi, I., Copat, C., Ferrante, M., 2020. Efficiency of wastewater treatment plants (WWTPs) for microplastic removal: a systematic review. International Journal of Environmental Research and Public Health 17 (21), 8014. https://doi.org/10.3390/ijerph17218014.

De Marco, G., Conti, G.O., Giannetto, A., Cappello, T., Galati, M., Iaria, C., Pulvirenti, E., Capparucci, F., Mauceri, A., Ferrante, M., Maisano, M., 2022. Embryotoxicity of polystyrene microplastics in zebrafish Daniorerio. Environmental Research 208, 112552. https://doi.org/10.1016/j.envres.2021.112552.

DEFRA, 2014. Energy from Waste A Guide to the Debate. https://assets.publishing.service.gov.uk/government/uploads/system/uploads/attachment_data/file/284612/pb14130-energy-waste-201402.pdf.

Domingo, J.L., Nadal, M., 2009. Domestic waste composting facilities: a review of human health risks. Environment International 35 (2), 382–389. https://doi.org/10.1016/j.envint.2008.07.004.

Domingo, J.L., Agramunt, M.C., Nadal, M., Schuhmacher, M., Corbella, J., 2002. Health risk assessment of PCDD/PCDF exposure for the population living in the vicinity of a municipal waste incinerator. Archives of Environmental Contamination and Toxicology 43 (4), 461–465. https://doi.org/10.1007/s00244-002-1280-6.

Ferrante, M., Cristaldi, A., Oliveri Conti, G., 2021. Oncogenic role of miRNA in environmental exposure to plasticizers: a systematic review. Journal of Personalized Medicine 11 (6), 500. https://doi.org/10.3390/jpm11060500.

Ferrante, M., Pietro, Z., Allegui, C., Maria, F., Antonio, C., Pulvirenti, E., Favara, C., Chiara, C., Grasso, A., Omayma, M., Gea, O.C., Banni, M., 2022. Microplastics in fillets of Mediterranean seafood. A risk assessment study. Environmental Research 204 (Pt C), 112247. https://doi.org/10.1016/j.envres.2021.112247.

Galise, I., Serinelli, M., Bisceglia, L., Assennato, G., 2012. Valutazione ex-ante dell'impatto sulla salute attribuibile all'inquinamento da inceneritore a Modugno (Bari) [Health impact assessment of pollution from incinerator in Modugno (Bari)]. Epidemiologia e Prevenzione 36 (1), 27–33.

Garrido Gamarro, E., Costanzo, V., 2022. Microplastics in Food Commodities – A Food Safety Review on Human Exposure through Dietary Sources. Food Safety and Quality Series No. 18. FAO, Rome. https://doi.org/10.4060/cc2392en.

Gatti, M.G., Bechtold, P., Campo, L., Barbieri, G., Quattrini, G., Ranzi, A., Sucato, S., Olgiati, L., Polledri, E., Romolo, M., Iacuzio, L., Carrozzi, G., Lauriola, P., Goldoni, C.A., Fustinoni, S., 2017. Human biomonitoring of polycyclic aromatic hydrocarbons and metals in the general population residing near the municipal solid waste incinerator of Modena, Italy. Chemosphere 186, 546–557. https://doi.org/10.1016/j.chemosphere.2017.07.122.

Harhay, M.O., Halpern, S.D., Harhay, J.S., Olliaro, P.L., 2009. Health care waste management: a neglected and growing public health problem worldwide. Tropical Medicine and International Health: TM & IH 14 (11), 1414–1417. https://doi.org/10.1111/j.1365-3156.2009.02386.x.

Jeon, J.H., Ahn, K.B., Kim, S.K., Im, J., Yun, C.H., Han, S.H., 2015. Bacterial flagellin induces IL-6 expression in human basophils. Molecular Immunology 65 (1), 168–176. https://doi.org/10.1016/j.molimm.2015.01.022.

Kirkland, D., Aardema, M.J., Battersby, R.V., Beevers, C., Burnett, K., Burzlaff, A., Czich, A., Donner, E.M., Fowler, P., Johnston, H.J., Krug, H.F., Pfuhler, S., Stankowski Jr., L.F., 2022. A weight of evidence review of the genotoxicity of titanium dioxide (TiO_2). Regulatory Toxicology and Pharmacology 136, 105263. https://doi.org/10.1016/j.yrtph.2022.105263.

Kozajda, A., Jeżak, K., Kapsa, A., 2019. Airborne *Staphylococcus aureus* in different environments-a review. Environmental Science and Pollution Research International 26 (34), 34741–34753. https://doi.org/10.1007/s11356-019-06557-1.

Laitinen, S., Rantio, T., 2014. Exposure to Dangerous Substances in the Waste Management Sector. European Agency for Safety and Health at Work.

Liang, Z., Yu, Y., Wang, X., Liao, W., Li, G., An, T., 2023. The exposure risks associated with pathogens and antibiotic resistance genes in bioaerosol from municipal landfill and surrounding area. Journal of Environmental Sciences (China) 129, 90–103. https://doi.org/10.1016/j.jes.2022.09.038.

Madsen, A.M., Raulf, M., Duquenne, P., Graff, P., Cyprowski, M., Beswick, A., Laitinen, S., Rasmussen, P. U., Hinker, M., Kolk, A., Górny, R.L., Oppliger, A., Crook, B., 2021. Review of biological risks associated with the collection of municipal wastes. Science of the Total Environment 791, 148287. https://doi.org/10.1016/j.scitotenv.2021.148287.

Najahi, H., Alessio, N., Squillaro, T., Conti, G.O., Ferrante, M., Di Bernardo, G., Galderisi, U., Messaoudi, I., Minucci, S., Banni, M., 2022. Environmental microplastics (EMPs) exposure alter the differentiation potential of mesenchymal stromal cells. Environmental Research 214 (Pt 4), 114088. https://doi.org/10.1016/j.envres.2022.114088.

Oliveri Conti, G., Ferrante, M., Banni, M., Favara, C., Nicolosi, I., Cristaldi, A., Fiore, M., Zuccarello, P., 2020. Micro- and nano-plastics in edible fruit and vegetables. The first diet risks assessment for the general population. Environmental Research 187, 109677. https://doi.org/10.1016/j.envres.2020.109677.

Piccardo, M.T., Geretto, M., Pulliero, A., Izzotti, A., 2022. Odor emissions: a public health concern for health risk perception. Environmental Research 204 (Pt B), 112121. https://doi.org/10.1016/j.envres.2021.112121.

Pulvirenti, E., Ferrante, M., Barbera, N., Favara, C., Aquilia, E., Palella, M., Cristaldi, A., Conti, G.O., Fiore, M., 2022. Effects of nano and microplastics on the inflammatory process: in vitro and in vivo studies systematic review. Frontiers in Bioscience 27 (10), 287. https://doi.org/10.31083/j.fbl2710287.

Reuters News, 2008. Afghanistan: Medical Waste Poses Health Risk in Urban Areas. Reuters News. http://www.alertnet.org/thenews/newsdesk/IRIN/fb0b02c252800fd18716034303f6dcc4.htm. Accessed 1 May 2009.

Signorelli, S.S., Oliveri Conti, G., Zanobetti, A., Baccarelli, A., Fiore, M., Ferrante, M., 2019. Effect of particulate matter-bound metals exposure on prothrombotic biomarkers: a systematic review. Environmental Research 177, 108573. https://doi.org/10.1016/j.envres.2019.108573.

Solberg, K.E., 2009. Trade in medical waste causes deaths in India. Lancet (London, England) 373 (9669), 1067. https://doi.org/10.1016/s0140-6736(09)60632-2.

Tait, P.W., Brew, J., Che, A., Costanzo, A., Danyluk, A., Davis, M., Khalaf, A., McMahon, K., Watson, A., Rowcliff, K., Bowles, D., 2020. The health impacts of waste incineration: a systematic review. Australian & New Zealand Journal of Public Health 44 (1), 40–48. https://doi.org/10.1111/1753-6405.12939.

Tchounwou, P.B., Yedjou, C.G., Patlolla, A.K., Sutton, D.J., 2012. Heavy metal toxicity and the environment. Experientia Supplementum 2012, 101, 133–164. https://doi.org/10.1007/978-3-7643-8340-4_6.

The European Parliament and the Council of the European Union, 2008/98/EC. Directive 2008/98/EC of the European Parliament and the Council on waste and repealing certain documents. Official Journal of the European Union L 312/3, 0003–0030.

UNSD, 1997. Glossary of Environment Statistics.

Wu, D., Feng, Y., Wang, R., Jiang, J., Guan, Q., Yang, X., Wei, H., Xia, Y., Luo, Y., 2023. Pigment microparticles and microplastics found in human thrombi based on Raman spectral evidence. Journal of Advanced Research 49, 141–150. https://doi.org/10.1016/j.jare.2022.09.004.

Yang, H., Ma, M., Thompson, J.R., Flower, R.J., 2018. Waste management, informal recycling, environmental pollution and public health. Journal of Epidemiology & Community Health 72 (3), 237–243. https://doi.org/10.1136/jech-2016-208597.

Yu, G.H., Wu, M.J., Wei, G.R., Luo, Y.H., Ran, W., Wang, B.R., Zhang, J.C., Shen, Q.R., 2012. Binding of organic ligands with Al(III) in dissolved organic matter from soil: implications for soil organic carbon storage. Environmental Science and Technology 46 (11), 6102–6109. https://doi.org/10.1021/es3002212.

Index

Note: 'Page numbers followed by "*f*" indicate figures and "*t*" indicate tables.'

Printed in the United States
by Baker & Taylor Publisher Services